최근 국제협약 개정에 따른

최신항해실습(I)

The Essential Guide to Navigation

윤귀호
조익순
배병덕
이윤석
김종성
전해동
공저

도서출판 상학당

欲爲大者 當爲人役

크고자 하거든 남을 섬기라

머리말

해기사의 직무능력에 대한 국제적인 최저 기준을 제시하는 국제해사기구(IMO)의 STCW 협약(선원의 훈련, 자격증명 및 당직근무의 기준에 관한 국제협약)은 1978년 채택된 이래로 1995년 대폭적인 협약개정을 거쳐 최근 2010년 마닐라 개정까지 많은 변화를 거쳐 왔다.

1995년 STCW 개정 협약에는 선원의 교육과 훈련목적 및 성취해야 할 능력의 향상, 해기사 시험 평가기준 설정 및 해기사 교육기관에 대한 내부 및 외부 감사제도 설정의 국내법 입법화 등이 추가되었으며, 특히 당직 해기사로서의 임무를 수행할 수 있는 능력의 평가를 강조하여 해기면허를 발급하는 주무관청이 승인한 훈련기록부에 실습생의 승선실습 내용을 기록하도록 하였다.

2010년 STCW 개정 협약에는 선교자원관리(BRTM), 전자해도(ECDIS), 해상교통관제(VTS), 조난신호 및 모르스 부호 해독능력(single letter), 선위통보제도, 탱커선 해기능력 및 시험제도, 선박보안기초교육 등의 내용이 기존 해기능력표에 추가되었다. 개정 협약은 2012년 1월 1일 발효되었으며, 2017년 1월 1일까지 경과규정을 두고 있으나, 모든 개정사항은 이미 해사대학 교과과정에 반영되어 진행 중이다.

따라서 항해계열 학생들은 최근 개정된 STCW 협약의 제II/1조 및 STCW 코드의 제A-II/1의 기준에 따라 교육 및 훈련을 이수해야 하며, 이러한 최근 국제협약의 동향을 반영한 적합한 항해실습 교재의 필요성이 대두됨에 따라 이 교재가 만들어지게 되었다.

이 교재의 주요 구성은 다음과 같다.

- 최신항해실습(Ⅰ) : 당직근무, 항해학, 해상안전
- 최신항해실습(II) : 항해기기, 해사 / 통신영어
- 최신항해실습(Ⅲ) : 선박운용, 선박적화, 선체구조 및 정비
- 최신항해실습(Ⅳ) : 선박보안 및 안전관리, PSC, 해양사고비상대응
- 최신항해실습(Ⅴ) : 선화운송, 탱커운송

이 교재는 최근 STCW 개정 협약을 반영한 항해계열 3학년 교과과정에 맞춰 구성되었으며, 정박 및 항해 중에 필요한 이론 및 실습의 내용에 대한 체계적인 정리를 통해 실습생이나 수업에 임하는 교수 모두에게 편리한 교재가 될 것으로 예상한다.

부족한 교재이지만 아무쪼록 이 교재가 항해계열 실습생의 교육 및 훈련에 도움이 되길 바라며, 앞으로도 지속적인 개정 및 보완을 할 것임을 다짐하는 바이다.

끝으로 이 교재의 편집과 출판에 많은 도움을 주신 도서출판 상학당 관계자에게도 감사의 말씀을 전하는 바이다.

2016. 3.

저자 일동

차 례

제1장

당직근무

1.1 선장복무지침서(Master's Standing Order)

1.1.1 Introduction(서론)

항해사에 있어서 당직근무(duty works, 當直勤務)는 항해당직 및 정박당직에 관계없이 가장 중요하고 핵심적인 사항이다. 선장복무지침서(Master's standing order)는 당직근무의 기본원칙을 정하여 항해사로 하여금 항상 숙지할 수 있도록 해야 할 것이다.

실무상 줄여 'Standing Order'라고도 하며, 이는 선박이 항해 중 항해당직사관이 선교에서 당직에 임할시 숙지해야 할 기본사항을 열거한 것으로써 회사의 항해당직방침에 기초를 두고 선박의 운항환경, 취급화물(특수화물, 예를 들면 Crude oil, Product, Chemical, LPG 및 LNG 등), 선박의 종류(일반화물선, 컨테이너선, 탱커, 자동차운반선 등), 선박의 주요 운항항로 및 특별한 환경(신조선 혹은 노후선박 등)을 비롯한 당해 선박 고유의 특성을 반영하고, 또한 당해 해역의 항해당직수행 시 본선 선장의 특별한 지시사항들을 반영하여 작성하는 당직수행지침서로써 항해당직사관 및 당직부원이 선교에서의 항해당직수행을 위하여 준수해야 할 선교에서의 지휘 및 통솔의 근간이 되는 지침이다.

이러한 선장의 복무지침서의 내용은 관련 국제규칙인 COLREG's(Convention on the International Regulations for Preventing Collisions at Sea, 1972), SOLAS(International Convention for the Safety of Life at Sea, 1974, 1983), STCW(International Convention on Standards of Training, Certification and Watchkeeping for Seafarers, 1978), ISM(International Safety Management) Code 및 회사가 제정한 선박의 안전관리시스템(Safety Management System, SMS) 등과 서로 모순이 되어서는 아니 된다.

따라서 선교에서 항해당직을 수행하는 모든 항해사는 당직수행 전 선장의 복무지침서의 내용을 철저히 숙지하여야 하며, 이러한 내용을 알고 있다는 표시로 선장복무지침서에 본인이 직접 직책과 서명을 해야 한다. 선장의 복무지침서의 내용은 선박에 따라 조금씩 상이한 내용도 있으나, 대부분은 국제규칙에서 권고하는 사항을 수용하여 작성되었기 때문에 해운회사와 각 선박마다 근본적인 내용은 유사하다고 말할 수 있을 것이다.

이 장에서는 국제해사기구(International Maritime Organization, IMO) 및 STCW 등의 국제규칙에서 권고하는 항해당직사관의 항해당직수행을 위한 권고사항을 영한대역(英韓

對譯)으로 표기하여 알아보고, 또한 현재 일반상선에서 사용하고 있는 선장복무지침서의 내용을 비교 검토하도록 할 것이다.

다음은 항해당직근무에 관련하여 IMO에서 권고하는 내용이다.

1.1.2 IMO Recommendation on Navigational Watch-keeping

1. SECTION Ⅰ

Basic principles to be Observed in Keeping a Navigational Watch
항해당직을 유지함에 있어서 지켜야 할 기본원칙

A. Introduction(서론)

1. The master of every ship is bound to ensure that watch-keeping arrangements are adequate for maintaining a safe navigational watch. Under his general direction, the officers of the watch are responsible for navigating the ship safely during their periods of duty when they will be particularly concerned to avoid collision and stranding.
   선장은 안전한 항해당직을 유지하도록 적절한 항해당직배치를 확보하여야 한다. 그의 일반적인 지휘하에 항해당직사관은 그의 당직시간동안 선박을 안전하게 운항할 책임이 있고 충돌과 좌초를 피한다는 의미에서는 특히 그러하다.
2. This Section includes the basic principles which shall at least be taken into account by all ships.
   이 절은 모든 선박이 고려하여야 할 최소한 기본원칙을 포함한다.

B. Watch arrangement(당직의 배치)

3. The composition of the watch, including the requirement for look-out(s), shall at all times be adequate and appropriate to the prevailing circumstances and conditions.
   당직의 구성은 경계에 대한 요건을 포함하여 모든 경우에 당시의 사정과 조건에 적합하고 알맞은 것이라야 한다.

4. When deciding the composition of the watch on the bridge the following points are among those to be taken into account.
선교당직자의 구성을 결정할 때에는 다음 사항을 고려해야 한다.
   (a) At no time shall the bridge be left unattended;
   어떠한 경우에도 선교를 비워서는 아니 된다.
   (b) The weather conditions, visibility and weather there is daylight or darkness;
   기상조건, 시정 및 주간 또는 야간의 구별에 따른 배려
   (c) The proximity of navigational hazards which may make it necessary for the officer in charge to carry out additional navigational duties;
   당직사관이 부가적으로 수행하여야 할 일을 만들어 줄 수 있는 항해장애물의 존재 여부
   (d) The use and operational condition of navigational aids such as radar or electronic position-indicating devices and any other equipment affecting the safe navigation of the ship;
   레이더 또는 전자적인 선위지시장치 및 선박의 안전항해에 영향을 미치는 기타 장비의 작동상태와 사용가능성
   (e) Whether the ship is fitted with automatic steering;
   자동조타장치의 유무
   (f) Any additional demands on the navigational watch that may arise as a result of special operational circumstances.
   특수한 운항사정 때문에 항해당직 중 생길 수 있는 부가적인 요구사항들

C. Fitness for duty(당직임무의 적합성)

5. The watch system shall be such that the efficiency of the watch-keeping members of the crew is not impaired by fatigue. Accordingly the duties shall be so organised that the first watch at the commencement of a voyage and the subsequent relieving watches are sufficiently rested and otherwise fit when going on duty.
당직조직은 당직자들의 근무능률이 피로 때문에 방해를 받지 아니하도록 짜여야 한다. 따라서 항해시작시에 첫 당직을 맡는 사람들과 당직을 인수받는 당직자들이

충분하게 휴식을 취할 수 있도록 그리고 당직시에 적합하도록 조직되어야 한다.

㉠ 항내에서 야간 하역근무를 계속한 일등항해사가 출항 즉시 당직사관 근무를 하게 되는 경우는 그가 잠시 쉬도록 하고 출항시 선장이 잠시 선교 근무를 시행하고 교대하여야 하는 것 등

D. Navigation(항해)

6. The intended voyage shall be planned in advance taking into consideration all pertinent information and any course laid down shall be checked.

항해계획은 모든 적절한 정보를 참작하여 사전에 수립되어야 하고 그어진 모든 침로는 다시 한 번 검토하여야 한다.

7. On taking over the watch the ship's estimated or true position, intended track, course and speed shall be confirmed; any navigational hazard expected to be encountered during the watch shall be noted.

당직인수자는 선박의 추정위치, 또는 진위치, 계획된 항로, 현재침로 및 속력을 확인하여야 하고, 자기 당직 중 조우될 것으로 예상되는 항해상의 장해물을 표시하여 두어야 한다.

8. During the watch the course steered, position and speed shall be checked at sufficiently frequent intervals using any available navigational aids necessary to ensure that the ship follows the planned course.

당직 중에는 모든 사용 가능한 항해계기를 이용하여 충분히 잦은 시간간격으로 조타침로, 위치와 속력을 조사하고 선박이 계획된 항로상을 따르고 있는가를 확인하여야 한다.

9. The safety and navigational equipment with which the ship is provided and the manner of the its operation shall be clearly understood; in addition its operational condition shall be fully taken into account.

선박에 설치되어 있는 안전장비와 항해장비 및 그들의 조작방법에 대하여는 명확하게 이해하고 있어야 하며, 추가하여 작동상태를 충분히 참작하여야 한다.

㉠ 고장난 계기에 대하여는 그 고장난 정도를 당직인계 시 다음 당직자에게 알려 주어야 한다.

10. Whoever is in charge of a navigational watch shall not be assigned or undertake

any duties which would interfere with the safe navigation of the ship.
누가 항해 당직을 맡든 간에 선박의 안전항해를 방해하는 임무를 그에게 주어서는 아니 되고 또한 그러한 일에 착수하여서도 아니 된다.

E. Look-out(경계)

11. Every ship shall at all times maintain a proper look-out by sight and hearing as wall as by all available means appropriate in the prevailing circumstances and condition so as to make a full appraisal of the situation and of the risk of collision, stranding and other hazards to navigation. Additionally, the duties of the look-out shall include the detection of ships or aircraft in distress, shipwrecked persons, wrecks and debris. In applying these principles the following shall be observed.
모든 선박은 선박이 처하여 있는 상황과 충돌, 좌초 및 기타 항해상의 장애사항을 충분히 평가할 수 있도록 시각 및 청각뿐만 아니라, 당시의 사정과 조건에 알맞은 모든 유용한 수단을 동원하여 항상 적당한 경계를 유지하여야 한다. 추가하여 경계의 임무는 조난을 당한 선박 또는 항공기, 조난자, 난파선 및 이의 잔해를 발견하는 것도 포함하여야 한다. 이러한 원칙들을 적용함에 있어서는 다음 사항을 준수하여야 한다.
    (a) Whoever is keeping a look-out must be able to give full attention to the task and no duties shall be assigned or undertaken which would interfere with the keeping of a proper look-out;
    누가 경계임무에 실시하든 간에 그는 경계임무에 열중하여야 하며, 경계를 방해하는 임무가 부가되거나 또는 그러한 일에 착수하여서는 아니 된다.
    (b) The duties of the person on look-out and helmsman are separate and the helmsman should not be considered the person on look-out while steering; except in small vessels where an unobstructed all round view is provided at the steering position and there is no impairment of night vision or other impediment to the keeping of a proper look-out;
    경계임무와 조타임무는 별개의 일이고 따라서 조타 중에 있는 조타수를 경계임무를 가진 사람으로 생각하여서는 아니 된다. 다만, 조타위치에서 사방을 볼 수 있고, 또한 적당한 경계를 방해하는 야경이나 다른 방해물이 없는 소형 선박의 경우에는 예외로 할 수 있다.

(c) There may be circumstances in which the officer of the watch can safely be the sole look-out in daylight. However, this practice shall only be followed after the situation has been carefully assessed on each occasion and it has been established without doubt that it is safe to do so. Full account shall be taken of all relevant factors including but not limited to the state of weather, conditions of visibility, traffic density, proximity of navigational hazards and if navigating in or near a traffic separation scheme (TSS). 'Assistance must be summoned to the bridge when any change in the situation necessitates this and such assistance must be immediately available'.

주간에는 당직사관 혼자만으로 충분한 경계를 할 수 있는 경우가 있을 수 있다. 그러나 이렇게 하는 것은 각 경우에 상황을 충분히 검토하고 이렇게 하여도 안전하다는 것이 확인된 후에만 따라야 하는 것이다. 이러한 일에는 모든 관계요소들을 고려해야 하며, 기상상태, 시정조건, 교통밀도, 항해장애물의 존재여부 및 통항분리방식의 부근 또는 그 내에서 항해하고 있을 때만 한정하여 생각할 문제는 아니다. '어떠한 사정의 변화 때문에 선교에서 협조자를 필요로 할 때에는 그를 즉시 소집하고, 그 협조자는 즉시 이용될 수 있어야 한다'.

F. Navigation with pilot embarked(도선사 승선 중 선박 운항)

12. Despite the duties and obligations of a pilot, his presence on board does not relieve the master or officer in charge of the watch from their duties and obligations for the safety of the ship. The master and the pilot shall exchange information regarding navigation procedures, local conditions and the ship's characteristics.

도선사의 임무와 책임에도 불구하고, 도선사 승선으로 선박의 안전을 위한 선장 또는 당직사관의 임무나 책임이 면제되는 것은 아니다. 선장과 도선사는 항해절차, 지역적인 조건 및 선박특성에 관한 정보를 서로 교환하여야 한다.

G. Protection of the marine environment(해양환경의 보호)

13. The master and officer in charge of the watch shall be aware of the serious effects of operational or accidental pollution of the marine environment and shall take all possible precautions to prevent such pollution international regulations.
선장과 당직사관은 운항과정에서 생기든 또는 과실사고에 의하여 생기든 간에 그로 인하여 생기는 해양환경오염의 중대한 영향을 알고 있어야 한다. 그리고 그들은 그러한 오염을 막을 수 있도록 사전에 모든 주의를 하여야 하며, 특히 현존하는 국제규칙의 범위 내에서는 그렇게 하여야 한다.

2. SECTION II

Operational Guidance for Officers in charge of a Navigational Watch
항해당직사관을 위한 운항지침

A. Introduction(서론)

1. This Section contains operational guidance of general application for officers in charge of a navigational watch, which masters are expected to supplement as appropriate. It is essential that officers of the watch appreciate that the efficient performance of their duties is necessary in the interest of safety of life and property at sea and the avoidance of pollution of the marine environment.
이 절은 항해당직사관을 위한 일반적인 운항지침을 포함하고 있으며 선장은 여기에 적당한 보충지침을 추가할 것으로 기대된다. 우선 당직사관이 그들의 효율적인 임무수행이 해상에 있어서의 인명 및 재산의 보호와 해양환경의 오염을 방지하는데 필요하다는 것을 인식하는 일이 중요하다.

B. General(일반원칙)

2. The officer of the watch is the master's representative and his primary responsibility at all times is the safe navigation of the vessel. He must at all times comply with the applicable regulations for preventing collisions at sea.

당직사관은 선장의 대행자이며 그의 최우선적 임무는 항상 선박의 안전운항이다. 그는 항상 해당 국제해상충돌방지규칙을 준수하여야 한다.

3. The officer of the watch should keep his watch on the bridge which he should in no circumstances leave until properly relieved. It is of especial importance that at all times the officer of the watch ensures that an efficient look-out is maintained. In a vessel with a separate chart room the officer of the watch may visit this, when essential, for a short period for the necessary performance of his navigational duties, but he should previously satisfy himself that it is safe to do so and ensure that an efficient look-out maintained.
당직사관은 선교에서 당직을 수행해야 하며 여하한 경우라도 적절히 교대를 한 후가 아니면 그 곳을 이탈하여서는 아니 된다. 항상 유효한 경계를 유지하도록 당직사관이 확인하는 일은 특별히 중요하다. 해도실이 별도로 되어 있는 선박에서는 당직사관은 그의 항해임무수행을 위하여 필요한 때에 잠깐 동안 그 곳에 갈 수 있다. 그러나 그는 그가 그 곳에 들어가기 전에 그렇게 하더라도 안전하다는 것을 사전에 확인하고 또한 유효한 경계를 확보한 후에 들어가야 한다.

4. There may be circumstances in which the officer of the watch can safely be the sole look-out in daylight. However, this practice shall only be followed after the situation has been carefully assessed on each occasion and it has been established without doubt that it is safe to do so. Full account shall be taken of all relevant factors including but not limited to the state of weather, conditions of visibility, traffic density, proximity of navigational hazards and if navigating in or near a traffic separation scheme. When the officer of the watch is acting as the sole look-out he must not hesitate to summon assistance to the bridge, and when for any reason he is unable to give his undivided attention to the look-out such assistance must be immediately available.
주간에 있어서는 당직사관의 경계만으로 충분한 경우가 있다. 그러나 이렇게 하는 것은 각 경우에 상황을 충분히 검토하고 이렇게 하여도 안전하다는 것이 확인된 후에만 따라야 하는 것이다. 이것을 결정함에 있어서는 모든 관련 요소들을 충분히 고려하여야 하며 기상상태, 시정조건, 교통밀도, 항해장해물의 존재여부 및 통항분리방식의 부근 또는 그 안에서 항행하고 있는가에만 국한하여 생각할 문제는 아니다. 당직사관이 혼자서 경계를 하고 있을 때는 선교에 보조자를 불러올리는 일에 주저하여서는 아니 된다. 그리고 어떠한 이유이든 간에 그는 그가 경계임무에 주력할 수 없을 때에는 즉시 보조자를 이용할 수 있어야 한다.

5. The officer of the watch should bear in mind that the engines are at his disposal and he should not hesitate to use them in case of need. However, timely notice of intended variations of engine speed should be given when possible. He should also keep prominently in mind the maneuvering capabilities of his ship including its stopping distance.
당직사관은 주기관의 사용은 그에게 맡겨져 있다는 것을 알고 필요할 때에는 그 사용을 주저하여서는 아니 된다. 그러나 가능할 때에는 적시에 속력의 변경의도를 알려주어야 한다. 그는 최단 정지거리를 포함하여 자선의 조종성능을 확실하게 파악하고 있어야 한다.

6. The officer of the watch should also bear in mind that the sound signalling apparatus is at his disposal and he should not hesitate to use it in accordance with the applicable regulations for preventing collisions at sea.
당직사관은 음향신호기구가 자기에게 맡겨져 있음을 알아야 하고 국제해상충돌방지규칙에 합치되도록 이것을 사용하는데 주저하여서는 아니 된다.

7. The officer of watch continues to be responsible for the safe navigation of the vessel despite the presence of master on the bridge until the master informs him specifically that he has assumed responsibility and this is mutually understood.
선장이 특히 책임을 지고 조선하겠다는 것을 당직사관에게 알리고 그리고 그것이 상호간에 이해되기까지는 선장이 선교에 있다고 하더라도 선박의 안전항해에 관한 당직사관의 책임은 계속된다.

C. Taking over the watch(당직의 인수)

8. The officer of the watch should not hand over the watch to the relieving officer if he has any reason to believe that the latter is apparently under any disability which would preclude him from carrying out his duties effectively. If in doubt,

the officer of the watch should ensure that members of his watch are apparently fully capable of performing their duties and in particular the adjustment to night vision.

당직사관은 당직을 인수받을 사관이 당직업무를 유효하게 수행할 수 없는 상태에 있음이 분명하고 믿을만한 이유가 있을 때에는 그에게 당직을 인계하여서는 아니 된다. 만약 의심스러운 경우에는 당직사관은 즉시 선장에게 그 취지를 보고하여야 한다. 당직을 인수하는 사관은 자기의 당직부원들이 명백히 그들의 임무를 수행할 수 있다는 것을 확인하여야 하고, 특히 야간에 시력조정을 행하는 단계에서는 그러하다.

9. The relieving officer should not take over the watch until his vision is fully adjusted to the light conditions and he has personally satisfied himself regarding.

당직인수사관은 그의 시각이 당시의 명암조건에 충분히 조정되고 다음 상황에 관하여 자신이 직접 확인할 때까지는 당직을 인수하여서는 아니 된다.

(a) Standing orders and other special instructions of the master relating to the navigation of the vessel;

선박의 운항에 관하여 선장이 지시한 당직근무수칙 및 기타 특별한 지시사항

(b) The position, course, speed and draught of the vessel;

선박의 위치, 침로, 속력 및 흘수

(c) Prevailing and predicted tides, currents, weather, visibility and the effect of these factors upon the course and speed;

현재 우세한 조석류, 해류, 기상, 시계 및 이들에 대한 예측, 그리고 그러한 요소들이 침로 및 속력에 미치는 영향

(d) The navigational situation including but not limited to the following:

다음 사항을 포함한 모든 항해상태

(i) The operational condition of all navigational and safety equipment being used or likely to be used during the watch;

현재 사용되고 있거나 당직 중 사용될 가능성이 있는 모든 항해 및 안전 장비의 작동상태

(ii) Errors of gyro and magnetic compasses;
자이로 및 자기 컴퍼스의 오차

(iii) The presence and movement of vessels in sight or known to be in the vicinity;
현재 보이고 있거나 부근에 있는 것으로 알려진 선박의 존재 및 그 동정

(iv) Conditions and hazards likely to be encountered during his watch;
그의 당직 중 조우될 것으로 예상되는 조건과 장애물

(v) The possible effects of heels, trim, water density and squat on under keel clearance.
횡·종경사, 물의 밀도 및 스커트 현상이 여유수심에 미칠 수 있는 영향

10. If at the time the officer of the watch is to be relieved a manoeuvre or other action to avoid any hazard is taking place, the relief of the officer should be deferred until action is completed.
만약 당직이 인계되어야 할 시간에 어떤 위험을 피하기 위하여 선박의 조종 또는 기타 동작이 취하여지고 있을 때에는 그러한 동작이 완료될 때까지 당직교대를 해서는 아니 된다.

D. Periodic checks of navigational equipment(항해장비의 주기적인 점검)

11. The officer of the watch should make regular checks to ensure that.
당직사관은 다음 사항을 확인하기 위하여 규칙적인 점검을 하여야 한다.

(a) The helmsman or the automatic pilot is steering the correct course;
조타수 또는 자동조타장치가 올바른 침로를 조타하고 있는지의 여부

(b) The standard compass error is established at least once a watch and when possible, after any major alteration of course. The standard and the gyro compasses should be frequently compared and repeaters should be synchronized with their master compass;
당직 중에 적어도 한 번은 Standard compass의 오차확인 및 가능하면 대각도 변침 후에도 확인할 것. Magnetic compasses와 Gyro compasses의 각도를 자주 비교할 것. Gyro repeater와 Main Gyro compasses의 동기상태

(c) The automatic pilot is tested in the manual position at least once a watch;
당직 중에 적어도 한 번은 자동조타를 수동으로 바꾸고 시험운전할 것

(d) The navigation and signal lights and other navigational equipment are functioning properly.
항해등 및 신호등과 기타 항해장비의 적당한 작동상태 확인

E. Automatic pilot(자동조타)

12. Officers of the watch should bear in mind the need to station the helmsman and to put the steering into manual control in good time to allow any potentially hazardous situation to be dealt with in a safe manner. With a vessel under automatic steering it is highly dangerous to allow a situation tp develop to the point where the officer of the look-out in order to take emergency action. The change over from automatic to manual steering and vice versa should be made by, or under the supervision of, a responsible officer.
당직사관은 어떤 잠재적인 위험을 안고 있는 경우가 안전한 방법으로 처리되기 위해서는 적시에 조타수를 배치하고 수동조타(manual steering)를 할 필요가 생긴다는 것을 마음에 새겨 두어야 한다. 자동조타를 하고 있는 선박에서 당직사관이 보조자가 없기 때문에 응급조치를 위하여 경계임무를 그만두지 아니할 수 없는 지경까지 사태가 발전한다는 것은 대단히 위험하다. 자동조타에서 수동조타로 또는 그와 반대되는 조치는 책임사관의 감독하에 또는 책임사관 자신에 의하여 행하여져야 한다.

F. Electronic navigational aids(전자항해보조계기)

13. The officer of the watch should be throughly familiar with the use of electronic navigational aids carried, including their capabilities and limitations.
당직사관은 성능과 한계를 포함하여 설치된 전자항해보조계기의 사용을 완전히 숙지하고 있어야 한다.

G. Echo-sounder(음향측심기)

14. The echo-sounder is a valuable navigational aid and should be used whenever

appropriate.

음향측심기는 매우 중요한 항해보조계기로 필요한 때에는 언제나 이것을 사용하여야 한다.

H. Navigational records(항해기록)

15. A proper record of the movements and activities of the vessel should be kept during the watch.
당직 중 선박의 항해과정과 운동상태를 올바르게 기록하여 두어야 한다.

I. Radar(레이더)

16. The officer of the watch should use the radar when appropriate and whenever restricted visibility is encountered or expected and at all times in congested waters having due regard to its limitations.
당직사관은 적절한 때, 시계가 제한될 때 또는 제한될 것이 예측될 때에는 언제나 레이더를 사용하여야 하고, 선박이 폭주하는 해역에서는 항상 레이더를 사용하여야 하며 레이더의 성능한계에 충분히 유의하여야 한다.

17. Whenever radar is in use, the officer of the watch should select an appropriate range scale, observe the display carefully and plot effectively.
레이더를 사용할 때에는 언제나 당직사관은 적절한 거리눈금을 선택하여 조작하고 영상을 주의깊게 관찰하여 효과적으로 플로팅(plotting)하여야 한다.

18. The officer of the watch should ensure that range scales employed are changed at sufficiently frequent intervals so that echoes are detected as early as possible and that small or poor echoes do not escape detection.
당직사관은 자주 자주 레이더의 거리눈금을 바꾸고 영상이 가능한 한 빨리 발견되고 작거나 또는 약한 영상도 놓치지 아니하도록 하여야 한다.

19. The officer of the watch should ensure that plotting or systematic analysis is commenced in ample time, remembering the sufficient time can be made available by reducing speed if necessary.
당직사관은 필요하다면 감속을 함으로써 충분한 시간을 얻을 수 있다는 것을 상기하면서 충분한 시간적인 여유를 두고 영상을 플로팅 혹은 체계적인 분석을 하여야 한다.

20. In clear weather, whenever possible, the officer of the watch should carry out radar practice.
당직사관은 청명한 날씨에는 가능한 레이더 연습을 실시하여야 한다.

J. Navigation in costal waters(연안 항해)

21. The largest scale chart on board, suitable for the area and corrected with the latest available information, should be used. Fixes should be taken at frequent intervals; whenever circumstances allow, fixing should be carried out by more than one method.
해도는 선내에 있는 가장 척도가 큰 해도로서 해당 지역에 적합하고 최신의 정보에 따라서 개정된 해도를 사용하여야 한다. 선위는 자주 측정하여야 하고 사정이 허락한다면 언제나 선위는 한 가지 이상의 방법에 의하여 결정하여야 한다.

22. The officer of the watch should positively identify all relevant navigation marks.
당직사관은 항해와 관련 있는 모든 항로표지를 적극적으로 확인하여야 한다.

K. Clear Weather(청명한 날씨)

23. The officer of the watch should take frequent and accurate compass bearings of a approaching vessel as a means of early detection of risk of collision; such risk may sometimes exist even when an appreciable bearing change is evident, particularly when approaching a very large vessel or a tow or when approaching a vessel at close range. He should also take early and position action in compliance with the applicable regulations for preventing collisions at sea and subsequently check that such action is having the desired effect.
당직사관은 충돌의 위험을 조기에 발견하는 방법으로서 접근하는 선박의 컴퍼스 방위를 자주 확인해야 한다. 충돌위험은 컴퍼스 방위에 상당한 변화가 생겨도 있을 수 있고, 특히 거대선박 또는 예인선열에 접근하거나 또는 지근거리에서 다른 선박에 접근할 때에는 그러하다. 당직사관은 해당 국제해상충돌방지규칙에 따라서 조기에 적극적인 동작을 취하여야 하고 동작을 취한 후에는 그러한 동작이 기대하였던 효과를 발생하고 있는지 검토하여야 한다.

L. Restricted visibility(제한시계)

24. When restricted visibility is encountered or suspected, the first responsibility of the officer of the watch is to comply with the relevant rules of the applicable regulation for preventing collisions at sea, with particular regard to the sounding of fog signals, proceeding at a moderate speed and he shall have the engines ready for immediate manoeuvrering. In addition, he should.
제한시계에 조우, 조우할 것으로 예측되면 당직사관의 일차적인 임무는 해상충돌방지규칙의 관계조항에 따르는 일이며, 특히 무중신호를 울리고 적당한 속력으로 진행하는 일에 유의하여야 하고 즉각적인 조종을 할 수 있도록 기관을 준비하여야 한다. 추가하여 그는 다음 일을 하여야 한다.

(a) Inform the master;(선장에게 보고)

(b) Post look-out(s) and helmsman and, in congested waters, revert to hand steering immediately;
폭주해역에서는 경계원과 조타수를 배치하고 수동조타로 변경

(c) Exhibit navigation light;(항해등 점등)

(d) Operate and use the radar. It is important that the officer of the watch should have the manoeuvring capabilities including the 'stopping distance' of his own vessel prominently in mind.
레이더를 작동하고 사용한다. 당직사관이 자선의 정지거리를 포함하여 제반 조종성능을 명백히 알고 있는 것이 대단히 중요하다.

M. Calling the master(선장에게 연락해야 할 경우)

25. The officer of the watch should notify the master immediately under the following circumstances.
다음과 같은 경우 당직사관은 즉시 선장에게 보고해야 한다.

(a) If restricted visibility is encountered or suspected;
제한시계의 조우 혹은 예상

(b) If the traffic conditions or the movements of other vessel are causing concern;
교통조건 또는 다른 선박들의 동정이 불안을 조성할 경우

(c) If difficulty is experienced in maintaining course;(침로유지가 곤란한 경우)

(d) On failure to sight land, a navigation mark or to obtain soundings by the expected time;

예정시간에 육지나 항로표지를 발견하지 못할 경우 또는 측심을 못했을 경우

(e) If land or a navigation mark is sighted or a charge in soundings occurs unexpectedly;

예측하지 아니한 육지나 또는 항로표지를 발견한 경우 또는 측심에 예측하지 아니한 변화가 생긴 경우

(f) On the breakdown of the engines, steering gear or any essential navigational equipment;

주기관, 조타장치 또는 기타 중요한 항해장비에 고장이 생겼을 경우

(g) In heavy weather if in any doubt about the possibility of weather damage;

악천후 속에서 악천우 때문에 생길 수 있는 손상에 대하여 의문이 있을 경우

(h) In any other emergency or situation in which he is in any doubt. Despite the requirement to notify the master immediately in the foregoing circumstances, the officer of the watch should in addition not hesitate to take immediate action for the safety of the ship, where circumstances so require.

비상시 또는 평상시를 막론하고 그에게 어떤 의문이 생겼을 경우. 위의 모든 경우에는 즉시 선장에게 보고하여야 함은 물론, 당직사관은 부가하여 사정이 그렇게 요구하는 경우에는 선박의 안전을 위하여 즉각적인 동작을 취하는데 주저하지 말아야 한다.

N. Navigation with pilot embarked(도선사가 승선 중 선박 운항)

26. Despite the duties and obligations of a pilot, his presence on board does not relive the officer of the watch from his duties and obligations for the safety of the ship. He should cooperate closely with the pilot and maintain an accurate check on the vessel's positions and movement. If he is in any doubt as to the pilot's actions or intentions, he should seek clarification from the pilot and if doubt still exists he should notify the master immediately and take

whatever action is necessary before the master arrives.

도선사의 임무와 책임에도 불구하고, 도선사의 승선은 선박의 안전을 위한 당직사관의 임무와 책임을 면제하여 주는 것은 아니다. 그는 도선사와 긴밀하게 협력하여 선박의 위치와 이동을 정확하게 확인하여 나가야 한다. 만약 그가 도선사의 행동과 의도에 어떤 의심을 가지게 되면, 그는 도선사에게 그 의심을 풀 수 있는 설명을 요구할 것이며, 그리하여도 의심이 남아 있으면 즉시 선장에게 알리고 선장이 선교에 도착하기 전이라도 필요한 조치가 있으면 그 조치를 취하여야 한다.

O. The watch-keeping personnel(당직부원)

27. The officer of the watch should give the watch-keeping personnel all appropriate instructions and information which will ensure the keeping of a safe watch including an appropriate look-out.

당직사관은 적절한 경계를 포함 안전당직을 유지하는 것을 확보할 수 있는 모든 지시와 정보를 당직부원에게 제공하여야 한다.

P. Ship at anchor(정박선)

28. If the master considers it necessary a continuous navigational watch should be maintained. In all circumstances, however, the officer of the watch should;

선장이 필요하다고 판단하면 정박 중이라도 항해당직을 배치하여야 한다. 그러나 모든 경우에 당직사관은 다음 사항을 이행하여야 한다.

(a) Determine and plot the ship's position on appropriate chart as soon as practicable and at sufficiently frequent intervals check when circumstances permit, by taking bearing of fixed navigational marks or readily identifiable shore objects, whether the ship is remaining securely at anchor;

실행 가능한 한 선위를 결정하여 적합한 해도에 표시하고 사정이 허락한다면 언제나 충분히 잦은 시간간격으로 고정항로표지 또는 쉽게 식별할 수 있는 육상물표의 방위를 검사하여 선박이 안전하게 정박하고 있는가를 확인한다.

(b) Ensure that and efficient look-out is maintained;

유효한 경계의 유지를 확보

(c) Ensure that inspection rounds of the vessel are made periodically;

일정한 시간간격으로 선내순시 점검을 실시하도록 한다.

(d) Observe meteorological and tidal conditions and the state of the sea;
기상, 조석류 및 해면상태를 관측한다.

(e) Notify the master and undertake all necessary measures if the vessel drags the anchor;
만약 주묘하면 선장에게 보고하고 필요한 모든 조치를 취한다.

(f) Ensure that the state of readiness of the main engines and other machinery is in accordance with the master's instruction;
주기관의 준비상태 및 다른 기계 장비들이 선장의 지시와 일치된 상태에 있는지 확인한다.

(g) If visibility deteriorates notify the master and comply with the applicable regulation for preventing collision's at sea;
만약 시계가 악화되면 선장에게 보고하고 해당 해상충돌방지규칙을 이행한다.

(h) Ensure that the vessel exhibits the appropriate lights and shapes and that appropriate sound signals are made at all times;
선박이 항상 적합한 등화와 형상물을 표시하고 적당한 음향신호를 울리고 있는가를 확인한다.

(i) Take measure to protect the environment from pollution by the ship and comply with the applicable pollution regulations.
선박에 의하여 일어날 수 있는 환경오염을 막도록 모든 조치를 다하고 해당 오염방지법규를 준수한다.

1.1.3 사용 중인 선교복무지침서(Standing Orders for Bridge Watch Officers)의 예

1. General Watch-keeping Requirements

The bridge watch officer is in charge of the vessel during his/her watch period and is responsible to me for the safety of the vessel, crew and cargo. It is his/her duty to see that all navigational laws, Rules of the Road, standing orders, etc. are complied with. He/she should be completely familiar with the vessel, her characteristics, operation, safety and fire-fighting equipment, bridge and navigation equipment, including portable equipment and signals. The watch officer is responsible for the conduct, actions, and performance of the personnel on his/her watch : instructing them in proper watch-standing duties; and ensuring that the instructions are carried out. The watch officer must never leave the bridge at any time unless relieved by me or by another licensed officer. The course and speed of the ship must not be changed without my authority, except as planned, or to avoid immediate danger in which case the watch officer must bear in mind that the engines are at his/her disposal and he/she should not hesitate to use them in case of need. In any event, any course and/or speed changes must be reported to me immediately. The watch officer must be familiar with, and periodically review, the vessel data information and maneuvering characteristics which are posted and available with these standing orders. All watch officers *must have a complete working knowledge of the Regulations for the Prevention of Collision at Sea and observe those rules at all times.* When taking any action as prescribed by the rules, be sure that the action is timely and sufficient, and that it has the desired result. Notwithstanding anything contained in these orders, all watches are to be kept in accordance with the recommendations of the IMO document *Keeping a Saft Navigational Watch.* A copy of the document is attached and it must be read by all watch-keeping officers.

2. Bridge Watch System

When under way, bridge watches shall be maintained as follows:

0000 to 0400 (1200 to 1600)________________ Second Officer

0400 to 0800 (1600 to 2000)________________ Chief Officer

0800 to 1200 (2000 to 2400)________________ Third Officer

The system may be temporarily modified, particularly on the occasion of the first watch when leaving port, to ensure that watch-keeping officers are not impaired by fatigue.

3. Calling the Master

Never fail to call me at any time if in any doubt whatsoever. Use any means at your disposal to contact me. Despite the requirements to notify me immediately in the following circumstances, the watch officer *should not hesitate to take immediate action for the safety of the ship where circumstances require.*

A. Under way

(a) If restricted visibility is encountered or expected.
(b) If the traffic conditions or the movements of other ships are causing concern.
(c) If difficulty is experienced in maintaining course.
(d) On failure to sight land or a navigation mark or to obtain soundings by the expected time.
(e) If, unexpectedly, land or a navigational mark is sighted or a change in sounding occurs.
(f) On the breakdown of engines, steering gear, or any essential navigation equipment.
(g) In heavy weather if in any doubt about the possibility of weather damage.
(h) If the shop meets any hazard to navigation.
(i) In any other emergency or situation in which you are in any doubt.

B. At Anchor

(a) If the movements of other ships are causing concern.
(b) If the vessel is dragging her anchor(s).
(c) If the state of the weather and sea is expected to deteriorate.
(d) If restricted visibility is encountered or expected.
(e) On the approach of any unidentified craft attempting to come alongside.
(f) In any other emergency or situation in which you are in any doubt.

4. Assuming the Watch

If is essential that watches are relieved punctually. The relieving watch officer must be on the bridge *at least 10minutes prior to the time he/she is to take over the watch.*

A. Assuming the Watch at Sea

Prior to assuming the watch at sea, the relieving watch officer must ensure that his/her vision is fully adjusted to the light conditions and he/she is personally satisfied regarding:

(a) Standing orders, night orders, and other special instruction relating to the navigation of the ship.

(b) The position of the ship and the depth of water.

(c) The true, magnetic, and gyro course being steered. The errors of the gyro and magnetic compasses. The direction and rate of prevailing and predicted tides, currents, and winds, and the amount of leeway and/or set being applied.

(d) The engine control mode, horsepower, or revolutions per minute (RPM) in use, and the speed made good and through the water and over the ground.

(e) The DR track for the watch with any proposed course changes. Landmarks or navigation of vessels, lights, or objects in sight, and status of bearings on them.

(g) Navigation equipment in use, its operational status, and an evaluation of information obtained from it.

(h) Conditions and hazards likely to be encountered during the watch.

(i) Publications required for reference and data.

(j) Proper setting of bridge clocks and course recorder; adjust as necessary for accuracy.

(k) Status of vessel for current and anticipated weather. Precautions taken and orders in effect.

(l) Weather information on Wind direction and force, sea and swell, and barometric tendency; indications from weather reports.

(m) Any other information necessary to ensure a thorough understanding of the vessel status and the existing situation.

B. Assuming the Watch at Anchor

Prior to assuming the watch at anchor, the relieving watch officer must ensure that his/her vision is fully adjusted to the light conditions and he/she is personally satisfied regarding :

(a) Standing orders, night orders, and any special port regulations.

(b) The position of the ship and the depth of water; the times and heights of high and low waters; the times the tidal current is due to change and whether there is sufficient swinging room.

(c) The anchor(s) in use, amount of cable out, arrangements for the slip of the cable, and status of anchor windlass.

(d) The state of the main engines and their availability for emergency use.

(e) The appropriate signals or lights are properly hoisted and exhibited.

(f) The appropriate VHF channel is available for use between the ship and port authorities.

(g) Presence and movement of vessels, lights, or objects in sight, and status of bearings on them.

(h) Navigation equipment in use, its operational status, and an evaluation of information obtained from it.

(i) Conditions and hazards likely to be encountered during the watch.

(j) Publications required for reference and data.

(k) Status of vessel for current and anticipated weather. Precautions taken and orders in effect.

(l) Weather information on wind direction and force, sea and swell, and barometric tendency; indications from weather reports.

5. Responsibility with a Pilot on Board

Attention is drawn to the following extract from IMO Resolution A 285(VIII) : *Despite the duties and obligations of a pilot, his presence on board does not relieve the watch officer from his duties and obligations for the safety of the ship. He should co-operate closely with the pilot and maintain an accurate check on the vessel's position and movement. If he is in any doubt as to the pilot's action or intentions, he should seek clarification from the pilot and if doubt still exists he should notify the maser immediately and take whatever action is necessary before the master arrives.*

6. Watch-keeping Duties at Sea

A. Proper Look-out

The watch officer must keep an, efficient look-out at all times except when essential for a short period for the necessary performance of his/her navigational duties, but he/she should satisfy him/herself that it is safe to do so and ensure that an efficient look-out is being maintained.

B. Helm Orders and Monitoring of Helmsman

Helm orders must be loud and clear and leave the helmsman in no doubt as to what he is required to do. The orders must be repeated in a similar manner by the helmsman.

C. Automatic Steering

The watch officer must supervise changes of steering mode from hand to auto and vice versa.

D. Monitoring of Position and Track

At all times when under way the vessel's progress must be monitored to ensure that the intended track of the vessel is maintained and the vessel is within that track's specified margins of margins of safety. The vessel's position must be fixed by the most reliable method available at the time. Fixing would include, but not be limited to, fixes.

E. Periodic Checks of Navigation Equipment

Operational tests of navigational equipment should be carried out as frequently and practicably as the circumstances permit. Where appropriate, the tests should also be recorded. In particular, the watch officer must make regular checks to ensure that :

(a) The helmsman or autopilot is steering the correct course.

(b) The gyro repeaters are synchronized with the master gyro.

(c) The magnetic and gyro compass errors are determined at least once a watch and after any alterations of course to a new heading. All details of compass errors must be entered in the compass error book.

(d) The autopilot is tested manually at least once a watch.

(e) The navigation and signal lights and other navigational equipment are functioning properly.

7. Watch Personnel

A. Helmsman

Watch officers are to ensure that helmsmen are sober, competent, alert, and properly dressed for duty.

B. Look-outs

Look-outs are to be posted at all times between sunset and sunrise, during reduced visibility, and when in heavy or congested traffic.

C. Stand-by

The standby is to be suitably dressed and available on call at all times.

8. Procedures in Restricted Visibility

When restricted visibility is encountered or expected, the watch officer is to comply with the relevant rules of the applicable Regulations for Preventing Collisions at Sea. In particular, the following actions must be taken:

(a) Place main engines on standby.

(b) Commence sounding appropriate fog signal ; switch on navigation lights.

(c) Post lookout (audio and visual).

(d) Advise master and engine room.

(e) Engage hand steering.

(f) Reduce to safe speed if necessary.

(g) Maintain listening watch on VHF channel 16 unless local regulations require another listening watch channel.

(h) Commence radar plot of approaching targets.

(i) Shut specified watertight doors.

9. Preparation for Departure

At least on hour prior to sailing the vessel is to be prepared for departure as follows using the departure checklist, a copy of which is attached.

(a) Check on necessary calls of personnel.

(b) Set bridge clocks. Give time check to engine room. Test engine order telegraph, telephone, and console alarms.

(c) Test whistles and general alarm.

(d) Check that rudder and propeller are clear. Test steering gear on each motor and leave on last side tested.

(e) Check master gyro compass heading and synchronize all gyro repeater headings with master gyro. Compare magnetic compass headings with gyro compass heading. Place azimuth circles on bridge wing repeaters and check lighting.

(f) Energize and check for proper operation : radars, GPS. Unlock satellite navigator and set. Check that all instrument lights are operative.

(g) Test navigation lights.

(h) Check transmission and reception on VHF channels 13 and 16. Set any third channel required to be monitored. Set all volume controls.

(i) Lay out binoculars, flashlights, and walkie-talkies. Have charts in order of use in chart table. Lay out chart instruments and necessary publications.

(j) Hoist necessary flag/light signals, and lay out others that may be required.

(k) Obtain crew muster check, and check that ship's papers and port clearance are aboard. Confirm that tanker safety checklist is completed and signed by master.

(l) Energize bow thrusters 20 minutes before sailing.

(m) Prepare the master/pilot information exchange form if using the services of a pilot.

(n) Ensure that the departure draft is obtained as soon as all cargo and trimming operations have been completed.

(o) Immediately prior to departure ensure that the steering gear compartment is manned by a licensed engineer and also by a qualified seaman.

(p) As soon as possible after the vessel has left the berth get a stowaway report from each department head. Recorded all tests and proceedings in the bell book. The bridge must not be left unattended once the gear has been tested.

10. Preparation for Arrival

At least on hour prior to arrival the vessel is to be prepared as follows using the arrival checklist, a copy of which is attached.

(a) Notify the engine room of one hour's notice of arrival. Synchronize bridge and engine room clocks. Request power on deck. Test engine order telegraph and steering gear.

(b) Check on necessary calls of personnel for arrival duties, clearing of hawse pipes, removing anchor lashings, etc.

(c) Check that all necessary navigation equipment is working and ready for use. Have necessary charts and publications ready for use. Monitor appropriate VHF distress and calling channels. If appropriate, contact pilot station, provide ETA, and confirm details of pilot embarkation.

(d) Prepare pilot embarkation gear and accommodation ladder.

(e) Hoist or prepare any necessary signal flags. Check and lay out walkie- talkies.
(f) Open bow thruster vent covers and clear controls. Energize when required.
(g) Prepare master / pilot information exchange form. Record all tests and proceedings in the bell book.

11. Handing over the Watch

The watch officer being relieved must ensure his relief is able and in condition to relieve, and thoroughly acquainted with the necessary facts before allowing himself to be relieved. The relief will indicate his assumption of the watch duties and responsibilities when he states, "I relieve you." If a maneuver or other action to avoid any hazard is taking place at the time the watch officer is to be relieved, *the relief of the officer should be deferred until such action has been completed.* 2. Deck log book.

1.2 야간지시록(Night Order Book)

1.2.1 정 의

통상 항해당직 중에는 선장복무지침서(Master's Standing Order)로써 항해당직을 수행하는데 충분하나, 특정해역(예를 들면 봄철 우리나라를 포함한 동지나해에는 짙은 안개가 자주 발생한다)이나 선박의 통행량이 폭주하는 싱가포르 해역, 도버해협 등을 항해하는 경우 선장복무지침서에 추가하여 특별한 상황을 위한 특별한 지시사항이 필요할 때가 있는데 이러한 때 선장은 선교에 당일의 당직항해사가 야간항해당직 수행시 주의해야 할 내용을 기록하게 되는데 이 기록을 야간지시록(夜間指示錄), 즉 'Night Order Book'이라고 한다. 실무상 야간지시록이라는 어구는 사용하지 않으며, 실습생들은 통상 영어로 Night Order Book으로 기억하는 것이 좋다.

1.2.2 야간지시록의 주요 내용

1. 선박 폭주수역 혹은 어선군(漁船群) 밀집지역 항해시 주의를 환기하라는 내용

① Keep a sharp lookout for crabs and fishing boats during the duty.

② Give all traffic a wide berth and maintain minimum CPA 2 miles.

③ Lookout for passing vessels and fishing vessels, especially for fishing vessels carrying no lights.

④ Ascertain the risk of collision by watching the compass bearing of an approaching ship.

⑤ Keep a sharp lookout.

2. 기상과 관련된 내용

① If wind direction or intensity changes significantly, call me.

② As soon as wind and sea subside enough to alter course to west, call me.

③ If wind or sea increase so vessel is spraying heavy water over forecastle head, call me.

④ Let me know if weather becomes thick or wind force increases.

⑤ When the visibility becomes poor, sound the whistle at regulated intervals and let me know it.

3. 기타 항해안전 조치에 관련된 내용

① Call me when Nam Hyung Je Do is @(abeam to stbd or port).

② Call me when you are in spot marked on chart 9048(BA).

③ In order to avoid immediate danger, reduced her RPM or stop the engine at once, if necessary, and then call me.

④ Call me at any time you are in doubt and at 0600 by telephone.

⑤ Observe the same orders as the previous day.

1.2.3 야간지시록 예문

① Observe the standing orders.

② Keep vessel on 261° track.

③ Keep helm in hand steering for night.

④ If wind direction or intensity changes significantly, call me.

⑤ No clock change tonight.

⑥ As soon as wind and sea subside enough to alter course to west, call me.

⑦ Keep a sharp lookout for crabs and fish boats.

⑧ Log. time and Lat. of meridian passage of 180°.

⑨ If wind or sea increase so vessel is spraying heavy water over forecastle head, call me.

1.2.4 야간지시록(Night Order Book) 예

Page

VOY. NO.		DATE	20 . . . HRS
FROM		TO(AT)	
MASTER'S SPECIAL INSTRUCTIONS			

WATCH OFFICER	TIME ACKNOWLEDGED	SIGNATURE	
			MASTER

[그림 1-1]

1.3 국제해상충돌방지규칙

1.3.1 국제해상충돌방지규칙의 개요

1. 국제해상충돌방지규칙의 연혁

1) 1863년의 영국의 충돌방지규칙

① 19세기 후반에 들어와 범선과 함께 기선의 출현과 선박교통량의 현격히 증가함

② 조선 및 항해술의 개발과 기상학의 발달로 좌초, 침몰의 해난은 감소하였으나, 해상교통량의 증가로 인한 충돌사고는 증가 : 충돌방지를 위한 해상교통통제의 필요성 제기함

③ 기존의 분산되어 있던 항해규정을 모아 새로운 해상 항해규정 제정함

④ 이 항해규정은 선박의 정의, 등화, 안개신호, 항행규칙 및 처벌 등의 내용을 포함한 해상교통규칙으로 30여 개국이 채택함으로써 국제해상교통법이 됨

2) 1889년 워싱턴 국제해사회의

① 해상에 있어서 선박의 충돌을 방지하기 위한 규칙(Regulation for Prevention Collision at Sea)에 관한 국제회의가 워싱턴에서 개최함

② '워싱턴 회의규칙'이라는 새로운 국제해상충돌방지규칙 체결함

3) 1913년 국제해상충돌방지규칙

① 1912년의 타이타닉(Titanic)호 사건으로 해상안전에 대한 경각심 부각으로 해상인명안전회의를 개최함

② 1913년 SOLAS(International Convention for the Safety of Life at Sea) 협약을 체결함

③ 제3장 (항해의 안전) 중에 충돌방지규칙에 관한 사항들 언급, 그 내용을 국내법으로 채용할 것을 약속 : 제1차 세계대전으로 실시하지 못함

4) 1929년 국제해상충돌방지규칙

제2회 국제해상인명안전회의가 개최됨으로 인해 국제해상충돌방지규칙을 협약의 부속서로 첨부하도록 함(제2차 세계대전으로 발효되지 못함)

5) 1948년 국제해상충돌방지규칙

① 2차 세계대전 이후 개최된 국제해상인명안전회의에서 채택
② 범세계적으로 채택된 최초의 통일규칙

6) 1972년 국제해상충돌방지규칙

IMO 회의에서 SOLAS 협약에서 분리시켜 새로운 충돌방지규칙 채택 : 현재에 사용되고 있는 규칙임

2. 내용 구성

① 제1장 총칙(General) : 이 법의 적용, 책임 및 용어정의 등에 대하여 기술
② 제2장 항행규정(Steering and sailing rules)
- 제1절 모든 시정상태에서의 선박의 조치
- 제2절 다른 선박이 시계내에 있을 때 선박의 조치
- 제3절 시정이 제한된 상태에서 선박의 조치

③ 제3장 등화 및 형상물(Lights and shapes)
④ 제4장 음향 및 등화신호(Sound and light signals)
⑤ 제5장 면제

1.3.2 국제해상충돌방지규칙의 내용

1. PART A - GENERAL(제1장 총칙)

A. RULE 1 Application(제1조 적용)

(a) These Rules shall apply to all vessels upon the high seas and in all waters connected therewith navigable by seagoing vessels.

이 규칙의 규정은 해양의 수면과 거기에 접속되어 항행선이 항행할 수 있는 전 수역 안에 있는 모든 선박에 적용된다.

(b) Nothing in these Rules shall interfere with the operation of special rules made by an appropriate authority for roadsteads, rivers, lakes or inland waterways connected with the high seas and navigable by seagoing vessels. Such special rules shall conform as closely as possible to these Rules.

이 규칙의 어느 규정도 해양에 접속하고 또 항행선이 항행할 수 있는 정박지, 항구, 하천, 호수 또는 내국수로에 관하여 권한이 있는 당국이 제정한 특별규칙의 시행을 방해하는 것은 아니다. 그 특별규칙은 될 수 있는대로 이 규칙의 규정에 따라야 한다.

(c) Nothing in these Rules shall interfere with the operation of any special rules made by the Government of any State with respect to additional station or signal lights or whistle signals for ships of war and vessels proceeding under convoy, or with respect to additional station or signal lights for fishing vessels engaged in fishing as a fleet. These additional station or signal lights or whistle signals shall, so far as possible, be such that they cannot be mistaken for any light or signal authorized elsewhere under these Rules.

이 규칙의 어느 규정도 복수의 함정 및 호송되어 항행하고 있는 선대에 증설하는 위치등, 신호등 혹은 기적신호에 관하여 또는 선단을 이루어 어로에 종사하고 있는 어선에 증설하는 위치등, 신호등에 관하여 각국의 정부가 제정한 특별규칙의 시행을 방해하는 것은 아니다. 이들 증설하는 위치등, 신호등 또는 기적신호는 가능한 한 이 규칙의 다른 규정에서 인정하는 등화나 신호와 오인되지 아니하는 것이라야 한다.

(d) Traffic separation schemes may be adopted by the Organization for the purpose of these Rules.

국제해사기구는 이 규칙의 적용에 있어서 해상교통분리제도를 채택할 수 있다.

(e) Whenever the Government concerned shall have determined that a vessel of special construction or purpose cannot comply fully with the provisions of any of these Rules with respect to the number, position, range or are of visibility of lights or shapes, as well as to the disposition and characteristics of sound-signalling appliances, without interfering with the special function of the vessel, such vessel shall comply with such other provisions in regard to the number, position, range or are of visibility of lights or shapes, as well as to the disposition and characteristics of sound-signalling appliances, as her Government shall have determined to be the closest possible compliance with these Rules in respect to that vessel.

특수한 구조 또는 목적을 가진 선박이 그의 특수한 기능을 방해받지 않고서는 등화 또는 형상물의 수, 위치, 가시거리 혹은 가시권 및 음향신호장치의 배치와 특성에 관한 이 규칙의 어느 규정을 완전히 지킬 수 없다고 관계국의 정부가 인정할 경우에는 당해 선박은 그 등화 또는 형상물의 수, 위치, 가시거리 혹은 가시권 및 음향신호장치의 배치와 특성에 관하여 자국의 정부가 이 규칙의 규정에 될 수 있는 대로 준한다고 인정하는 특례규정에 따라야 한다.

B. RULE 2 Responsibility(제2조 책임)

(a) Nothing in these Rules shall exonerate any vessel, or the owner, master or crew thereof, from the consequences of any neglect to comply with these Rules or of the neglect of any precaution which may be required by the ordinary practice of seamen, or by the special circumstances of the case.

이 규칙의 어느 규정도 이 규칙의 규정을 지키는 일의 태만 또는 선원의 상무로서 혹은 그 경우의 특수한 사정에 의하여 요구되는 주의의 태만으로 인하여 생긴 결과에 대하여 선박, 선박소유자, 선장 또는 해원의 책임을 면제하는 것은 아니다.

(b) In construing and complying with these Rules due regard shall be had to all dangers of navigation and collision and to any special circumstances, including the limitations of the vessels involved, which may make a departure from these Rules necessary to avoid immediate danger.

이 규칙의 규정을 해석하고 이행함에 있어서는 모든 항행의 위험과 충돌의 위험에 대하여 상당한 주의를 하여야 하며, 또 급박한 위험을 피하기 위하여 이 규칙의 규정에서 벗어나게 할 필요가 있는 모든 특수한 사정 및 관련된 선박의 성능상의 제한을 포함하여서 상당한 주의를 다하여야 한다.

C. RULE 3 General Definitions(제3조 일반적 정의)

For the purpose of these Rules, except where the context otherwise requires;
문맥상 달리로 해석하는 경우를 제외하고, 이 규칙들의 목적상

(a) The word "vessel" includes every description of water craft, including non-displacement craft, WIG craft and seaplanes, used or capable of being used as a means of transportation on water.
"선박"이라 함은 무배수량선, WIG선 및 수상비행기를 포함하여 수상의 운송수단으로 사용하거나 또는 사용할 수 있는 모든 수상선류를 말한다.

(b) The term "power-driven vessel" means any vessel propelled by machinery.
"동력선"이라 함은 기계로 추진하는 모든 선박을 말한다.

(c) The term "sailing vessel" means any vessel under sail provided that propelling machinery, if fitted, is not being used.
"범선"이라 함은 추진기계를 장치하고 있을지라도 이를 사용하고 있지 아니하는 것을 조건으로 하여 돛으로 진행하는 모든 선박을 말한다.

(d) The term "vessel engaged in fishing" means any vessel fishing with nets, lines, trawls or other fishing apparatus which restrict manoeuvrability, but does not include a vessel fishing with trolling lines or other fishing apparatus which do not restrict manoeuvrability.
"어로에 종사하고 있는 선박"이라 함은 조종성능을 제한하는 어망, 낚싯줄, 트롤망 또는 기타의 어구를 사용하여 어로하고 있는 선박을 말하며, 조정성능을 제한하지 아니하는 예승 또는 기타의 어구를 사용하여 어로하고 있는 선박을 포함하지 아니한다.

(e) The word "seaplane" includes any aircraft designed to manoeuvre on the water.
"수상비행기"라 함은 수면에서 이동할 수 있도록 설계한 항공기를 말한다.

(f) The term "vessel not under command" means a vessel which through some exceptional circumstance is unable to manoeuvre as required by these Rules and is therefore unable to keep out of the way of another vessel.
"조종이 자유롭지 못한 상태에 있는 선박"이라 함은 어떠한 예외적인 사정으로 인하여 이 규칙의 규정에 따라서 조종할 수 없고 또 그 때문에 다른 선박의 진로를

피할 수 없는 선박을 말한다.

(g) The term "vessel restricted in her ability to manoeuvre" means a vessel which from the nature of her work is restricted in her ability to manoeuvre as required by these Rules and is therefore unable to keep out of the way of another vessel.
"조종능력이 제한되어 있는 선박"이라 함은 그 선박의 작업의 성질상 이 규칙의 규정에 따라서 조종하는 능력이 제한되고 또 그 때문에 다른 선박의 진로를 피할 수 없는 선박을 말한다.
The following vessels shall be regarded as vessels restricted in their ability to manoeuvre:
다음 각 호의 선박은 조종능력이 제한되어 있는 선박으로 본다.
  (i) A vessel engaged in laying, servicing or picking up a navigation mark, submarine cable or pipeline;
  항로표지, 해저전선 또는 파이프라인의 부설, 보수 또는 인양작업에 종사하고 있는 선박
  (ii) A vessel engaged in dredging, surveying or underwater operations;
  준설, 측량 또는 수중작업에 종사하고 있는 선박
  (iii) A vessel engaged in replenishment or transferring persons, provisions or cargo while underway;
  항행 중에 보급에 종사하거나 또는 사람, 식량 혹은 화물을 옮겨 싣는데 종사하고 있는 선박
  (iv) A vessel engaged in the launching or recovery of aircraft;
  항공기의 발전 또는 회수작업에 종사하고 있는 선박
  (v) A vessel engaged in mine sweeping operations;
  소해작업에 종사하고 있는 선박
  (vi) A vessel engaged in a towing operation such as severely restricts the towing vessel and her tow in their ability to deviate from their course.
  예선과 피예인선의 침로 이탈능력을 심히 제한하는 그러한 예항작업에 종사하고 있는 선박

(h) The term "vessel constrained by her draught" means a power-driven vessel which because of her draught in relation to the available depth of water is severely restricted in her ability to deviate from the course she is following.

"흘수 때문에 제약을 받고 있는 선박"이라 함은 이용할 수 있는 수심과 자선의 흘수와의 관계로 침로에서 벗어날 수 있는 능력이 심히 제한되고 있는 동력선을 말한다.

(i) The word "underway" means that a vessel is not at anchor, or made fast to the shore, or aground.

"항행 중"이라 함은 선박이 정박하거나 육지에 계류하거나 또는 얹혀 있는 것이 아닌 상태를 말한다.

(j) The words "length" and "breadth" of a vessel mean her length overall and greatest breadth.

"선박의 길이와 폭"이라 함은 그 선박의 전장과 최대의 폭을 말한다.

(k) Vessels shall be deemed to be in sight of one another only when one can be observed visually from the other.

한 선박이 다른 선박을 시각에 의하여 관찰할 수 있는 경우에 한하여 선박이 서로 시야에 있다고 본다.

(l) The term "restricted visibility" means any condition in which visibility is restricted by fog, mist, falling snow, heavy rainstorms, sandstorms or any other similar causes.

"제한된 시정"이라 함은 안개, 엷은 안개, 강설, 폭풍우, 모래폭풍 또는 기타 유사한 원인에 의하여 시정이 제한된 상태를 말한다.

(m) The term 'Wing-In-Ground(WIG) craft' means a multimodal craft which, in its main operational mode, flies in close proximity to the surface by utilizing surface-effect action.

"WIG선"이라 함은 복합운항형태를 가지는 선박으로 주 운항 모드에서는 표면효과 작용을 이용하여 해면에 가깝게 나타나는 선박을 말한다.

2. PART B - STEERING AND SAILING RULES(제2장 항법규정)

1) SECTION I Conduct of Vessels in any Condition of Visibility

제1절 모든 시정상태에 있어서의 선박의 행동

A. RULE 4 Application(제4조 적용)

Rules in this Section apply in any condition of visibility.
이 절의 규정은 모든 시정상태에 있어서 이를 적용한다.

B. RULE 5 Look-out(제5조 경계)

Every vessel shall at all times maintain a proper look-out by sight and hearing as well as by all available means appropriate in the prevailing circumstances and conditions so as to make a full appraisal of the situation and of the risk of collision.

모든 선박은 처지 및 충돌의 위험성을 충분히 판단할 수 있도록 시각과 청각에 의할 뿐만 아니라 당시의 사정과 조건에 알맞은 모든 이용할 수 있는 수단에 의하여 언제나 적당한 경계를 유지하여야 한다.

C. RULE 6 Safe Speed(제6조 안전속력)

Every vessel shall at all times proceed at a safe speed so that she can take proper and effective action to avoid collision and be stopped within a distance appropriate to the prevailing circumstances and conditions. In determining a safe speed the following factors shall be among those taken into account:

모든 선박은 충돌을 피하기 위하여 적절하고 유효한 동작을 취할 수 있고 또 당시의 사정과 조건에 적합한 거리에서 정선할 수 있도록 언제나 안전한 속력으로 항행하여야 한다. 안전한 속력을 결정하는데 있어서는 다음과 같은 요인을 고려하여야 한다.

(a) By all vessels:(모든 선박의 경우에는)

(i) The state of visibility;(시정상태)

(ii) The traffic density including concentrations of fishing vessels or any other vessels;

어선 혹은 기타의 선박의 집중을 포함하는 교통의 밀도

(iii) The manoeuvrability of the vessel with special reference to stopping distance and turning ability in the prevailing conditions;

당시 조건하에서 특히 정지거리와 선회성능에 관계되는 선박의 조정성능

(iv) At night the presence of background light such as from shore lights or from back scatter of her own lights;

야간에 육지의 등화 또는 자선의 등화의 반사광등에서 생기는 배경광의 존재

(v) The state of wind, sea and current, and the proximity of navigational hazards;

바람, 해면 및 해류의 상태의 항행상의 위험의 근접

(vi) The draught in relation to the available depth of water.

이용할 수 있는 수심과 흘수와의 관계

(b) Additionally, by vessels with operational radar:

추가적으로, 레이더를 사용하는 선박의 경우에는

(i) The characteristics, efficiency and limitations of the radar equipment;

레이더 장치의 성질, 능력 및 한계

(ii) Any constraints imposed by the radar range scale in use;

사용하고 있는 레이더의 레인지 스케일(거리눈금)에서 생기는 제약

(iii) The effect on radar detection of the sea state, weather and other sources of interference;

해면상태, 기상 및 기타의 방해원이 레이더의 탐지에 미치는 영향

(iv) The possibility that small vessels, ice and other floating objects may not be detected by radar at an adequate range;

소형선박, 얼음 및 기타의 떠도는 물체는 충분한 레인지(거리)에서는 레이더로 탐지되지 않을 수도 있다는 가능성

(v) The number, location and movement of vessels detected by radar;

레이더로 탐지한 선박의 척수, 위치 및 움직임

(vi) The more exact assessment of the visibility that may be possible when radar is used to determine the range of vessels or other objects in the vicinity.

부근에 있는 선박 또는 기타의 물체의 거리를 측정하기 위하여 레이더를 사용할 경우에는 시정의 더욱 정확한 사정이 가능하다는 사실

D. RULE 7 Risk of Collision(제7조 충돌의 위험성)

(a) Every vessel shall use all available means appropriate to the prevailing circumstances and conditions to determine if risk of collision exists. If there is any doubt such risk shall be deemed to exist.
모든 선박은 충돌의 위험성이 있는지의 여부를 판정하기 위하여 당시의 사정과 조건에 적합한 모든 이용할 수 있는 수단을 활용하여야 한다. 의심스러운 경우에는 충돌의 위험성이 있다고 보아야 한다.

(b) Proper use shall be made of radar equipment if fitted and operational, including long-range scanning to obtain early warning of risk of collision and radar plotting or equivalent systematic observation of detected objects.
레이더 장치가 설비되어 사용할 수 있는 상태에 있다면 이를 적절하게 이용하여야 하는 바, 그 중에는 충돌의 위험성에 대한 조기의 경보를 얻기 위한 장거리의 주사와 탐지된 물체에 대한 레이더 플로팅 또는 이와 대등한 기타의 계통적인 관찰 등이 포함되어야 한다.

(c) Assumptions shall not be made on the basis of scanty information, especially scanty radar information.
불충분한 정보, 특히 불충분한 레이더 정보에 근거를 둔 억측을 하여서는 아니 된다.

(d) In determining if risk of collision exists the following considerations shall be among those taken into account:
충돌의 위험성이 있는지의 여부를 판정함에 있어서는 다음과 같은 유의사항을 고려하여야 한다.

(i) Such risk shall be deemed to exist if the compass bearing of an approaching vessel does not appreciably change;
접근하여 오는 다른 선박의 컴퍼스 방위가 감지할 수 있을 정도로 변경하지 않을 때에는 충돌의 위험성이 있다고 보아야 한다.

(ii) Such risk may sometimes exist even when an appreciable bearing change is evident, particularly when approaching a very large vessel or a tow or when approaching a vessel at close range.

접근하여 오는 다른 선박의 컴퍼스 방위에 감지할만한 변화가 나타나는 경우일지라도 특히 초대형선 혹은 피예인선열에 접근하거나 또는 다른 선박에 가까운 거리로 접근하는 때에는 간혹 충돌의 위험성이 있을 수 있다.

E. RULE 8 Action to Avoid Collision(제8조 충돌을 피하기 위한 동작)

(a) Any action taken to avoid collision shall, if the circumstances of the case admit, be positive, made in ample time and with due regard to the observance of good seamanship.

충돌을 피하기 위한 모든 동작은 그 경우의 사정이 허락하는 한 명확하여야 하고 충분히 여유 있는 시각에 적당한 선박운용술에 따르도록 유의해서 취하여야 한다.

(b) Any alteration of course and / or speed to avoid collision shall, if the circumstances of the case admit, be large enough to be readily apparent to another vessel observing visually or by radar; a succession of small alterations of course and/or speed should be avoided.

충돌을 피하기 위한 침로나 속력 또는 그 양자의 변경은 그 경우의 사정이 허락하는 한, 시각 또는 레이더에 의하여 관찰하고 있는 다른 선박이 용이하게 알아볼 수 있을 만큼 충분히 크게 하여야 한다. 조금씩 침로나 속력 또는 그 양자를 연속적으로 변경하는 것은 피하여야 한다.

(c) If there is sufficient sea room, alteration of course alone may be the most effective action to avoid a close quarters situation provided that it is made in good time, is substantial and does not result in another close-quarters situation.

충분히 넓은 조선수역이 있는 경우에는 침로의 변경만으로도 박근상태를 피하기 위한 가장 유효한 동작이 될 수 있다. 다만, 침로의 변경은 적당한 시각에 대각도로 행하여 다른 박근상태를 초래하지 않도록 하여야 한다.

(d) Action taken to avoid collision with another vessel shall be such as to result in passing at a safe distance. The effectiveness of the action shall be carefully checked until the other vessel is finally past and clear.

다른 선박과의 충돌을 피하기 위하여 취하는 동작은 안전한 거리를 두고 통과할 수 있도록 하는 것이라야 한다. 그 동작의 효과는 다른 선박이 완전히 지나가 버릴 때까지 주의깊게 확인하여야 한다.

(e) If necessary to avoid collision or allow more time to assess the situation, a vessel shall slacken her speed or take all way off by stopping or reversing her means of propulsion.

선박이 충돌을 피하기 위하여 또는 상황을 판단하기 위한 시간적 여유를 얻기 위하여 필요하다면 자선의 속력을 늦추거나 또는 추진수단을 정지 혹은 역전하여 선박의 전진타력을 완전히 멈추어야 한다.

(f) (i) A vessel which, by any of these Rules, is required not to impede the passage or safe passage of another vessel shall, when required by the circumstances of the case, take early action to allow sufficient sea room for the safe passage of the other vessel.

이 규칙의 어느 규정에 의하여 다른 선박의 통항 또는 안전한 통항을 방해하지 아니하도록 요구된 선박은 그 당시의 사정이 요구된다면, 다른 선박이 안전하게 통항할 수 있는 충분하게 넓은 수역을 허용하기 위하여 빨리 동작을 취하여야 한다.

(ii) A vessel required not to impede the passage or safe passage of another vessel is not relieved of this obligation if approaching the other vessel so as to involve risk of collision and shall, when taking action, have full regard to the action which may be required by the Rules of this part.

다른 선박의 통항 또는 안전통항을 방해하지 아니하도록 요구된 선박은 다른 선박에 접근하여 충돌의 위험성이 생긴 경우에도 이 의무를 면하지 아니한다. 그리고 만일 피항동작을 취하는 경우, 이 장의 규정에서 요구된 동작에 충분하게 주의하여야 한다.

(iii) A vessel the passage of which is not to be impeded remains fully obliged to comply with the Rules of this part when the two vessels are approaching one another so as to involve risk of collision.

통항의 방해를 받지 아니하는 선박은 다른 선박과 서로 접근하여 충돌의 위험성이 생긴 경우에 이 장의 규정을 충실하게 지킬 의무가 있다.

F. RULE 9 Narrow Channels(제9조 좁은 수로)

(a) A vessel proceeding along the course of a narrow channel or fairway shall keep as near to the outer limit of the channel or fairway which lies on her starboard side as is safe and practicable.

좁은 수로 또는 항로를 따라 진행하고 있는 선박은 안전하고 또 실행 가능한 한 자선의 우현쪽에 있는 수로 또는 항로의 외측한계에 접근하여 항행하여야 한다.

(b) A vessel of less than 20 metres in length or a sailing vessel shall not impede the passage of a vessel which can safely navigate only within a narrow channel or fairway.

길이 20미터 미만의 선박 또는 범선은 좁은 수로 또는 항로 안쪽에서만 안전하게 항행할 수 있는 선박의 통항을 방해하여서는 아니 된다.

(c) A vessel engaged in fishing shall not impede the passage of any other vessel navigating within a narrow channel or fairway.

어로에 종사하고 있는 선박은 좁은 수로 또는 항로안에서 항행하고 있는 다른 선박의 통항을 방해하여서는 아니 된다.

(d) A vessel shall not cross a narrow channel or fairway if such crossing impedes the passage of a vessel which can safely navigate only within such channel or fairway. The latter vessel may use the sound signal prescribed in Rule 34 (d) if in doubt as to the intention of the crossing vessel.

선박은 좁은 수로 또는 항로 안쪽에서만 안전하게 항행할 수 있는 다른 선박의 통항을 방해하는 경우에는 좁은 수로 또는 항로를 횡단하여서는 아니 된다. 후자의 선박은 횡단하고 있는 선박의 의도에 대하여 의심이 있는 경우에는 제34조 (d)항에 규정하는 음향신호를 이용할 수 있다.

(e) (i) In a narrow channel or fairway when overtaking can take place only if the vessel to be overtaken has to take action to permit safe passing, the vessel intending to overtake shall indicate her intention by sounding the appropriate signal prescribed in Rule 34(c)(i).

The vessel to be overtaken shall, if in agreement, sound the appropriate signal prescribed in Rule 34(c) (ii) and take steps to permit safe passing.

If in doubt she may sound the signals prescribed in Rule 34(d).

좁은 수로 또는 항로에서 추월당하게 될 선박이 다른 선박을 안전하게 통과시키기 위한 동작을 취하여야만 추월이 가능한 경우에는 추월하려는 선박은 제34조 (c)항 (1)호에 규정하는 적당한 음향신호에 의하여 그 의사를 표시하여야 한다. 추월당하게 될 선박이 이에 동의한다면 제34조 (c)항 (2)호에 규정하는 적당한 신호를 울리고 또 안전하게 통과시킬 수 있도록 조치하여야 한다. 만약 의심이 있으면 추월당하게 될 선박은 제34조 (d)항에 규정하는 신호를 울릴 수 있다.

(ii) This Rule does not relieve the overtaking vessel of her obligation under Rule 13.

이 조는 추월하는 선박에게 제13조에 규정하는 의무를 면제하는 것은 아니다.

(f) A vessel nearing a bend or an area of a narrow channel or fairway where other vessels may be obscured by an intervening obstruction shall navigate with particular alertness and caution and shall sound the appropriate signal prescribed in Rule 34(e).

좁은 수로 또는 항로에서 개재하는 장애물 때문에 다른 선박을 볼 수 없는 굴곡부 또는 수역에 접근하고 있는 선박은 특별한 경계와 주의를 다하여 항행하여야 하며 또 제34조(e)항에 규정하는 적당한 신호를 울려야 한다.

(g) Any vessel shall, if the circumstances of the case admit, avoid anchoring in a narrow channel.

모든 선박은 그 경우의 사정이 허락하는 한 좁은 수로에 정박하지 않도록 하여야 한다.

G. RULE 10 Traffic Separation Schemes(제10조 해상교통분리제도)

(a) This Rule applies to traffic separation schemes adopted by the Organization and does not relieve any vessel of her obligation under any other rule.

이 조의 규정은 국제해사기구가 채택한 해상교통분리제도에 이를 적용하며 또한 어떠한 선박에게도 다른 규정에 의거한 의무를 면제시키지 아니한다.

(b) A vessel using a traffic separation scheme shall:

해상교통분리제도를 이용하는 선박은

(i) Proceed in the appropriate traffic lane in the general direction of traffic flow for that lane;

적당한 교통로 안에서 그 통로에 대한 교통류의 일반적인 방향으로 진행하여야 한다.

(ii) So far as practicable keep clear of a traffic separation line or separation zone;

가능한 한 교통분리선 또는 분리대에 접근하지 않도록 하여야 한다.

(iii) Normally join or leave a traffic lane at the termination of the lane, but when joining or leaving from the side shall do so at as small an angle to the general direction of traffic flow as practicable.

일반적으로 항행로로 들어가거나 나올 때에는 항행로의 말단에서 출입하여야 한다. 그러나 항행로 측면에서 출입할 때에는 실행 가능한 한 교통흐름의 일반적인 방향에 대하여 소각도로 출입하여야 한다.

(c) A vessel shall, so far as practicable, avoid crossing traffic lanes, but if obliged to do so shall cross on her heading as nearly as practicable at right angles to the general direction of traffic flow.

선박은 가능한 한 항행로를 횡단하지 않도록 하여야 한다. 그러나 부득이 횡단하여야 할 경우에는 교통흐름의 일반적인 방향에 대하여 선수방향이 가능한 한 직각으로 횡단하여야 한다.

(d) (i) "A vessel shall not use an inshore traffic lane within the adjacent traffic separation scheme. However, vessel of less than 20metres in length, sailing vessels and vessels engaged in fishing may use the inshore traffic zone."

"인근 교통분리수역 안의 적당한 항행로를 안전하게 사용할 수 있는 선박은 연안항로대를 이용할 수 없다. 다만 길이 20미터 미만의 선박, 범선 및 어로 중인 선박은 연안항로대를 이용할 수 있다."

(ii) Notwithstanding subparagraph (d)(i), a vessel may use an inshore traffic zone when en rute to or from a port, offshore installation or structure, pilot station or any other place situated within the inshore traffic zone, or to avoid immediate danger.

제4항 (1)의 규정에도 불구하고, 어떠한 선박이라도 항, 해양구조물, 도선지점 또는 연안항로대 안에 위치한 기타 장소에 출입항하기 위하여 또한 긴박한 위험을 피하기 위하여 연안항로대를 이용할 수 있다.

(e) A vessel, other than a crossing vessel, shall not normally enter a separation zone or cross a separation line except:
항행로에 들어가거나 나오기 위한 선박 또는 횡단하는 선박이 아니면, 다음 각 호의 경우를 제외하고는 통상적으로 분리대로 들어가거나 분리선을 횡단하여서는 아니 된다.

(i) In cases of emergency to avoid immediate danger;
급박한 위험을 피하기 위한 경우

(ii) To engage in fishing within a separation zone.
분리대 안에서 어로에 종사하기 위한 경우

(f) A vessel navigating in areas near the terminations of traffic separation schemes shall do so with particular caution.
해상교통분리수역의 끝 부근의 수역을 항행하는 선박은 특별한 주의를 기울여 항행하여야 한다.

(g) A vessel shall so far as practicable avoid anchoring in a traffic separation scheme or in areas near its terminations.
선박은 가능한 한 교통분리수역 안 또는 그 끝 부근의 수역에서 정박하지 않아야 한다.

(h) A vessel not using a traffic separation scheme shall avoid it by as wide a margin as is practicable.
교통분리수역을 이용하지 아니하는 선박은 실행 가능한 한 충분한 여지를 두고 교통분리수역에서부터 떨어져야 한다.

(i) A vessel engaged in fishing shall not impede the passage of any vessel following a traffic lane.
어로에 종사하고 있는 선박은 항행로를 따라서 항행하고 있는 다른 선박의 통항을 방해하여서는 아니 된다.

(j) A vessel of less than 20 metres in length or a sailing vessel shall not impede

the safe passage of a power-driven vessel following a traffic lane.
길이 20미터 미만의 선박 또는 범선은 항행로를 따라서 항행하고 있는 동력선의 안전한 통항을 방해하여서는 아니 된다.

(k) A vessel restricted in her ability to maneuver when engaged in an operation for the maintenance of safety of navigation in a traffic separation scheme is exempted from complying with this Rule to the extent necessary to carry out the operation.
해상교통분리수역에서 항해안전설비의 유지작업에 종사하고 있는 경우에 있어서 조종성능이 제한된 선박은 그 작업을 이행하기 위하여 필요한 범위까지 이 규정의 적용을 면제받는다.

(l) A vessel restricted in her ability to maneuver when engaged in an operation for the laying, servicing, or picking up of a submarine cable, within a traffic separation scheme, is exempted from complying with this Rule to the extent necessary to carry out the operation.
해상교통분리수역에서 해저전선을 설치하거나, 관리하거나 또는 인양하는 작업에 종사하는 경우에 있어서 조종성능이 제한된 선박은 그 작업을 이행하기 위하여 필요한 범위까지 이 규정의 적용을 면제받는다.

2) SECTION II Conduct of Vessels in Sight of one Another
제2절 서로 시야내에 있는 선박의 행동

A. RULE 11 Application(제11조 적용)

Rules in this Section apply to vessels in sight of one another.
이 절의 규정은 서로 시야내에 있는 선박에 이를 적용한다.

B. RULE 12 Sailing Vessels(제12조 범선)

(a) When two sailing vessels are approaching one another, so as to involve risk of collision, one of them shall keep out of the way of the other as follows:
2척의 범선이 서로 접근할 때에 충돌의 위험성을 내포하는 경우에는 그 중의 한 범선이 다른 범선의 진로를 다음과 같이 피하여야 한다.

(i) When each has the wind on a different side, the vessel which has the wind on the port side shall keep out of the way of the other;
각 선박이 다른 현에 바람을 받고 있는 경우에는 좌현에 바람을 받고 있는 선박이 다른 선박의 진로를 피하여야 한다.

(ii) When both have the wind on the same side, the vessel which is to windward shall keep out of the way of the vessel which is to leeward;
양 선박이 같은 현에 바람을 받고 있는 경우에는 바람 불어오는 쪽의 선박이 바람 불어가는 쪽의 선박의 진로를 피하여야 한다.

(iii) If a vessel with the wind on the port side sees a vessel to windward and cannot determine with certainty whether the other vessel has the wind on the port or on the starboard side, she shall keep out of the way of the other.
좌현에 바람을 받고 있는 선박은 바람 불어오는 쪽에 다른 선박을 보고 그 선박이 바람을 좌현에 받고 있는지, 또는 우현에 받고 있는지 확인할 수 없는 경우에는 그 다른 선박의 진로를 피하여야 한다.

(b) For the purposes of this Rule the windward side shall be deemed to be the side opposite to that on which the mainsail is carried or, in the case of a square-rigged vessel, the side opposite to that on which the largest fore-and-aft sail is carried.
이 조의 적용에 있어서는 주범을 달고 있는 쪽의 반대측 또는 횡범선의 경우에는 최대의 종범을 달고 있는 쪽의 반대측을 바람 불어오는 쪽이라고 본다.

C. RULE 13 Overtaking(제13조 추월)

(a) Notwithstanding anything contained in the Rules of this Section any vessel overtaking any other shall keep out of the way of the vessel being overtaken.
다른 선박을 추월하고 있는 모든 선박은 이 절의 다른 규정에 불구하고 추월당하고 있는 선박의 진로를 피하여야 한다.

(b) A vessel shall be deemed to be overtaking when coming up with another vessel from a direction more than 22.5degrees abaft her beam, that is, in such a position with reference to the vessel she is overtaking, that at night she would

be able to see only the stern light of that vessel but neither of her sidelights.
선박은 다른 선박의 정횡 22.5도를 넘는 후방으로부터, 즉 추월당하고 있는 선박에 관하여는 야간에 그 선박의 선미등만 볼 수 있고 어느 쪽 현등도 볼 수 없는 위치로부터 다른 선박을 앞지르고 있는 경우에는 추월하는 선박이라고 본다.

(c) When a vessel is in any doubt as to whether she is overtaking another she shall assume that this is the case and act accordingly.
선박은 자선이 다른 선박을 추월하고 있는지의 여부에 대하여 의심이 있는 경우에는 다른 선박을 추월하고 있다고 가정하고 그것에 합당한 동작을 취하여야 한다.

(d) Any subsequent alteration of the bearing between the two vessels shall not make the overtaking vessel a crossing vessel within the meaning of these Rules or relieve her of the duty of keeping clear of the overtaken vessel until she is finally past and clear.
그 후에 두 선박 사이의 방위에 어떠한 변경이 있더라도 추월하는 선박은 이 규칙에서 의미하는 바는 횡단하는 선박으로 되는 것이 아니고, 또 추월당하는 선박을 완전히 앞질러 지나갈 때까지 그 선박을 피하여야 할 의무를 면제받는 것이 아니다.

D. RULE 14 Head-on Situation(제14조 마주치는 상태)

(a) When two power-driven vessels are meeting on reciprocal or nearly reciprocal courses so as to involve risk of collision each shall alter her course to starboard so that each shall pass on the port side of the other.
2척의 동력선이 상반되는 침로 또는 거의 상반되는 침로에서 마주치는 경우에 충돌의 위험성을 내포하는 때에는 각 선박이 서로 다른 선박의 좌현쪽을 통과할 수 있도록 각기 침로를 우현쪽으로 변경하여야 한다.

(b) Such a situation shall be deemed to exist when a vessel sees the other ahead or nearly ahead and by night she could see the masthead lights of the other in a line or nearly in a line and/or both sidelights and by day she observes the corresponding aspect of the other vessel.
선박이 다른 선박을 선수방향 또는 거의 선수방향에 보는 경우로서, 야간에는 다른 선박의 마스트 정부등들을 일직선 또는 거의 일직선에 볼 수 있는 때 또는 양측의 현등을 볼 수 있는 때 및 그 양자의 상태를 동시에 볼 수 있는 때, 그리고 주간에는 다른 선박을 이와 마찬가지 방향에서 관찰하는 때에는 마주치는 상태에 있다고 보아야 한다.

(c) When a vessel is in any doubt as to whether such a situation exists she shall assume that it does exist and act accordingly.
선박은 마주치는 상태에 있는지의 여부에 대하여 의심이 있는 경우에는 마주치는 상태에 있다고 가정하고 그것에 합당한 동작을 취하여야 한다.

E. RULE 15 Crossing Situation(제15조 횡단상태)

When two power-driven vessels are crossing so as to involve risk of collision, the vessel which has the other on her own starboard side shall keep out of the way and shall, if the circumstances of the case admit, avoid crossing ahead of the other vessel.

2척의 동력선이 서로 진로를 횡단할 경우에 충돌의 위험성을 내포하는 때에는 다른 선박을 자선의 우현쪽에 두고 있는 선박이 다른 선박의 진로를 피하여야 하며 또 그 경우의 사정이 허락하는 한 다른 선박의 선수를 횡단하여서는 아니 된다.

F. RULE 16 Action by Give-Way Vessel(제16조 피항선의 동작)

Every vessel which is directed to keep out of the way of another vessel shall, so far as possible, take early and substantial action to keep well clear.

다른 선박의 진로를 피하여야 할 모든 선박은 다른 선박으로부터 충분히 떨어지도록 가능한 한 조기에 큰 동작을 취하여야 한다.

G. RULE 17 Action by Stand-on Vessel(제17조 유지선의 동작)

(a) (i) Where one of two vessels is to keep out of the way the other shall keep her course and speed.
두척의 선박 중에서 한 척이 다른 선박의 진로를 피하여야 할 경우에 그 다른 선박은 그의 침로와 속력을 유지하여야 한다.

(ii) The latter vessel may however take action to avoid collision by her manoeuvre alone, as soon as it becomes apparent to her that the vessel required to keep out of the way is not taking appropriate action in compliance with these Rules.

그럼에도 불구하고 후자의 선박은 진로를 피하여야 할 선박이 이 규칙의 규정에 따른 적당한 동작을 취하지 않고 있음이 판명되면 곧 자선의 조종만으로 충돌을 피하기 위한 동작을 취할 수 있다.

(b) When, from any cause, the vessel required to keep her course and speed finds herself so close that collision cannot be avoided by the action of the give-way vessel alone, she shall take such action as will best aid to avoid collision.

침로와 속력을 유지하여야 할 선박은 어떠한 원인으로 다른 선박에 아주 가까이 접근하였기 때문에 진로를 피하여야 할 선박의 동작만으로는 충돌을 피할 수 없다고 인정할 때에는 충돌을 피하는데 도움이 될 최선의 동작을 취하여야 한다.

(c) A power-driven vessel which takes action in a crossing situation in accordance with sub-paragraph(a)(ii) of this Rule to avoid collision with another power-driven vessel shall, if the circumstances of the case admit, not alter course to port for a vessel on her own port side.

동력선은 횡단하는 상태에 있어서 다른 동력선과의 충돌을 피하기 위하여 이 조의 (a)항(2)호의 규정에 따라서 동작을 취하는 때에는 그 경우의 사정이 허락한다면 자선의 좌현쪽에 있는 선박을 향하여 좌현쪽으로 침로를 변경하여서는 아니 된다.

(d) This Rule does not relieve the give-way vessel of her obligation to keep out of the way.

이 조의 규정은 피항선으로 하여금 진로를 피하여야 할 의무를 면제시키는 것은 아니다.

H. RULE 18 Responsibilities between Vessels(제18조 선박 상호간의 책임)

Except where Rules 9, 10 and 13 otherwise require:

제9조, 제10조 및 제13조에서 이와 달리 규정하고 있는 경우를 제외하고는

(a) A power-driven vessel underway shall keep out of the way of:

항행 중인 동력선은 다음 각 호의 선박의 진로를 피하여야 한다.

(i) A vessel not under command;

조종이 자유롭지 못한 상태에 있는 선박

(ii) A vessel restricted in her ability to manoeuvre;

조종능력이 제한되어 있는 선박

(iii) A vessel engaged in fishing;

어로에 종사하고 있는 선박

(iv) A sailing vessel.(범선)

(b) A sailing vessel underway shall keep out of the way of:

항행 중인 범선은 다음 각 호의 선박의 진로를 피하여야 한다.

(i) A vessel not under command

조종이 자유롭지 못한 상태에 있는 선박

(ii) A vessel restricted in her ability to manoeuvre;

조종능력이 제한되어 있는 선박

(iii) A vessel engaged in fishing;

어로에 종사하고 있는 선박

(c) A vessel engaged in fishing when underway shall, so far as possible, keep out of the way of:

어로에 종사하고 있는 항행 중의 선박은 가능한 한 다음 각 호의 선박의 진로를 피하여야 한다.

(i) A vessel not under command;

조종이 자유롭지 못한 상태에 있는 선박

(ii) A vessel restricted in her ability to manoeuvre.

조종능력이 제한되어 있는 선박

(d) (i) Any vessel other than a vessel not under command or a vessel restricted in her ability to manoeuvre shall, if the circumstances of the case admit, avoid impeding the safe passage of a vessel constrained by her draught, exhibiting the signals in Rule 28.

조종이 자유롭지 못한 상태에 있는 선박 또는 조종능력이 제한되어 있는 선박이 아닌 그 밖의 선박은 그 경우의 사정이 허락한다면 제28조에 규정하는 신호를 표시하고 있는 흘수 때문에 제약을 받고 있는 선박의 안전한 통항을 방해하지 않도록 하여야 한다.

(ii) A vessel constrained by her draught shall navigate with particular caution having full regard to her special condition.

흘수 때문에 제약을 받고 있는 선박은 그의 특수한 조건을 충분히 고려하여 특히 신중하게 항행하여야 한다.

(e) A seaplane on the water shall, in general, keep well clear of all vessels and avoid impeding their navigation. In circumstances, however, where risk of collision exists, she shall comply with the Rules of this Part.

수면에 있는 수상비행기는 원칙적으로 모든 선박으로부터 충분히 떨어져서 그들 선박의 항행을 방해하지 않도록 하여야 한다. 그러나 충돌의 위험성이 있는 상황에서는 이 장의 규정을 지켜야 한다.

(f) (i) A WIG craft, when taking off, landing and in flight near the surface, keep well clear of all other vessels avoid impeding their navigation :

WIG선은 수면에서 이수하거나 착수할 때 그리고 비행할 때 다른 선박으로부터 충분히 떨어져서 그들 선박의 항행을 방해하지 않도록 하여야 한다.

(ii) A WIG craft operating on the water surface shall comply with the Rules of these Part as a power-driven vessel.

수면 위에서 항진 중인 WIG선은 동력선으로서 이 장의 규정을 지켜야 한다.

3) SECTION III Conduct of Vessels in Restricted Visibility

제3절 제한된 시정에 있어서의 선박의 행동

A. RULE 19 Conduct of Vessels in Restricted Visibility

제19조 제한된 시정에 있어서의 선박의 행동

(a) This Rule applies to vessels not in sight of one another when navigating in or near an area of restricted visibility.

이 조의 규정은 시정이 제한된 수역 또는 그 부근에서 항행하고 있는 선박이 서로 시야 내에 있지 아니하는 경우에 적용한다.

(b) Every vessel shall proceed at a safe speed adapted to the prevailing circumstances and conditions of restricted visibility. A power-driven vessel shall have her engines ready for immediate manoeuvre.

모든 선박은 시정이 제한된 그 당시의 사정과 조건에 적응하는 안전한 속력으로 항행하여야 한다. 동력선은 기관을 당장 사용할 수 있도록 준비해 두어야 한다.

(c) Every vessel shall have due regard to the prevailing circumstances and conditions of restricted visibility when complying with the Rules of Section I of this Part.

모든 선박은 이 장의 제1절의 규정에 따라서 행동하고 있는 때에는 시정이 제한된 그 당시의 사정과 조건을 충분히 고려하여야 한다.

(d) A vessel which detects by radar alone the presence of another vessel shall determine if a close-quarters situation is developing and/or risk of collision exists. If so, she shall take avoiding action in ample time, provided that when such action consists of an alteration of course, so far as possible the following shall be avoided:

다른 선박이 존재함을 레이더만으로 탐지한 선박은 박근상태로 되고 있는지 혹은 충돌의 위험성이 있는지 또는 그 양자의 상태에 있는지의 여부를 판단하여야 한다. 그러한 상태에 있다면 충분히 여유 있는 시기에 피항동작을 취하여야 한다. 다만, 그 동작이 침로의 변경만으로 이루어질 경우에는 가능한 한, 다음 각 호의 동작을 피하여야 한다.

(i) An alteration of course to port for a vessel forward of the beam, other than for a vessel being overtaken;

추월당하고 있는 선박에 대한 경우를 제외하고, 자선의 정횡 전방에 있는 선박에 대하여 좌현쪽으로 침로를 변경하는 일

(ii) An alteration of course towards a vessel abeam or abaft the beam.

정횡 또는 정횡 후방에 있는 선박의 방향으로 침로를 변경하는 일

(e) Except where it has been determined that a risk of collision does not exist, every vessel which hears apparently forward of her beam the fog signal of another vessel, or which cannot avoid a close quarters situation with another vessel forward of her beam, shall reduce her speed to the minimum at which she can be kept on her course. She shall if necessary take all her way off and in any event navigate with extreme caution until danger of collision is over.

충돌의 위험성이 없다고 판단한 경우를 제외하고, 자선의 정횡 전방이라고 생각되

는 곳에서 다른 선박의 무중신호를 듣는 선박 또는 자선의 정횡 전방에 있는 다른 선박과의 박근상태를 피할 수 없는 모든 선박은 자선의 침로를 유지할 수 있는 최소의 속력으로 감속하여야 한다. 당해 선박은 필요하다면 자선의 진행을 완전히 멈추어야 하고 또 어떠한 경우에도 충돌의 위험성이 사라질 때까지 극히 조심하여 항행하여야 한다.

3. PART C－LIGHTS AND SHAPES(제3장 등화와 형상물)

A. RULE 20 Application(제20조 적용)

(a) Rules in this Part shall be complied with in all weathers.

이 장의 규정은 모든 천후에 있어서 이를 지켜야 한다.

(b) The Rules concerning lights shall be complied with from sunset to sunrise, and during such times no other lights shall be exhibited, except such lights as cannot be mistaken for the lights specified in these Rules or do not impair their visibility or distinctive character, or interfere with keeping of a proper look-out.

등화에 관한 규정은 일몰시로부터 일출시까지의 사이에 이를 지켜야 하며, 또 이 시간 내에는 이 규칙에 규정하는 등화로 오인되는 등화, 이 규칙에 규정하는 등화의 가시도 혹은 그 특성의 식별을 감손하는 등화 또는 적당한 경계의 이행을 방해하는 등화를 제외한 그 밖의 등화를 표시하여야 한다.

(c) The lights prescribed by these Rules shall, if carried, also be exhibited from sunrise to sunset in restricted visibility and may be exhibited in all other circumstances when it is deemed necessary.

이 규칙에 규정하는 등화는 이를 설치하고 있다면 일출시로부터 일몰시까지의 사이에도 제한된 시정에 있어서는 역시 표시하여야 하며 또 필요하다고 인정되는 그 밖의 모든 경우에 표시할 수 있다.

(d) The Rules concerning shapes shall be complied with by day.

형상물에 관한 규정은 주간에 이를 지켜야 한다.

(e) The lights and shapes specified in these Rules shall comply with the provisions of Annex Ⅰ to these Regulations.

이 규칙에 규정하는 등화와 형상물은 이 규칙은 부속서 Ⅰ의 규정에 따라야 한다.

B. RULE 21 Definitions(제21조 정의)

(a) "Masthead light" means a white light placed over the fore and aft centreline of the vessel showing an unbroken light over an arc of the horizon of 225 degrees and so fixed as to show the light from right ahead to 22.5 degrees abaft the beam on either side of the vessel.
"마스트 정부등"이라 함은 선수미의 중심선상에 설치되어 온전한 불빛의 225도에 이르는 수평의 호를 비추고, 또 그 불빛이 정선수 방향으로부터 각 현의 정횡후방 22.5도까지 비출 수 있도록 장치한 백색등을 말한다.

(b) "Sidelights" means a green light on the starboard side and a red light on the port side each showing an unbroken light over an arc of the horizon of 112.5 degrees and so fixed as to show the light from right ahead to 22.5 degrees abaft the beam on its respective side. In a vessel of less than 20 metres in length the sidelights may be combined in one lantern carried on the fore and aft centreline of the vessel.
"현등"이라 함은 온전한 불빛이 각기 112.5도에 이르는 수평의 호를 비추고, 또 그 불빛이 정선수 방향으로부터 각기 좌우현의 정횡후 22.5도까지 비출 수 있도록 장치한 우현쪽의 녹색등과 좌현쪽의 홍색등을 말한다. 길이 20미터 미만의 선박에 있어서는 현등을 선수미의 중심선상에 달고 있는 하나의 랜턴속에 결합할 수 있다.

(c) "Sternlight" means a white light placed as nearly as practicable at the stern showing an unbroken light over an arc of the horizon of 135 degrees and so fixed as to show the light 67.5 degree from right aft on each side of the vessel.
"선미등"이라 함은 실행 가능한 한 선미에 가까이 설치되어 온전한 불빛이 135도에 이르는 수평의 호를 비추고, 또 그 불빛이 정선미 방향으로부터 각 현의 67.5도까지 비출 수 있도록 장치한 백색등을 말한다.

(d) "Towing light" means a yellow light having the same characteristics as the "Sternlight" defined in paragraph (c) of this Rule.
"예선등"이라 함은 이 조의 (3)항에 정의한 선미등과 동일한 특성을 가진 황색등을 말한다.

(e) "All round light" means a light showing an unbroken light over an arc of the horizon of 360 degrees.

"전주등"이라 함은 온전한 불빛이 360도에 이르는 수평의 호를 비추는 등화를 말한다.

(f) "Flashing light" means a light flashing at regular intervals at a frequency of 120 flashes or more per minute.

"섬광등"이라 함은 매분에 120회 이상의 주기로 규칙적인 간격을 두고 섬광을 발하는 등화를 말한다.

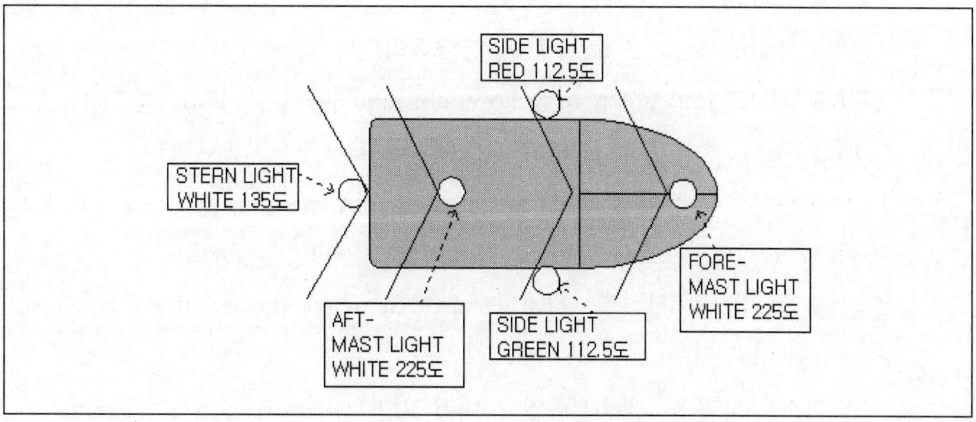

[그림 1-2] 각종 등화의 표시

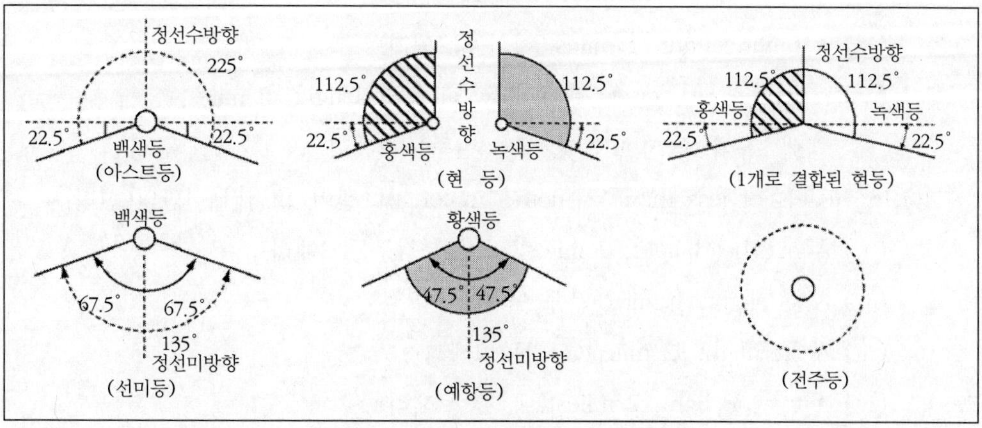

[그림 1-3] 각종 등화의 발광범위

C. RULE 22 Visibility of Lights(제22조 등화의 가시거리)

The lights prescribed in these Rules shall have an intensity as specified in Section 8 of Annex I to these Regulations so as to be visible at the following minimum ranges:

이 규칙에 규정하는 등화는 다음 각 항에 규정하는 최소거리에서 볼 수 있도록 부속서 I 의 제8항에 규정하는 광도를 가져야 한다.

(a) In vessels of 50 metres or more in length : (길이 50미터 이상의 선박)

(i) A masthead light, 6miles;(마스트 정부등 6해리)

(ii) A sidelight, 3 miles;(현등 3해리)

(iii) A sternlight, 3 miles;(선미등 3해리)

(iv) A towing light, 3 miles;(예선등 3해리)

(v) A white, red, green or yellow all-round light, 3 miles.
백색, 홍색, 녹색 또는 황색의 전주등 3해리

(b) In vessels of 12 metres or more in length but less than 50 metres in length:
길이 12미터 이상으로서 길이 50미터 미만의 선박

(i) A masthead light, 5 miles; except that where the length of the vessel in less than 20 metres, 3 miles;
마스트 정부등 5해리, 다만 20미터 미만의 선박에 있어서는 3해리

(ii) A sidelight, 2 miles;(현등 2해리)

(iii) A sternlight, 2 miles;(선미등 2해리)

(iv) A towing light, 2 miles;(예선등 2해리)

(v) A white, red, green or yellow all-round light, 2 miles.
백색, 홍색, 녹색 또는 황색의 전주등 2해리

(c) In vessels of less than 12 metres in length:(길이 12미터 미만의 선박)

(i) A masthead light, 2 miles;(마스트 정부등 2해리)

(ii) A sidelight, 1 mile;(현등 1해리)

(iii) A sternlight, 2 miles;(선미등 2해리)

(iv) A towing light, 2 miles;(예선등 2해리)

(v) A white, red, green or yellow all-round light, 2 miles.
백색, 홍색, 녹색 또는 황색의 전주등 2해리

〈표 1-1〉 등화의 광도 규정

	50미터 이상 선박	12~50미터 미만 선박	12미터 미만 선박
마스트등	6마일	5마일	2마일
선미등	3〃	2〃	1〃
현등	3〃	2〃	2〃
예선등	3〃	2〃	2〃
전주/섬광등	3〃	2〃	2〃

D. RULE 23 Power-Driven Vessels Underway(제23조 항행 중인 동력선)

(a) A power-driven vessel underway shall exhibit:

항행 중인 동력선은 다음 각 호에 규정하는 등화를 표시하여야 한다.

(i) A masthead light forward;(전부에 마스트 정부등 1개)

(ii) A second masthead light abaft of and higher than the forward one; except that a vessel of less than 50 metres in length shall not be obliged to exhibit such light but may do so;

전부 마스트 정부등의 후방으로 그보다 높은 위치에 제2마스트 정부등 1개, 다만, 길이 50미터 미만의 선박은 이를 표시하여야 할 의무는 없으나 표시하여도 무방하다.

(iii) Sidelights;(현등)

(iv) A sternlight.(선미등)

(b) An air-cushion vessel when operating in the non-displacement mode shall, in addition to the lights prescribed in paragraph (a) of this Rule, exhibit an all-round flashing yellow light.

에어 쿠션선이 무배수량의 상태로 가동하고 있을 때에는 이 조의 (a)항에 규정하는 등화에 부가하여 전주위에서 볼 수 있는 황색의 섬광등 1개를 표시해야 한다.

(c) A WIG craft only when taking off, landing and in flight near the surface shall, in addition to the lights prescribed in paragraph (a) of this Rule, exhibit a high-intensity all-round flashing red light.

위그선이 해수면 부근에서 비행, 이륙 · 착륙시에는 이 조의 (a)항에 규정하는 등화에 부가하여 고광도의 전주위에서 볼 수 있는 홍색의 섬광등 1개를 표시해야 한다.

(d) (i) A power-driven vessel of less than 12 metres in length may, in lieu of the lights prescribed in paragraph (a) of this Rule, exhibit an all-round white light and sidelights;

길이 12미터 미만의 동력선은 이 조의 (a)항에 규정하는 등화에 대신하여 1개의 백색 전주등과 현등을 표시할 수 있다.

(ii) A power-driven vessel of less than 7 metres in length whose maximum speed does not exceed 7 knots may, in lieu of the lights prescribed in paragraph (a) of this Rule, exhibit an all-round white light and shall, if practicable, also exhibit sidelights;

길이가 7미터 미만이고 최대속력이 7노트를 넘지 아니하는 동력선은 이 조의 (a)항에 규정하는 등화에 대신하여 백색의 전주등 1개를 표시할 수 있다. 그 선박은 실행이 가능하다면 현등 역시 표시하여야 한다.

(iii) The masthead light or all-round white light on a power-driven vessel of less than 12 metres in length may be displaced from the fore and aft centre line of the vessel if centreline fitting is not practicable, provided that the sidelights are combined in one lantern which shall be carried on the fore and aft centre line of the vessel or located as nearly as practicable in the same fore and aft line as the masthead light or the all-round white light.

길이 12미터 미만인 동력선의 마스트등과 백색 전주등은 선체의 중심선상에 설치되어야 한다. 만일 중심선상에 설치할 수가 없는 경우 마스트등 또는 백색 전주등과 거의 같은 위치에 설치하거나 선체 중심선상에서 설치할 수 있는 복합 랜턴을 달아야 한다.

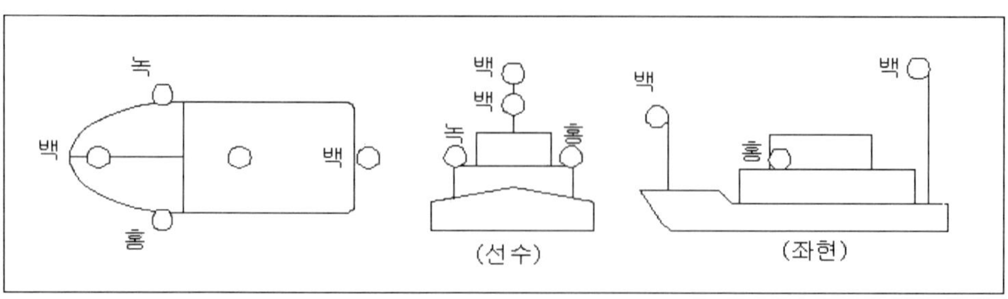

[그림 1-4] 길이 50미터 이상의 항행 중인 동력선

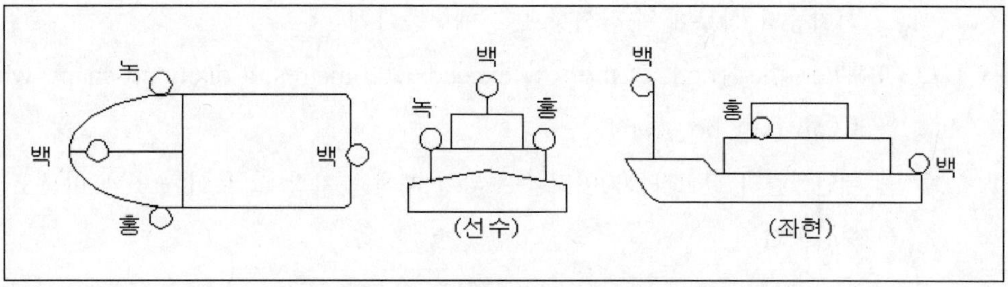

[그림 1-5] 길이 50미터 미만의 항행 중인 동력선

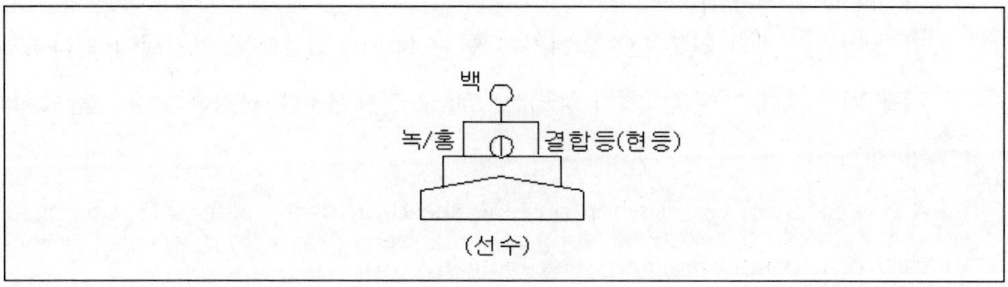

[그림 1-6] 길이 20미터 미만의 항행 중인 모든 선박(범선도 포함)

E. RULE 24 Towing and Pushing(제24조 선박의 끌기와 밀기)

(a) A power-driven vessel when towing shall exhibit:
동력선은 다른 선박을 끌고 있는 경우에는 다음 각 호에 규정하는 등화 또는 형상물을 표시하여야 한다.

(i) Instead of the light prescribed in Rule 23 (a) (i), two masthead lights forward in a vertical line. When the length of the tow, measuring from the stern of the towing vessel to the after end of the tow exceeds 200 metres, three such lights in a vertical line;
제23조 (a)항 (1)호에 규정하는 등화의 대신으로 전부에 수직선상으로 마스트 정부등 2개, 예선의 선미로부터 끌려가고 있는 선박의 후단까지 측정한 예인선열의 길이가 200미터를 초과할 경우에는 수직선상에 마스트 정부등 3개

(ii) Sidelights;(현등)

(iii) A sternlight;(선미등)

(iv) A towing light in a vertical line above the sternlight;

선미등의 상부에 수직선상으로 예선등 1개

(v) When the length of the tow exceeds 200 metres, a diamond shape where it can best be seen.

예인선열의 길이가 200미터를 초과할 때에는 가장 잘 보이는 곳에 마름모꼴의 형상물 1개

(b) When a pushing vessel and a vessel being pushed ahead are rigidly connected in a composite unit they shall be regarded as a power-driven vessel and exhibit the lights prescribed in Rule 23.

밀고 있는 선박과 앞쪽으로 밀려가고 있는 선박이 견고하게 연결되어 하나의 복합체를 이룬 때에는 이를 1척의 동력선으로 보고 제23조에 규정하는 등화를 표시하여야 한다.

(c) A power-driven vessel when pushing ahead or towing alongside, except in the case of a composite unit, shall exhibit:

동력선은 다른 선박을 앞으로 밀고 있거나 또는 뱃전에 대고 끌고 있는 때에는 복합체를 이룬 경우를 제외하고는 다음 각 호에 규정하는 등화를 표시해야 한다.

(i) Instead of the light prescribed in Rule 23 (a) (i), two masthead lights forward in a vertical line;

제23조 (a)항 (1)호에 규정하는 등화의 대신으로 전부에 수직선상으로 마스트 정부등 2개

(ii) Sidelights;(현등)

(iii) A sternlight.(선미등)

(d) A power-driven vessel to which paragraphs (a) and (c) of this Rule apply shall also comply with Rule 23 (a) (ii).

이 조의 (1)항과 (3)항의 규정을 적용받는 동력선은 제23조 (a)항 (2)호의 규정 역시 지켜야 한다.

(e) A vessel or object being towed, other than those mentioned in paragraph (g) of this Rule, shall exhibit:

이 규칙의 (g)항에서 언급된 규정 이외에 끌려가고 있는 선박 또는 물체는 다음 각 호의 등화 또는 형상물을 표시하여야 한다.

(i) Sidelights;(현등)

(ii) A sternlight;(선미등)

(iii) When the length of the tow exceeds 200 metres, a diamond shape where it can best be seen.

예인선열의 길이가 200미터를 초과할 경우에는 가장 잘 보이는 곳에 마름모꼴의 형상물 1개

(f) Provided that any number of vessels being towed alongside or pushed in a group shall be lighted as one vessel,

2척 이상의 선박이 일단이 되어 뱃전에 붙어서 끌려가고 있거나 밀려가고 있는 경우에는 하나의 선박처럼 불이 켜져야 하는 조건으로,

(i) A vessel being pushed ahead, not being part of a composite unit, shall exhibit at the forward end, sidelights;

앞쪽으로 밀려가고 있는 선박은 복합체의 일부가 아닌 경우에는 그의 전단에 양 현등을 표시하여야 한다.

(ii) A vessel being towed alongside shall exhibit a sternlight and at the forward end, sidelights.

뱃전에 붙어서 끌려가고 있는 선박은 선미등 1개를 표시하여야 하고 또 그이 전단에 양 현등을 표시하여야 한다.

(g) An inconspicuous, partly submerged vessel or object, or combination of such vessels or objects being towed, shall exhibit:

현저하게 부분적으로 잠겨있는 선박 또는 물체 혹은 이러한 피예선 혹는 피예물체의 결합체는 다음을 표시하여야 한다.

(i) If it is less than 25 metres in breadth, one all-round white light at or near the forward end and one at or near the after end except that dracones need not exhibit a light at or near the forward end;

폭이 25미터 미만인 경우, 앞쪽 끝 또는 앞쪽 및 뒤쪽 끝 또는 뒤쪽 부근에 백색 전주등 1개, 다만 드래콘은 앞쪽 끝 또는 앞쪽에 등을 표시할 필요가 없다.

(ii) If it is 25 metres or more in breadth, two additional all-round white lights at or near the extremities of its breadth;

폭이 25미터 이상인 경우, 양쪽 폭의 끝에 또는 근처에 2개의 추가 백색 전주등 1개

(iii) If it exceeds 100 metres in length, additional all-round white lights between the lights prescribed in sub-paragraphs (i) and (ii) so that the distance between the lights shall not exceed 100 metres;

길이가 100미터를 초과하는 경우, 등화사이의 거리가 100미터를 초과하지 않도록 (1), (2)항에서 규정한 등화사이에 추가 백색 전주등 1개

(iv) A diamond shape at or near the aftermost extremity of the last vessel or object being towed and if the length of the tow exceeds 200 metres an additional diamond shape where it can best be seen and located as far forward as is practicable.

최후단 선박 또는 예항되고 있는 물체의 맨 끝에 또는 가까이에 마름모꼴 형상 그리고 예인되고 있는 선박의 길이가 200미터를 초과하는 경우, 가장 잘 보이는 곳에 위치한 추가 마름모꼴의 형상물은 잘 보이는 것이어야 한다.

(h) Where from any sufficient cause it is impracticable for a vessel or object being towed to exhibit the lights prescribed in paragraph (e) of this Rule, all possible measures shall be taken to light the vessel or object towed or at least to indicate the presence of the unlighted vessel or object.

어떠한 충분한 원인으로 인하여 끌려가고 있는 선박 또는 물체에 이 조의 (e)항에 규정하는 등화를 표시하기가 실행 불가능한 경우에는 끌려가고 있는 선박 혹은 물체를 조명하기 위하여, 또는 적어도 등화를 표시하지 아니하는 그 선박 혹은 물체의 존재를 가리키기 위하여 모든 가능한 조치를 취하여야 한다.

(i) Where from any sufficient cause it is impracticable for a vessel not normally engaged in towing operations to display the lights prescribed in paragraph (a) or (c) of this Rule, such vessel shall not be required to exhibit those lights when engaged in towing another vessel in distress or otherwise in need of assistance. All possible measures shall be taken to indicate the nature of the relationship between the towing vessel and the vessel being towed as authorized by Rule

36, in particular by illuminating the towline.

어떠한 충분한 원인으로 인하여 이 조의 (a)항 또는 (c)항에서 규정하는 등화를 비정상적으로 예항작업에 종사하는 선박이 표시하기가 불가능한 경우, 즉 조난이나 도움을 필요로 하는 다른 선박을 예항하는데 종사할 때에는 그 선박은 예항등화를 표시할 필요가 없다. 제36조에 의하여 인정된 예항되는 선박과 예선 사이의 연결상태 특히 예선 밧줄의 비춤을 표시하기 위하여 가능한 모든 조치를 취하여야 한다.

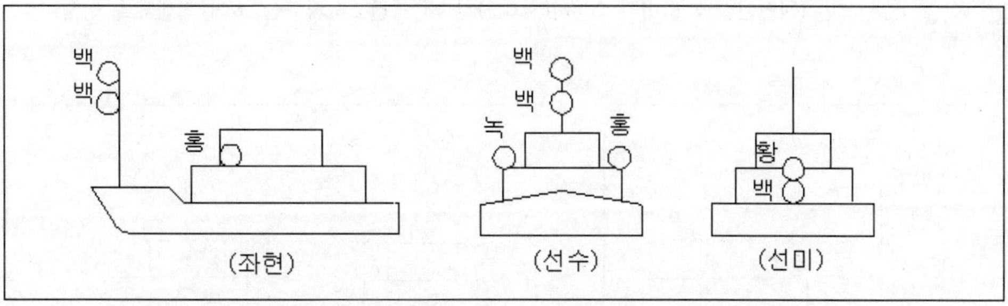

[그림 1-7] 길이 50미터 미만의 동력선이 타선을 선미에 끌고 있는 경우
(예인선열의 길이가 200미터를 초과하지 않는 경우)

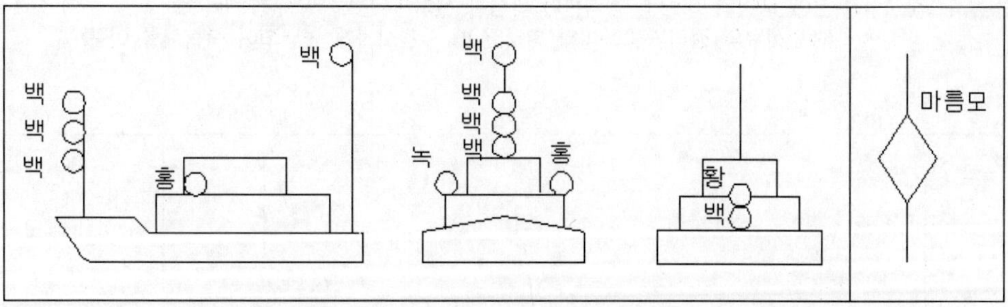

[그림 1-8] 길이 50미터 이상의 동력선이 타선을 선미에 끌고 있는 경우(a)
(예인선열의 길이가 200미터를 초과하는 경우)

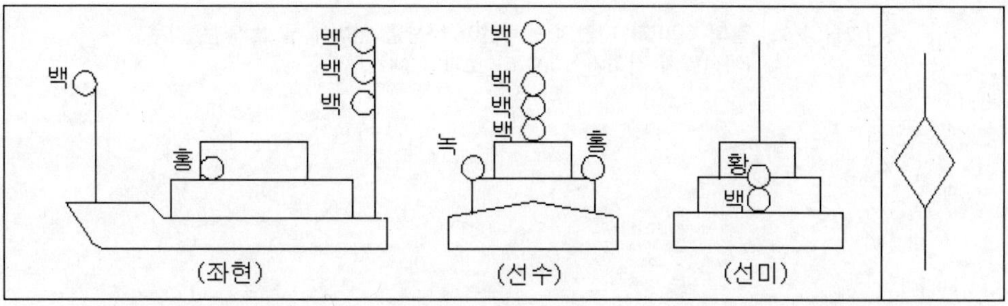

[그림 1-9] 길이 50미터 이상의 동력선이 타선을 선미에 끌고 있는 경우(b)
(예인선열의 길이가 200미터를 초과하는 경우)

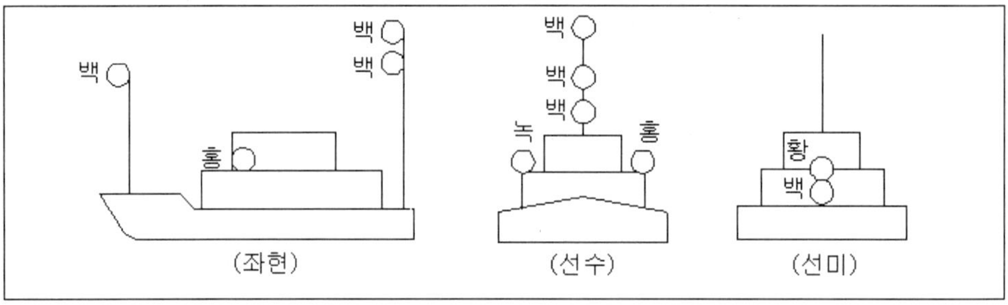

[그림 1-10] 길이 50미터 이상의 동력선이 타선을 선미에 끌고 있는 경우(a)
(예인선열의 길이가 200미터 이하일 때 - 견고하게 연결되어 복합체를 이룸)

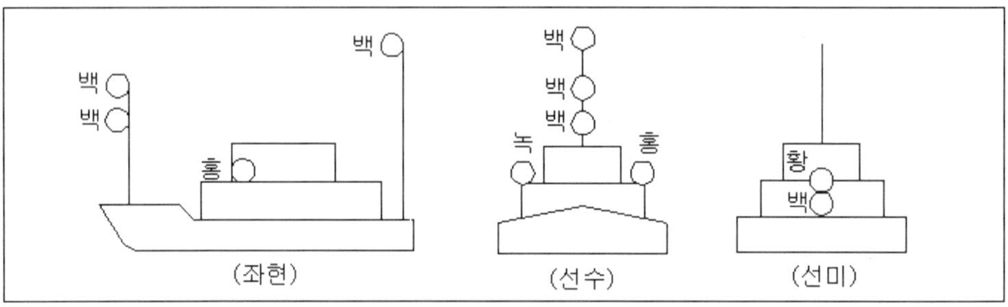

[그림 1-11] 길이 50미터 이상의 동력선이 타선을 선미에 끌고 있는 경우(b)
(예인선열의 길이가 200미터 이하일 때 - 견고하게 연결되어 복합체를 이룸)

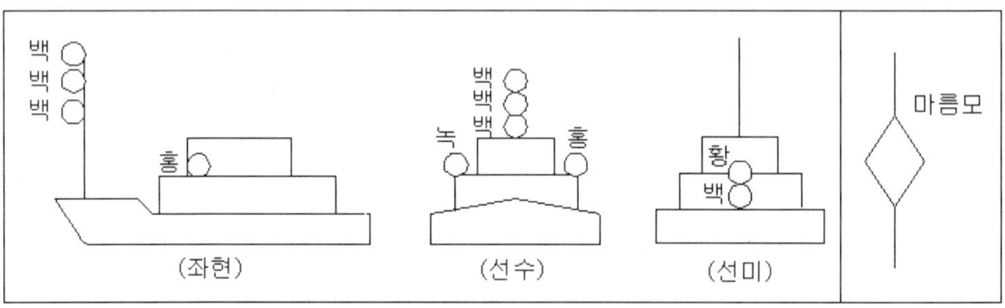

[그림 1-12] 길이 50미터 미만의 동력선이 타선을 선미에 끌고 있는 경우
(예선열의 길이가 200미터 초과일 때)

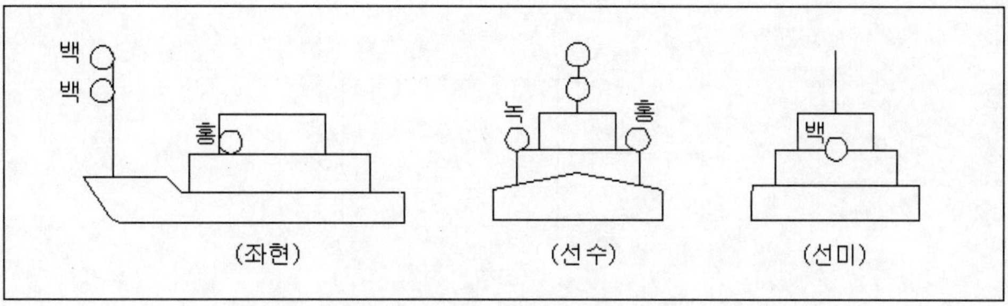

[그림 1-13] 타선을 앞으로 밀거나 뱃전에 끌고 있는 동력선

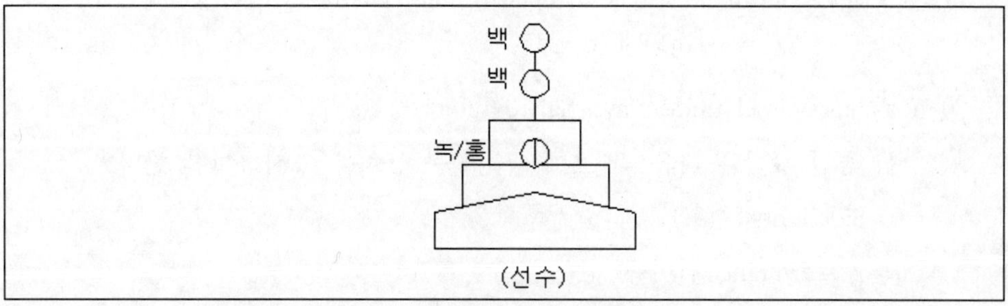

[그림 1-14] 타선을 끌고, 밀고 혹은 뱃전에 대고 끌고 있는 길이 20미터 미만의 선박

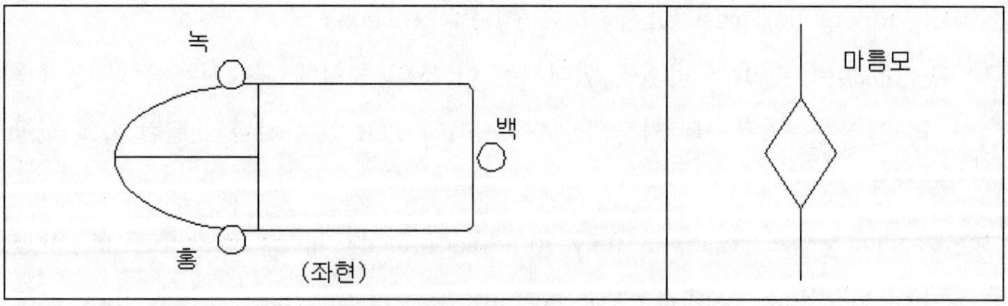

[그림 1-15] 끌려가고 있는 선박 또는 물체

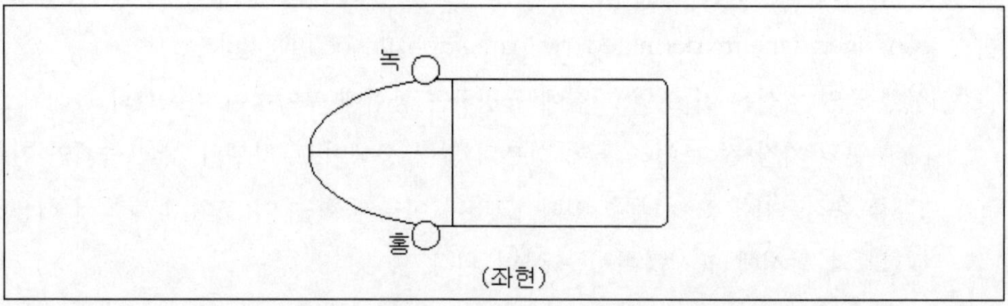

[그림 1-16] 앞으로 밀려가고 있는 선박의 전단에 현등(복합체가 아님)

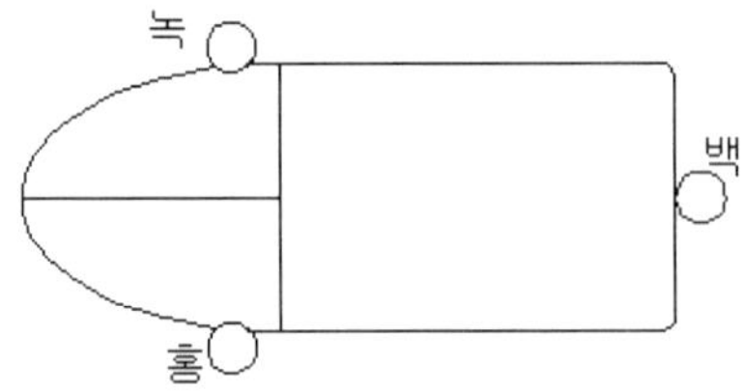

[그림 1-17] 뱃전에 붙어 끌려가고 있는 선박

F. RULE 25 Sailing Vessels Underway and Vessels Under Oars
(제25조 항행 중인 범선과 노도선)

(a) A sailing vessel underway shall exhibit:
항행 중인 선박은 다음 각 호에 규정하는 등화를 표시하여야 한다.
(i) Sidelights;(현등)
(ii) A sternlight.(선미등)

(b) In a sailing vessel of less than 12 metres in length the lights prescribed in paragraph (a) of this Rule may be combined in one lantern carried at or near the top of the mast where it can best be seen.
길이 12미터 미만의 범선에 있어서는 이 조의 (a)항에 규정하는 등화를 마스트의 꼭대기 또는 그 부근의 가장 잘 보이는 곳에 달고 있는 하나의 랜턴 속에 결합시킬 수 있다.

(c) A sailing vessel underway may, in addition to the lights prescribed in paragraph (a) of this Rule, exhibit at or near the top of the mast, where they can best be seen, two all-round lights in a vertical line, the upper being red and the lower green, but these lights shall not be exhibited in conjunction with the combined lantern permitted by paragraph (b) of this Rule.
항행 중인 범선은 이 조의 (a)항에 규정하는 등화에 부가하여 마스트의 꼭대기 또는 그 부근의 가장 잘 보이는 곳에 상부의 것이 홍색이고, 하부의 것이 녹색인 전주등 2개를 수직선상에 표시할 수 있다. 그러나 이들 등화는 이 조의 (b)항에서 허용하는 결합등과 동시에 표시하여서는 아니 된다.

(d) (i) A sailing vessel of less than 7 metres in length shall, if practicable, exhibit the lights prescribed in paragraph (a) or (b) of this Rule, but if she does not, she shall have ready at hand an electric torch or lighted lantern showing a white light which shall be exhibited in sufficient time to prevent collision.

길이 7미터 미만의 범선은 실행가능하다면 이 조의 (a)항 또는 (b)항에 규정하는 등화를 표시하여야 한다. 그러나 이를 표시하지 아니할 경우에는 백색의 불빛을 비추는 휴대용 전등 또는 점화한 랜턴 1개를 즉시 사용할 수 있도록 준비하여 놓고 충돌을 방지하기 위하여 충분한 시간을 두고 표시하여야 한다.

(ii) A vessel under oars may exhibit the lights prescribed in this Rule for sailing vessels, but if she does not, she shall have ready at hand an electric torch or lighted lantern showing a white light which shall be exhibited in sufficient time to prevent collision.

노도선은 이 조문에 규정하는 범선의 등화를 표시할 수 있다. 그러나 이를 표시하지 아니할 경우에는 백색의 불빛을 비추는 휴대용 전등 또는 점화한 랜턴 1개를 즉시 사용할 수 있도록 준비하여 놓고 충돌을 방지하기 위하여 충분한 시간을 두고 표시하여야 한다.

(e) A vessel proceeding under sail when also being propelled by machinery shall exhibit forward where it can best be seen a conical shape, apex downwards.

돛을 사용하는 선박이 기관을 병용하여 진행하고 있는 경우에는 전부의 가장 잘 보이는 곳에 원추형 형상물 1개를 그의 정점을 아래로 두고 표시하여야 한다.

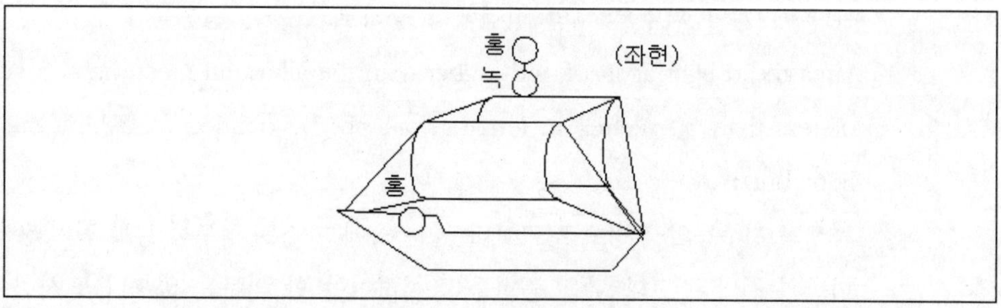

[그림 1-18] 항행 중인 범선

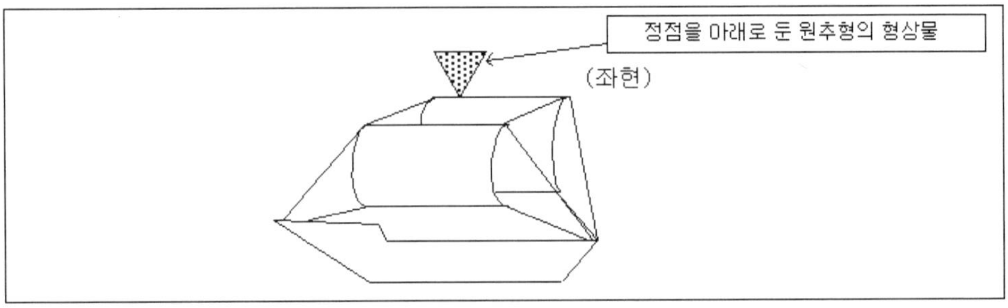

[그림 1-19] 범선이 돛과 기관을 동시 사용하여 진행할 때(야간-동력선 등화)

G. RULE 26 Fishing Vessels(제26조 어선)

(a) A vessel engaged in fishing, whether underway or at anchor, shall exhibit only the lights and shapes prescribed in this Rule.
어로에 종사하고 있는 선박은 항행 중이든 혹은 정박 중이이라도 이 조에 규정하는 등화 또는 형상물만을 표시하여야 한다.

(b) A vessel when engaged in trawling, by which is meant the dragging through the water of age net or other apparatus used as a fishing appliance, shall exhibit:
형망 또는 기타의 어구를 수중에서 끄는 것을 의미하는 트롤망 어로에 종사하고 있는 선박은 다음 각 호에 규정하는 등화 또는 형상물을 표시하여야 한다.

(i) Two all-round lights in a vertical line, the upper being green and the lower white of a shape consisting of two cones with their apexes together in a vertical line one above the other;
수직선상에 상부의 것이 녹색이고 하부의 것이 백색인 전주등 2개 또는 수직선상에 2개의 원추를 그의 정점에서 상하로 결합한 형상물 1개

(ii) A masthead light abaft of and higher than the all-round green light; a vessel of less than 50 metres in length shall not be obliged to exhibit such a light but may do so;
녹색의 전주등의 후방으로 그보다 높은 곳에 마스트 정부등 1개, 길이 50미터 미만의 선박은 그와 같은 등화를 표시하여야 할 의무는 없으나 표시하여도 무방하다.

(iii) When making way through the water, in addition to the lights prescribed in this paragraph, sidelights and a sternlight.

대수속력이 있는 경우에는 이 항에 규정하는 등화에 부가하여 양 현등과 선미등 1개

(c) A vessel engaged in fishing, other than trawling, shall exhibit:
트롤망 어로 이외의 어로에 종사하고 있는 선박은 다음 각 호에 규정하는 등화 또는 형상물을 표시하여야 한다.

(i) Two all-round lights in a vertical line, the upper being red and the lower white, or a shape consisting of two cones with apexes together in a vertical line one above the other;
수직선상에 상부의 것이 홍색이고 하부의 것이 백색인 전주등 2개 또는 수직선상에 2개의 원추를 그의 정점에서 상하로 결합한 형상물 1개

(ii) When there is outlying gear extending more than 150 metres horizontally from the vessel, an all-round white light or a cone apex upwards in the direction of the gear;
수평거리로 150미터가 넘는 어구를 선외로 내고 있는 경우에는 어구를 내고 있는 방향으로 백색의 전주등 1개 또는 그의 정점을 위로 한 원추형의 형상물 1개

(iii) When making way through the water, in addition to the lights prescribed in this paragraph, sidelights and a sternlight.
대수속력이 있는 경우에는 이 항에 규정하는 등화에 부가하여 양 현등과 선미등 1개

(d) The additional signals described in Annex II to these Regulations apply to a vessel engaged in fishing in close proximity to other vessels engaged in fishing.
이 규칙의 부속서 II에 규정된 부가신호는 어로에 종사하고 있는 다른 선박의 가까운 부근에서 어로 중인 선박에 적용하여야 한다.

(e) A vessel when not engaged in fishing shall not exhibit the lights or shapes prescribed in this Rule, but only those prescribed for a vessel of her length.
어로에 종사하고 있지 아니하는 선박은 이 조에 규정하는 등화 또는 형상물을 표시하여서는 아니 되고, 그 선박과 동일한 길이의 선박에 대하여 규정하는 등화 또는 형상물만을 표시하여야 한다.

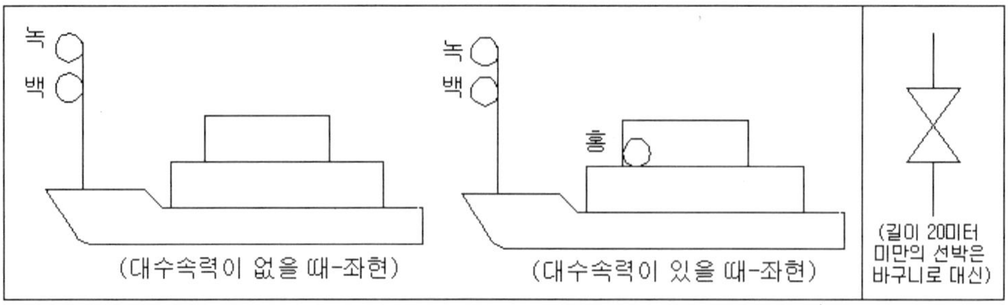

[그림 1-20] 트롤망 어로에 종사하고 있는 선박

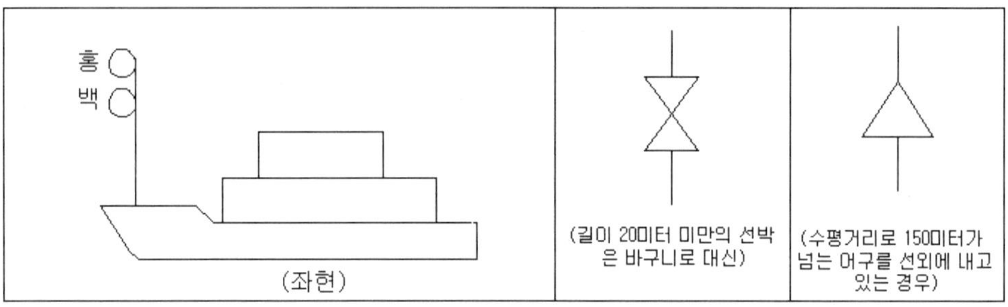

[그림 1-21] 트롤망 어로 이외의 어로에 종사하고 있는 선박

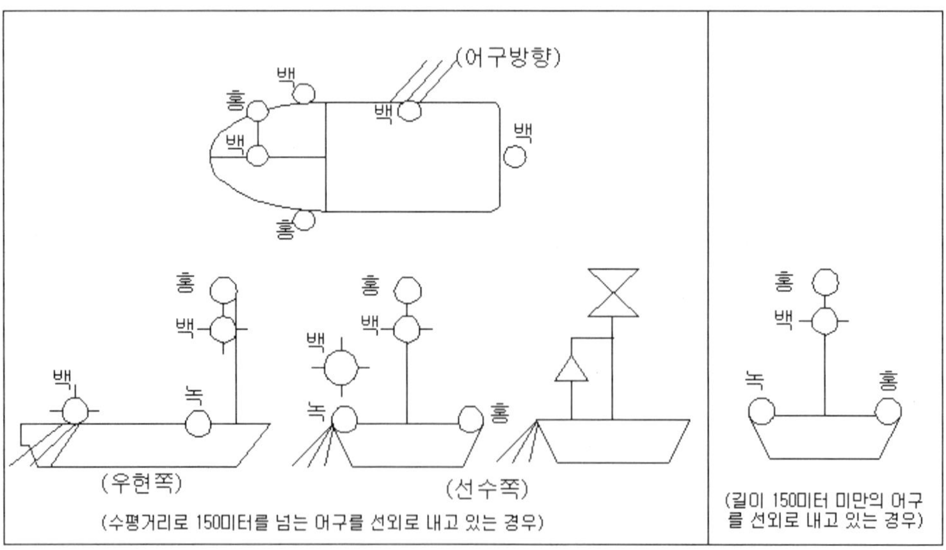

[그림 1-22] 트롤 이외의 어선이 어구를 선외로 내고 있을 때의 등화

H. RULE 27 Vessels not under Command or Restricted in Their Ability to Manoeuvre
(제27조 조종불능선 또는 조종능력이 제한되어 있는 선박)

(a) A vessel not under command shall exhibit:
조종불능선박은 다음 각 호에 규정하는 등화 또는 형상물을 표시하여야 한다:

(i) Two all-round red lights in a vertical line where they can best be seen;
가장 잘 보이는 곳에 수직선상으로 홍색의 전주등 2개

(ii) Two balls or similar shapes in a vertical line where they can best be seen;
가장 잘 보이는 곳에 수직선상으로 구형 또는 그와 유사한 형상물 2개

(iii) When making way through the water, in addition to the lights prescribed in this paragraph, sidelights and a sternlight.
대수속력이 있는 경우에는 이 항에 규정하는 등화에 부가하여 양 현등과 선미등 1개

(b) A vessel restricted in her ability to manoeuvre, except a vessel engaged in mine sweeping operations, shall exhibit:
조종능력이 제한되어 있는 선박은 소해작업에 종사하고 있는 선박을 제외하고는 다음 각 호에 규정하는 등화 또는 형상물을 표시하여야 한다.

(i) Three all-round lights in a vertical line where they can best be seen. The highest and lowest of these lights shall be red and the middle light shall be white;
가장 잘 보이는 곳에 수직선상으로 전주등 3개, 이들 등화 중에서 상단과 하단의 등화는 홍색, 중간의 등화는 백색이라야 한다.

(ii) Three shapes in a vertical line where they can best be seen. The highest and lowest of these shapes shall be balls and the middle one a diamond;
가장 잘 보이는 곳에 수직선상으로 형상물 3개, 이들 형상물 가운데서 상단과 하단의 형상물은 구형, 중간의 형상물은 마름모꼴이라야 한다.

(iii) When making way through the water, masthead lights, sidelights and a sternlight, in addition to the lights prescribed in subparagraph (i);
대수속력이 있는 경우에는 (1)호에 규정하는 등화에 부가하여 마스트 정부등, 양 현등 및 선미등 1개

(iv) When at anchor, in addition to the lights or shapes prescribed in subparagraph (i) and (ii), the light, lights or shape prescribed in Rule 30.

정박 중에는 (1)호와 (2)호에 규정하는 등화 또는 형상물에 부가하여 제30조에 규정하는 등화 또는 형상물

(c) A vessel engaged in a towing operation such as renders her unable to deviate from her course shall, in addition to the lights or shapes prescribed in subparagraph (b) (i) and (ii) of this Rule, exhibit the lights or shape prescribed in Rule 24(a).

선박으로 하여금 그의 침로를 벗어날 수 없게 만드는 그러한 예항작업에 종사하고 있는 선박은 이 조의 (b)항 (1)호와 (2)호에 규정하는 등화 또는 형상물에 부가하여 제24조 (a)항에 규정하는 등화 또는 형상물을 표시하여야 한다.

(d) A vessel engaged in dredging or underwater operations, when restricted in her ability to manoeuvre, shall exhibit the lights and shapes prescribed in paragraph (b) of this Rule and shall in addition, when an obstruction exists, exhibit:

준설 또는 수중작업에 종사하고 있는 선박은 조종능력이 제한되어 있는 경우에는 이 조의 (b)항에 규정하는 등화 또는 형상물을 표시하여야 하며 또 장애물이 있는 경우에는 이에 부가하여 다음 각 호에 규정하는 등화 또는 형상물을 표시하여야 한다.

(i) Two all-round red lights or two balls in a vertical line to indicate the side on which the obstruction exists;

장애물이 있는 쪽을 가리키기 위하여 수직선상에 홍색의 전주등 2개 또는 구형의 형상물 2개

(ii) Two all-round green lights or two diamonds in a vertical line to indicate the side on which another vessel may pass;

다른 선박이 통과할 수 있는 쪽을 가리키기 위하여 수직선상에 녹색의 전주등 2개 또는 마름모꼴의 형상물 2개

(iii) When making way through the water, in addition to the lights prescribed in this paragraph, masthead lights, sidelights and a sterniight;

대수속력이 있는 경우에는 이 항에 규정하는 등화에 부가하여 마스트 정부등, 양 현등 및 선미등 1개

(iv) A vessel to which this paragraph applies when at anchor shall exhibit the lights or shapes prescribed in sub-paragraphs (i) and (ii) instead of the lights or shape prescribed in Rule 30.

이 항의 적용을 받는 선박은 정박 중인 때에는 제30조에 규정하는 등화 또는 형상물에 대신하여 (1)호와 (2)호에 규정하는 등화 또는 형상물을 표시하여야 한다.

(e) Whenever the size of a vessel engaged in diving operations makes it impracticable to exhibit the shapes prescribed in paragraph (d) of this Rule, a rigid replica of the International Code flag "A" not less than 1 metre in height shall be exhibited. Measures shall be taken to ensure all-round visibility.

잠수작업에 종사하고 있는 선박은 그의 크기로 인하여 이 조의 (d)항에 규정하는 형상물을 표시할 수 없는 경우에는 높이가 1미터 이상인 국제신호서의 "A"기의 모사판을 표시해야 한다. 이 모사판은 이를 전주위에서 볼 수 있도록 조치해야 한다.

(f) A vessel engaged in mine sweeping operations shall, in addition to the lights prescribed for a power-driven vessel in Rule 23, exhibit three all-round green lights or three balls. One of these lights or shapes shall be exhibited at or near the foremast head and one at each end of the fore yard. These lights or shapes indicate that it is dangerous for another vessel to approach closer than 1,000 metres astern or 500 metres on either side of the mine sweeper.

종사하고 있는 선박은 제23조에 규정하는 동력선에 대한 등화에 부가하여 녹색의 전주등 3개 또는 구형의 형상물 3개를 표시하여야 한다. 이들 등화 또는 형상물 중에서 1개는 전부 마스트의 꼭대기 또는 그 부근에 표시하고 또 전부 돛 가름대의 양단에 1개씩 표시하여야 한다. 이들 등화 또는 형상물은 다른 선박이 소해선의 선미로 1,000미터 또는 그의 양현측으로 500미터 이내에 접근하면 위험하다는 것을 가리킨다.

(g) Vessels of less than 7 metres in length shall not be required to exhibit the lights prescribed in this Rule.

길이 7미터 미만의 선박은 이 조에 규정하는 등화를 표시하는 것을 요하지 아니한다.

(h) The signals prescribed in this Rule are not signals of vessels in distress and requiring assistance. Such signals are contained in Annex IV to these Regulations.

이 조에 규정하는 신호는 조난하여 구원을 요구하는 선박의 신호가 아니다. 그러한 신호는 이 규칙의 부속서 Ⅳ에 들어 있다.

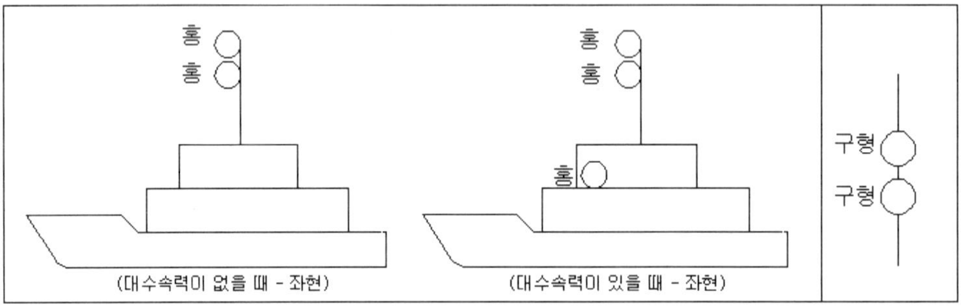

[그림 1-23] 조종불능선박(vessel not under command)

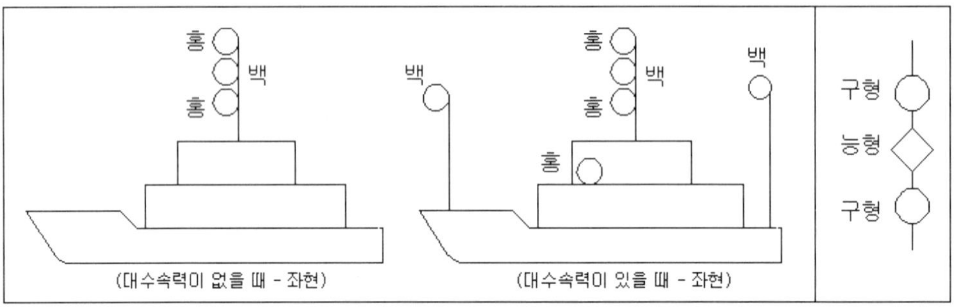

[그림 1-24] 조종능력 제한선박(vessel restricted in their ability to manoeuvre)

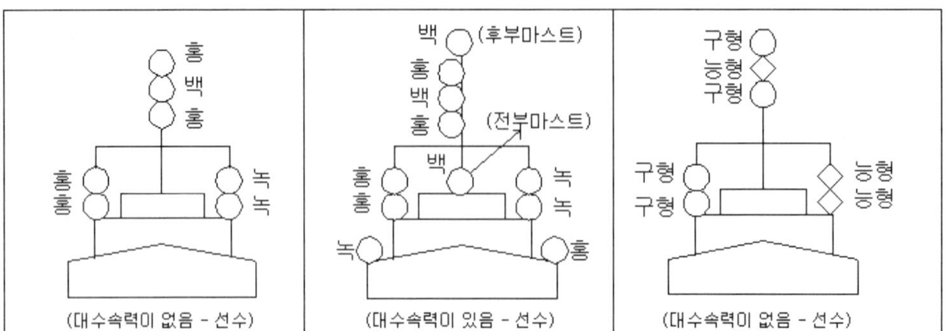

[그림 1-25] 준설 또는 수중 작업선

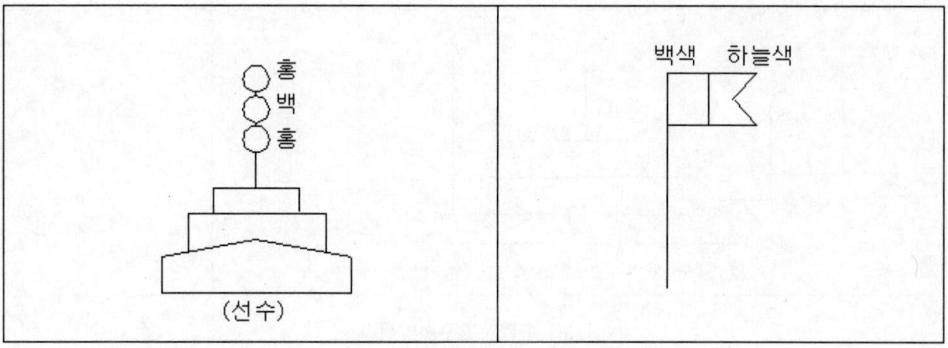

[그림 1-26] 잠수작업 종사선

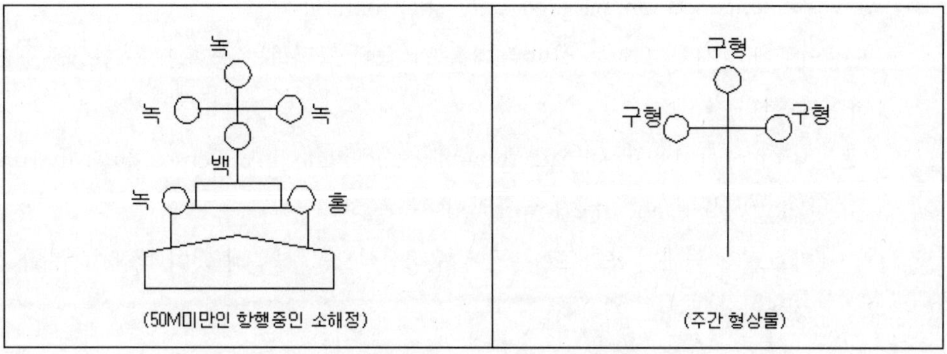

[그림 1-27] 기뢰제거작업 중인 선박 소해작업에 종사하는 선박

I. RULE 28 Vessels Constrained by Their Draught
(제28조 흘수 때문에 제약을 받고 있는 선박)

A vessel constrained by her draught may, in addition to the lights prescribed for power-driven vessels in Rule 23, exhibit where they can best be seen three all-round red lights in a vertical line, or a cylinder.

흘수 때문에 제약을 받고 있는 선박은 제23조에 규정하는 동력선에 대한 등화에 부가하여 가장 잘 보이는 곳에 홍색의 전주등 3개를 수직선상으로 표시하거나 또는 원통형의 형상물 1개를 표시할 수 있다.

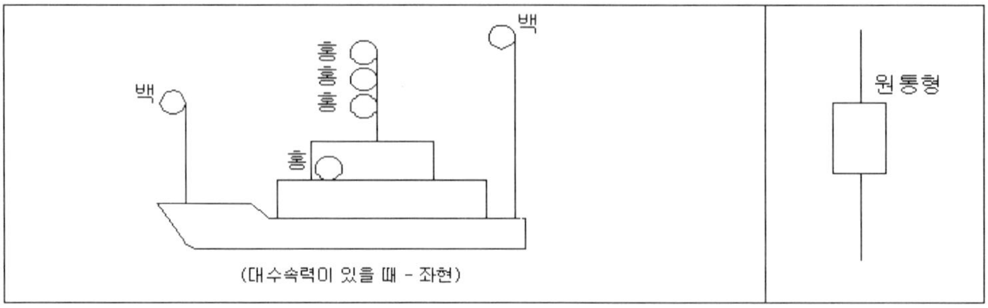

[그림 1-28] 흘수제약선박

J. RULE 29 Pilot Vessels(제29조 도선선)

(a) A vessel engaged on pilotage duty shall exhibit:
도선업무에 종사하고 있는 선박은 다음 각 호에 규정하는 등화 또는 형상물을 표시하여야 한다.

(i) At or near the masthead, two all-round lights in a vertical line, the upper being white and the lower red;
마스트의 꼭대기 또는 그 부근에 수직선상으로 상부의 것이 백색, 하부의 것이 홍색인 전주등 2개

(ii) When underway, in addition, sidelights and a stern-light;
항행 중에는 이에 부가하여 양 현등과 선미등 1개

(iii) When at anchor, in addition to the lights prescribed in sub-paragraph (i) the anchor light, lights or shape.
정박 중에는 (1)호에 규정하는 등화에 부가하여 정박등 또는 정박형상물

(b) A pilot vessel when not engaged on pilotage duty shall exhibit the lights or shapes prescribed for a similar vessel of her length.
도선업무에 종사하고 있지 아니하는 도선선은 그 선박과 동일한 길이의 유사한 선박에 대하여 규정하는 등화 또는 형상물을 표시하여야 한다.

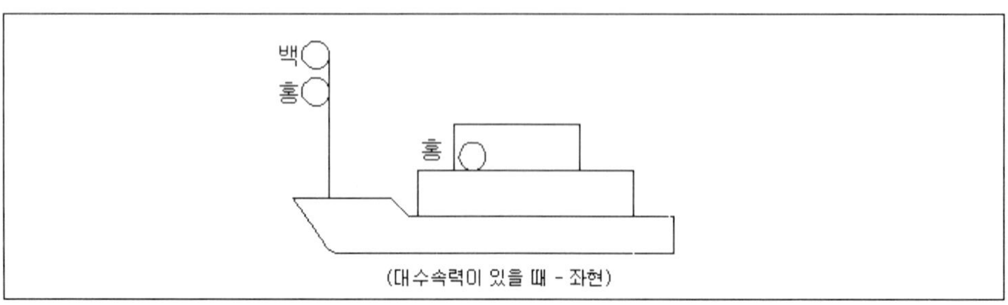

[그림 1-29] 항행 중인 도선선

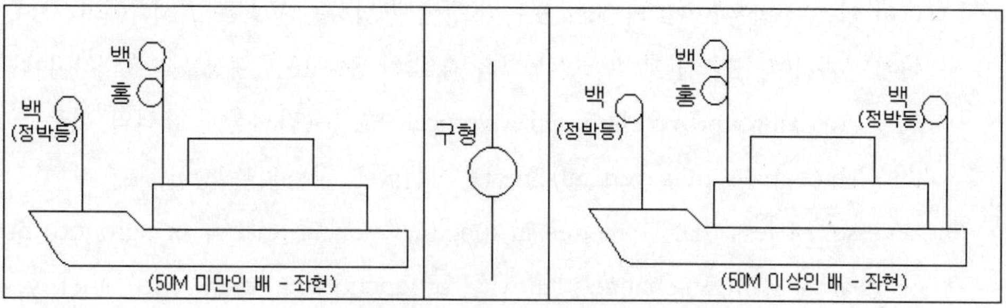

[그림 1-30] 정박 중인 도선선

K. RULE 30 Anchored Vessels and Vessels Aground
(제30조 정박선과 얹혀 있는 선박)

(a) A vessel at anchor shall exhibit where it can best be seen:
정박하고 있는 선박은 가장 잘 보이는 곳에 다음 각 호에 규정하는 등화 또는 형상물을 표시하여야 한다.

(i) In the fore part, an all-round white light or one ball;
전부에, 백색의 전주등 1개 또는 구형의 형상물 1개

(ii) At or near the stern and at a lower level than the light prescribed in sub-paragraph (i) an all-round white light.
선미 또는 그 부근에 (1)호에 규정하는 등화보다 낮은 위치로 백색의 전주등 1개

(b) A vessel of less than 50 metres in length may exhibit an all-round white light where it can best be seen instead of the lights prescribed in paragraph (a) of this Rule.
길이 50미터 미만의 선박은 이 조의 (a)항에 규정하는 등화에 대신하여 가장 잘 보이는 곳에 백색의 전주등 1개를 표시할 수 있다.

(c) A vessel at anchor may, and a vessel of 100 metres and more in length shall, also use the available working or equivalent lights to illuminate her decks.
또한, 정박하고 있는 선박은 자선의 갑판을 조명하기 위하여 이용할 수 있는 작업등 또는 이와 대등한 등화를 사용할 수 있으며, 또 길이 100미터 이상의 선박은 이들 등화를 사용하여야 한다.

(d) A vessel aground shall exhibit the lights prescribed in paragraph (a) or (b) of this Rule and in addition, where they can best be seen:

얹혀 있는 선박은 이 조의 (a)항 또는 (b)항에 규정하는 등화를 표시하여야 하며, 또 이에 부가하여 가장 잘 보이는 곳에 다음 각 호의 등화 또는 형상물을 표시하여야 한다.

(i) Two all-round red lights in a vertical line;(수직선상에 홍색의 전주등 2개)

(ii) Three balls in a vertical line.(수직선상에 구형의 형상물 3개)

(e) A vessel of less than 7 metres in length, when at anchor or aground, not in or near a narrow channel, fairway or anchorage, or where other vessels normally navigate, shall not be required to exhibit the lights or shapes prescribed in paragraphs (a), (b) or (d) of this Rule.

길이 7미터 미만의 선박은 좁은 수로, 항로 또는 묘지의 안 혹은 이들 부근 또는 다른 선박이 통상적으로 항행하는 곳 이외의 장소에 정박하거나 또는 얹혀 있는 때에는 이 조의 (a)항, (b)항 또는 (d)항에 규정하는 등화 또는 형상물을 표시하는 것을 요하지 아니한다.

(f) A vessel of less than 12 metres in length, when aground, shall not be required to exhibit the lights or shapes prescribed in sub-paragraphs (d) (i) and (ii) of this Rule.

길이 12미터 미만의 선박이 얹혀 있는 때에는 이 조의 (d)항 1호와 2호에 규정하는 등화 또는 형상물을 표시하는 것을 요하지 아니한다.

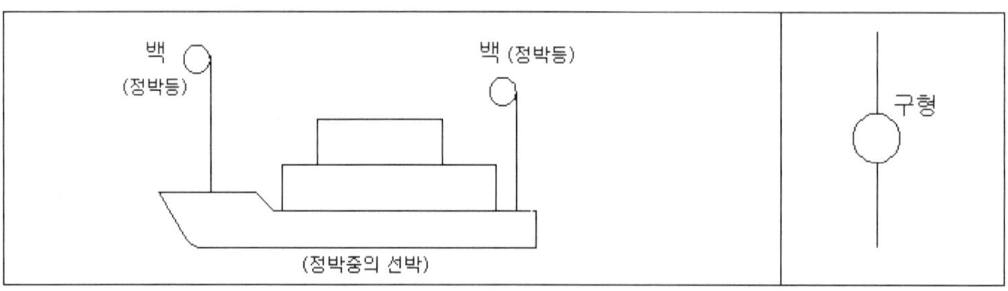

[그림 1-31] 정박 중인 선박(anchored vessels)

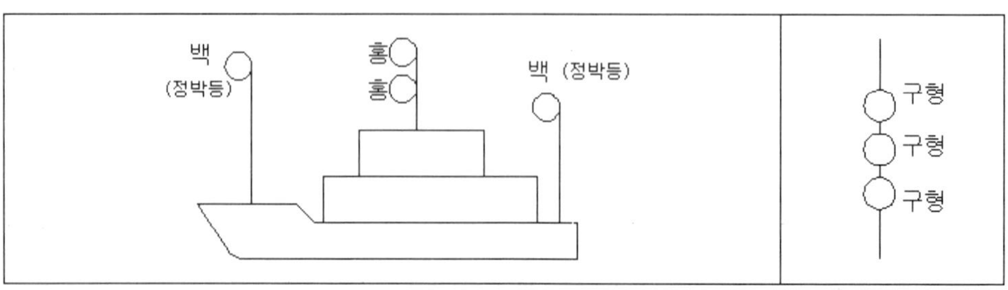

[그림 1-32] 얹혀 있는 선박(vessels aground)

L. RULE 31 Seaplanes(제31조 수상비행기)

Where it is impracticable for a seaplane or a WIG craft to exhibit lights and shapes of the characteristics or in the positions prescribed in the Rules of this Part she shall exhibit lights and shapes as closely similar in characteristics and position as is possible.

수상비행기나 WIG선이 등화와 형상물을 이 장의 제조에 규정하는 특성을 가진 것으로 표시할 수 없거나, 규정하는 위치에 표시할 수 없는 경우에는 특성과 위치에 있어서 가능한 한 이 장에 규정하는 것과 유사한 등화 또는 형상물을 표시하여야 한다.

4. PART D - SOUND AND LIGHT SIGNALS(제4장 음향신호와 발광신호)

A. RULE 32 Definitions(제32조 정의)

(a) The "whistle" means any sound signalling appliance capable of producing the prescribed blasts and which complies with the specifications in Annex III to these Regulations.

"기적"이라 함은 이 규칙에 규정하는 단음과 장음을 낼 수 있고, 또 이 규칙의 부속서 III에 규정하는 규격에 적합한 음향신호기구를 말한다.

(b) The term "short blast" means a blast of about one second's duration.

"단음"이라 함은, 약 1초 동안 계속하는 고동소리를 말한다.

(c) The term "prolonged blast" means a blast of from four to six seconds' duration.

"장음"이라 함은, 4초 내지 6초 동안 계속하는 고동소리를 말한다.

B. RULE 33 Equipment for Sound Signals(제33조 음향신호장치)

(a) A vessel of 12 metres or more in length shall be provided with a whistle and a bell and a vessel of 100 metres or more in length shall, in addition, be provided with a gong, the tone and sound of which cannot be confused with that of the bell. The whistle, bell and gong shall comply with the specifications in Annex III to these Regulations. The bell or gong or both may be replaced by other equipment having the same respective sound characteristics, provided that manual sounding of the required signals shall always be possible.

길이 12미터 이상의 선박은 기적 1개와 호종 1개를 비치하여야 하며, 길이 100미터 이상의 선박은 이에 부가하여 호종과 혼동되지 아니하는 음조와 소리를 가진 동라 1개를 비치하여야 한다. 기적, 호종 및 동라는 이 규칙의 부속서 III에 규정하는 규격에 적합하여야 한다. 호종 혹은 동라 또는 그 양자는 각기 이와 동일한 음색을 가진 다른 장치로 대치할 수 있다. 다만 이들은 필요한 신호음을 수동으로 항시 낼 수 있어야 한다.

(b) A vessel of less than 12 metres in length shall not be obliged to carry the sound signalling appliances prescribed in paragraph (a) of this Rule, but if she does not, she shall be provided with some other means of making an efficient sound signal.

길이 12미터 미만의 선박은 이 조의 (a)항에 규정하는 음향신호기구를 비치할 의무는 없다. 그러나 이를 비치하지 아니하는 경우에는 유효한 음향신호를 낼 수 있는 다른 수단을 갖추어야 한다.

C. RULE 34 Manoeuvring and Warning Signals(제34조 조종신호와 경고신호)

(a) When vessels are in sight of one another, a power-driven vessel underway, when manoeuvring as authorized or required by these Rules, shall indicate that manoeuvre by the following signals on her whistle:

선박이 서로 시야 내에 있는 경우에 항행 중인 동력선이 이 규칙의 규정에 의하여 인정되거나 또는 요구되는 조종을 할 때에는 기적을 사용한 다음 각 호의 신호에 의하여 그 조종을 표시하여야 한다.

(i) One short blast to mean "I am altering my course to starboard";

단음 1회 : "나는 침로를 우현쪽으로 변경하고 있다."

(ii) Two short blasts to mean "I am altering my course to port";

단음 2회 : "나는 침로를 좌현쪽으로 변경하고 있다."

(iii) Three short blasts to mean "I am operating astern propulsion".

단음 3회 : "나는 기관을 뒤로 추진하고 있다."

(b) Any vessel may supplement the whistle signals prescribed in paragraph (a) of this Rule by light signals, repeated as appropriate, whilst the manoeuvre is being carried out:

모든 선박은 당해 조종이 이루어지고 있는 동안에 다음 각 호의 발광신호를 적절히 반복하여 이 조의 1항에 규정하는 기적신호를 보충할 수 있다.

(i) These light signals shall have the following significance:

이들 발광신호는 다음과 같은 의미를 가진다.

- One flash to mean "I am altering my course to starboard";

  섬광 1회 : "나는 침로를 우현쪽으로 변경하고 있다."

- Two flashes to mean "I am altering my course to port";

  섬광 2회 : "나는 침로를 좌현쪽으로 변경하고 있다."

- Three flashes to mean "I am operating astern propulsion";

  섬광 3회 : "나는 기관을 뒤로 추진하고 있다."

(ii) The duration of each flash be about one second, the interval between flashes shall be about one second, and the interval between successive signals shall be not less than ten seconds;

각 섬광의 지속시간은 약 1초 동안이고, 섬광 사이의 간격은 약 1초이며 또 반복되는 신호 사이의 간격은 10초 이상이어야 한다.

(iii) The light used for this signal shall, if fitted, be an all-round white light, visible at a minimum range of 5 miles, and shall comply with the provisions of Annex I.

이 발광신호에 사용하는 등화는 설치되어 있다면 5해리의 최소 거리에서 볼 수 있는 백색의 전주등 1개이어야 하고 부속서 I 의 규정에 적합한 것이어야 한다.

(c) When in sight of one another in a narrow channel or fairway;

선박이 좁은 수로 또는 항로에 있어서 서로 시야내에 있는 경우에는

(i) A vessel intending to overtake another shall in compliance with Rule 9(e)

(i) Indicate her intention by the following signals on her whistle:

다른 선박을 추월하고자 하는 선박은 제9조 (e)항 (1)호의 규정에 따라서 기적을 사용한 다음과 같은 신호에 의하여 그의 의도를 표시하여야 한다.

- Two prolonged blasts followed by one short blast to mean "I intend to overtake you on your starboard side";

  장음 2회, 단음 1회 : "나는 그대의 우현쪽으로 추월하고자 한다."

- Two prolonged blasts followed by two short blasts to mean "I intend to overtake you on your port side."
  장음 2회, 단음 2회 : "나는 그대의 좌현쪽으로 추월하고자 한다."

(ii) A vessel about to be overtaken when acting in accordance with Rule 9(e)(i) shall indicate her agreement by the following signal on her whistle:
추월당하게 될 선박이 제9조 (e)항 (1)호의 규정에 따라서 다른 선박의 추월에 동의하는 경우에는 기적을 사용한 다음과 같은 신호에 의하여 그의 동의를 표시하여야 한다.

- One prolonged, one short, one prolonged and one short blast, in that order.
  장음 1회, 단음 1회, 장음 1회 및 단음 1회의 순서

(d) When vessels in sight of one another are approaching each other and from any cause either vessel fails to understand the intentions or actions of the other, or is in doubt whether sufficient action is being taken by the other to avoid collision, the vessel in doubt shall immediately indicate such doubt by giving at least five short and rapid blasts on the whistle. Such signal may be supplemented by a light signal of at least five short and rapid flashes.
서로 시야 내에 있는 선박이 접근하고 있을 경우에 어떠한 원인으로 인하여 한 선박이 다른 선박의 의도 또는 동작을 이해할 수 없는 때 또는 다른 선박이 충돌을 피하기 위하여 충분한 동작을 취하고 있는지의 여부가 의심스러운 때에는 의심을 가진 선박은 곧 기적을 사용하여 급속히 단음을 5회 이상 울려 그 의문을 표시하여야 한다. 그 신호는 5회 이상의 짧고 급속한 섬광을 발하는 발광신호로써 보충할 수 있다.

(e) A vessel nearing a bend or an area of a channel or fairway where other vessels may be obscured by an intervening obstruction shall sound one prolonged blast. Such signal shall be answered with a prolonged blast by any approaching vessel that may be within hearing around the bend or behind the intervening obstruction.
수로 또는 항로에 있어서 개재하는 장애물 때문에 다른 선박을 볼 수 없는 굴곡부 혹은 수역에 접근하고 있는 선박은 장음을 1회 울려야 한다. 접근하며 오는 선박이

굴곡부의 주위 또는 개재하는 장애물의 배후에서 이 신호를 들었을 때에는 장음 1회를 울려 당해 신호에 응답하여야 한다.

(f) If whistles are fitted on a vessel at a distance apart of more than 100 metres, one whistle only shall be used for giving manoeuvring and warning signals.
복수의 기적이 100미터 이상의 거리를 두고 선내에 설치되어 있는 경우에는 그 중에서 하나의 기적만을 조종신호와 경고신호를 울리는데 사용하여야 한다.

〈표 1-2〉 음향 및 발광신호

	음향신호		발광신호(간격은 10초 이상)	
변침 중 신호	오른쪽	•	오른쪽	*
	왼 쪽	• •	왼 쪽	* *
	후진 중	• • •	후진 중	* * *
좁은 수도의 추월신호	오른쪽 추월	— —•		
	왼쪽 추월	— —••		
	추월동의	—•—•		
	굴곡부 경고신호	—		
의문신호	•••••(단음 5회 이상)		* * * * * (5회 이상)	

D. RULE 35 Sound Signals in Restricted Visibility
(제35조 제한된 시정에 있어서의 음향신호)

In or near an area of restricted visibility, whether by day or night, the signals prescribed in this Rule shall be used as follows:
시정이 제한된 수역내 또는 그 부근에 있어서는 주야를 불문하고, 이 조에 규정하는 신호를 다음 각 호와 같이 사용하여야 한다.

(a) A power-driven vessel making way through the water shall sound at intervals of not more than 2 minutes one prolonged blast.
동력선은 대수속력이 있는 경우에는 2분간을 넘지 아니하는 간격으로 장음 1회를 울려야 한다.

(b) A power-driven vessel underway but stopped and making no way through the water shall sound at intervals of not more than 2 minutes two prolonged blasts in succession with an interval of about 2 seconds between them.

항행 중인 동력선은 정지하여 대수속력이 없는 경우에는 장음사이의 간격이 약 2초인 연속한 장음 2회를 2분간을 넘지 아니하는 간격으로 울려야 한다.

(c) A vessel not under command, a vessel restricted in her ability to manoeuvre, a vessel constrained by her draught, a sailing vessel, a vessel engaged in fishing and a vessel engaged in towing or pushing another vessel shall, instead of the signals prescribed in paragraphs (a) or (b) of this Rule, sound at intervals of not more than 2 minutes three blasts in succession, namely one prolonged followed by two short blasts.

조종이 자유롭지 못한 상태에 있는 선박, 조종능력이 제한되어 있는 선박, 흘수 때문에 제약을 받고 있는 선박, 범선, 어로에 종사하고 있는 선박 및 다른 선박을 끌고 있거나 또는 밀고 있는 선박은 이 조의 (a)항 또는 (b)항에 규정하는 신호에 대신하여 2분간을 넘지 아니하는 간격으로 연속한 3회의 기적, 즉 장음 1회에 이어 단음 2회를 울려야 한다.

(d) A vessel engaged in fishing, when at anchor, and a vessel restricted in her ability to manoeuvre when carrying out her work at anchor, shall instead of the signals prescribed in paragraph (g) of this Rule sound the signal prescribed in paragraph (c) of this Rule.

어로에 종사하고 있는 선박이 정박할 때 그리고 조종능력이 제한된 선박이 정박작업을 수행할 때 이조의 (g)항에 규정하는 신호대신 (c)항에 규정하는 신호를 울려야 한다.

(e) A vessel towed or if more than one vessel is towed the last vessel of the tow, if manned, shall at intervals of not more than 2 minutes sound four blasts in succession, namely one prolonged followed by three short blasts. When practicable, this signal shall be made immediately after the signal made by the towing vessel.

끌려가고 있는 선박 또는 2척 이상의 선박이 끌려가고 있을 때의 최후미의 선박은 승무원이 있을 경우에는 2분간을 넘지 아니하는 간격으로 연속한 4회의 기적, 즉 장음 1회에 이어 단음 3회를 울려야 한다. 이 신호는 실행이 가능한 한 끌고 있는 선박이 행하는 신호의 직후에 울려야 한다.

(f) When a pushing vessel and a vessel being pushed ahead are rigidly connected in a composite unit they shall be regarded as a power-driven vessel and shall give the signals prescribed in paragraphs (a) or (b) of this Rule.

밀고 있는 선박과 앞쪽으로 밀려가고 있는 선박이 견고하게 연결되어 하나의 복합체를 이룬 경우에는 이를 한 척의 동력선으로 보고 이 조의 (a)항 또는 (b)항에 규정하는 신호를 하여야 한다.

(g) A vessel at anchor shall at intervals of not more than one minute ring the bell rapidly for about 5 seconds. In a vessel of 100 metres or more in length bell shall be sounded in the forepart of the vessel and immediately after the ringing of the bell the gong shall be sounded rapidly for about 5 seconds in the after part of the vessel. A vessel at anchor may in addition sound three blasts in succession, namely one short, one prolonged and one short blast, to give warning of her position and of the possibility of collision to an approaching vessel.

정박하고 있는 선박은 1분간을 넘지 아니하는 간격으로 약 5초 동안 급속히 호종을 울려야 한다. 길이 100미터 이상의 선박에 있어서는 호종을 선박의 전부에서 울려야 하고, 또 호종을 울린 직후에 선박의 후부에서 동라를 약 5초 동안 급속히 울려야 한다. 정박하고 있는 선박은 접근하여 오는 선박에 대하여 자선의 위치와 충돌의 가능성을 경고하기 위해서는 이에 부가하여 연 3회의 기적 즉 단음 1회, 장음 1회 및 단음 1회를 울릴 수 있다.

(h) A vessel aground shall give the bell signal and if required the gong signal prescribed in paragraph (g) of this Rule and shall, in addition, give three separate and distinct strokes on the bell immediately before and after the rapid ringing of the bell. A vessel aground may in addition sound an appropriate whistle signal.

얹혀 있는 선박은 호종신호를 하여야 하고 또 필요하다면, 이 조의 (g)항에 규정하는 동라신호를 하고 이에 부가하여 급속히 울리는 호종신호의 직전과 직후에 호종을 명확하게 분리하여 3회 타종하여야 한다. 얹혀 있는 선박은 이에 부가하여 적절한 기적신호를 할 수 있다.

(i) A vessel of 12 metres or more but less than 20 metres in length shall not be obliged to give the bell signals prescribed in paragraphs (g) and (h) of this Rule. However, if she does not, she shall make some other efficient sound signal at intervals of not more than 2 minutes.

길이 12미터 이상 20미터 미만의 선박은 이 규칙의 (g)항과 (h)항에 규정된 호종신호를 울려서는 아니 된다. 그러나 그렇지 않을 경우 2분을 넘지 않은 간격으로 여타 유효한 다른 음향신호를 울려야 한다.

(j) A vessel of less than 12 metres in length shall not be obliged to give the above-mentioned signals but, if she does not, shall make some other efficient sound signal at intervals of not more than 2 minutes.

길이 12미터 미만의 선박은 전술한 신호를 하여야 할 의무는 없으나, 이 신호를 하지 않을 경우에는 2분간을 넘지 아니하는 간격으로 기타의 유효한 음향신호를 하여야 한다.

(k) A pilot vessel when engaged on pilotage duty may in addition to the signals prescribed in paragraphs (a), (b) or (f) of this Rule sound an identity signal consisting of four short blasts.

도선선이 도선업무에 종사하고 있는 경우에는 이 조의 (a)항, (b)항 또는 (f)항에 규정하는 신호에 부가하여 단음 4회로 이루어지는 식별신호를 울릴 수 있다.

〈표 1-3〉 제한된 시정에서의 음향신호

①	항해 중의 동력선(③과 ④는 제외)	—	
②	항행 중인 동력선이 대수속력이 없는 때	— —	
③	범선, 어로 중인 어선, 조종불능선, 조종성능 제한선, 흘수제한선 등	—••	
④	피예선(마지막 선)	—•••	
⑤	정박선	선수부	선미부
	길이 100m 이상의 선박	(종 약 5초)	(꽹과리 약 5초)
	길이 100m 미만의 선박	(종 약 5초)	
	경고신호	•—•	
⑥	얹힌 선박	선수부	선미부
	길이 100m 이상의 선박	(종 3점타)(종약 5초)(종 3점타)	(꽹과리 약 5초)
	길이 100m 미만의 선박	(종 3점타)(종약 5초)(종 3점타)	
⑦	도선선	••••(항행 중, 정박 중의 신호에 추가)	

E. RULE 36 Signals to Attract Attention(제36조 주의환기신호)

If necessary to attract the attention of another vessel any vessel may make light or sound signals that cannot be mistaken for any signal authorized elsewhere in these Rules, or may direct the beam of her searchlight in the direction of the danger, in such a way as not to embarrass and vessel.

모든 선박은 다른 선박의 주의를 환기하기 위하여 필요한 경우에는 이 규칙의 다른 조항에서 인정하는 어떠한 신호와도 오인되지 아니할 등화신호 또는 음향신호를 할 수 있고, 또 다른 선박을 당황하게 하지 아니하는 방법으로 위험이 있는 방향에 탐조등을 비출 수 있다.

F. RULE 37 Distress Signals(제37조 조난신호)

When a vessel is in distress and requires assistance she shall use or exhibit the signals prescribed in Annex IV to these Regulations.

선박이 조난하여 구원을 요청할 경우에는 이 규칙의 부속서 IV에 규정하는 신호를 사용하거나 또는 표시하여야 한다.

〈표 1-4〉 조난신호방법

	조난신호방법
(1)	약 1분간의 간격으로 행하는 1회의 발포 기타의 폭발에 의한 신호
(2)	무중신호기구에 의한 연속된 음향신호
(3)	단시간의 간격으로 1회에 1개씩 발사되어 별모양의 붉은 불꽃을 발하는 로켓 또는 유탄에 의한 신호
(4)	무선전신 또는 기타의 신호방법에 의한 "모르스" 부호「··· — — —···」(SOS)의 신호
(5)	무선전화에 의한 "메이데이"라는 말의 신호
(6)	국제기류신호에 의한 NC의 조난신호
(7)	방형기로서 그위 또는 아래에 공 또는 이와 유사한 것 1개를 붙여 이루어지는 신호
(8)	선상에서의 발염(타르통, 기름통 등의 연소로 생기는) 신호
(9)	낙하산이 달린 적색의 염화 로켓 또는 적색의 수중 염화에 의한 신호
(10)	오렌지색의 연기를 발하는 발연신호
(11)	좌우로 벌린 팔을 천천히 반복하여 올렸다 내렸다 하는 신호
(12)	무전전신 경보신호
(13)	무선전화 경보신호
(14)	비상용 위치지시 무선표지에 의하여 발신하는 신호

5. PART E - EXEMPTIONS(제5장 면제)

A. RULE 38 Exemptions(제38조 적용의 면제)

Any vessel (or class of vessels) provided that she complies with the requirements of the International Regulations for Preventing Collisions at Sea, 1960, the keel of which is laid or which is at a corresponding stage of construction before the entry into force of these Regulations may be exempted from compliance therewith as follows:

이 규칙이 발효하기 이전에 용골이 놓였거나 또는 이와 대등한 건조단계에 있는 선박(또는 선박의 종류)은 1960년의 국제해상충돌예방규칙의 규정을 지킬 것을 조건으로, 다음과 같이 이 규칙의 적용을 면제받을 수 있다:

(a) The installation of lights with ranges prescribed in Rule 22, until four years after the date of entry into force of these Regulations.

제22조에 규정하는 가시거리를 가진 등화의 설치는 이 규칙의 효력발생일로부터 4년간

(b) The installation of lights with colour specifications as prescribed in Section 7 of Annex I to these Regulations, until four years after the date of entry into force of these Regulations.

이 규칙의 부속서 I 의 제7항에 규정하는 색도명세를 가진 등화의 설치는 이 규칙의 효력발행일로부터 4년간

(c) The repositioning of lights as a result of conversion from Imperial to metric units and rounding off measurement figures, permanent exemption.

영국식 단위에서 미터식 단위로 환산하고 남은 우수리(단수) 때문에 생기는 등화의 위치의 변경은 영구히 면제

(d) (i) The repositioning of masthead lights on vessels of less than 150 metres in length, resulting from the prescriptions of Section 3(a) of Annex I, permanent exemption;

부속서 I 의 제3(a)항의 규정에 의한 길이 150미터 미만인 선박의 마스트 정부 등의 위치의 변경은 영구히 면제

(ii) The repositioning of masthead lights on vessels of 150 metres or more in length, resulting from the prescriptions of Section 3(a) of Annex I to these Regulations, until nine years after the date of entry into force of these Regulations.

이 규칙의 부속서 Ⅰ의 제3(a)항의 규정에 의한 길이 150미터 이상인 선박의 마스트 정부 등의 위치의 변경은 이 규칙의 효력발생일로부터 9년간

(e) The repositioning of masthead light resulting from the prescriptions of Section 2(b) of Annex I, until nine years after the date of entry into force of these Regulations.

부속서 Ⅰ의 제2(b)항의 규정에 의한 마스트 정부 등의 위치의 변경은 이 규칙의 효력발생일로부터 9년간

(f) The repositioning of sidelights resulting from the prescriptions of Sections 2(g) and 3(b) of Annex I, until nine years after the date of entry into force of these Regulations.

부속서 Ⅰ의 제2(g)항과 제3(b)항의 규정에 의한 현등의 위치의 변경은 이 규칙의 효력발생일로부터 9년간

(g) The requirements for sound signal appliances prescribed in Annex III, until nine years after the date of entry into force of these Regulations.

부속서 Ⅲ에 규정하는 음향신호기구에 대한 요건은 이 규칙의 효력발생일로부터 9년간

(h) The repositioning of all-round lights resulting from the prescriptions of Section 9(b) of Annex I to these Regulations, permanent exemption.

부속서 Ⅰ의 제9(b)항의 규정에 의한 전주등의 위치변경은 영구히 면제

1.4 해사안전법

1.4.1 의 의

해사안전법은 선박의 안전운항을 위한 안전관리체계를 확립하여 선박항행과 관련된 모든 위험과 장해를 제거함으로써 해사안전(海事安全) 증진과 선박의 원활한 교통에 이바지함을 목적으로 선박의 안전운항을 위한 안전관리체제를 확립하고, 해상에서 일어나는 선박항행상의 모든 위험과 장해를 제거함으로써 해상에서의 안전 및 원활한 교통을 확보함을 목적으로 하는 교통공법이다.

이 법은 우리나라의 영해 이내의 해역에 있어서 국적에 관계없이 해상교통의 목적에 사용하는 모든 종류의 선박에 관하여 해상교통 질서를 유지하는데 그 입법목적을 두고 있으며, 1972년 국제해상충돌방지규칙 제정과 그 배경과 맥을 같이 하며 한국 연안에서도 선박의 대형화, 고속화 그리고 특정해역에서의 해상교통량 확대로 인하여 예상되는 해양사고의 증가를 방지하기 위하여 우리나라 연안에 있어서의 특수한 지리적 조건과 특성에 맞추어 입법된 기본적인 해상교통질서법이다.

또한 과거에 입법되었던 해상충돌예방법과는 그 입법목적을 달리하는데, 즉 해상충돌예방법은 해상에 있어서 선박의 충돌과 항해의 위험을 방지하기 위한 행위에 대하여 선박 스스로의 판단과 단독적 실행만을 요구하며 또 그 위반효과에 대해서 구체적인 책임규정이 없는 소극적 해상교통법이었다. 반면에 해사안전법은 1986년 제정되었으며, 2013년 전면적인 개정으로 정부와 사용자의 의무를 구체적으로 추가, 해사안전관리(제2장), 수역안전관리(제3장), 해상교통안전관리(제4장) 등의 규정이 신설 혹은 보완되어 현재에 이르고 있다.

이 법은 해상활동에 있어서 이용자에게 교통질서의 유지의무와 교통시설과 항로표지의 설치와 관리에 대하여 정부에게 관리자 의무를 규정하여 적극적인 해상교통법으로써의 성질을 가지고 있다.

1.4.2 적용범위

1. 적용대상

대한민국의 영해, 내수(해상항행선박이 항행을 계속할 수 없는 하천 · 호수 · 늪 등은 제외한다. 이하 같다)에 있는 선박이나 해양시설. 다만, 대한민국선박이 아닌 선박(외국선박) 중

① 대한민국의 항(港)과 항 사이만을 항행하는 선박

② 국적의 취득을 조건으로 하여 선체용선(船體傭船)으로 차용한 선박

③ 대한민국의 영해 및 내수를 제외한 해역에 있는 대한민국선박

④ 대한민국의 배타적 경제수역에서 항행장애물을 발생시킨 선박

⑤ 대한민국의 배타적 경제수역 또는 대륙붕에 있는 해양시설

적용선박의 종류는 물에서 항행수단으로 사용하거나 사용할 수 있는 모든 종류의 배(물 위에서 이동할 수 있는 수상항공기와 수면비행선박을 포함한다)를 말한다. (동법 제2조(정의) 2항)

2. 장소적 적용범위

장소적 적용범위에 있어서 해사안전법 제3조 제1항은 이 법은 대한민국의 영해 및 내수(해상 항행선박이 항행을 계속할 수 없는 하천 · 湖沼 등을 제외한다)에 있는 모든 선박과 대한민국의 영해나 내수가 아닌 해역에 있는 대한민국 선박에 대하여 이를 적용한다고 하여 그 적용범위에 대하여 정하고 있다. 즉 동법은 기본적으로 대한민국의 영해 및 내수에 있는 대한민국의 선박은 물론 외국적 선박에 대하여도 적용하고, 대한민국의 선박에 대하여는 공해에서도 적용한다. 대한민국의 선박이란 선박법상의 선박*을 말한다.

그러므로 영해 내에서 뿐만 아니라 공해에서 우리나라 국적선박끼리 조우할 때 지켜야 할 법은 해사안전법이 된다. 영해 내에서 우리나라 국적선박과 외국적 선박이 충돌한 경우에도 해사안전법이 준거법이 된다.

---

* 선박법 제1조의2 이 법에서 '선박'이라 함은 수상 또는 수중에서 항행용으로 사용하거나 사용될 수 있는 배종류를 말한다.

① 기선 : 기관을 사용하여 추진하는 선박(기관과 돛을 모두 사용하는 경우로서 주로 기관을 사용하는 것을 포함한다)

② 범선 : 돛을 사용하여 추진하는 선박(기관과 돛을 모두 사용하는 경우로서 주로 돛을 사용하는 것을 포함한다)

③ 부선 : 자력항행능력이 없어 다른 선박에 의하여 끌리거나 밀려서 항행되는 선박

반면 공해상에서 우리나라 선박과 외국선박이 충돌하면 섭외사건이 되어 만약 우리나라가 관할권을 가지면 가해선박의 해상교통법이 준거법이 된다. 해상항행선박이 항행을 계속할 수 없는 하천 및 호소(湖沼) 등을 제외한다고 되어 있으므로, 국제규칙에서와 마찬가지로 해사안전법은 한강이나 충주호 등에는 적용되지 아니한다.

3. 인적 적용범위

인적 범위에 있어서는 이 법 또는 시행령 중 선박소유자에 관한 규정은 선박공유의 경우에 선박관리인을 임명한 때에는 이를 선박관리인에게, 선박임대차의 경우에는 이를 선박임차인에게 각각 적용하고 선장에 관한 규정은 선장에 갈음하여 그 직무를 행하는 자에게도 이를 적용한다. 즉 선박의 운항지휘자로서 선장뿐만 아니라 선장의 지휘, 감독 아래 해상교통의 안전의무를 지고 있는 항해사와 도선사를 포함한다. 또한 해양시설의 소유자에 관한 규정은 해양시설임대차의 경우에는 이를 해양시설의 임차인에게 적용하도록 하고 있다.

1.4.3 구 성

제1장은 총칙, 제2장은 해사안전관리계획, 제3장의1은 수역안전관리(해양시설의 보호수역 설정 및 관리), 제3장의2는 교통안전특정해역 등의 설정과 관리, 제3장의3은 유조선통항금지해역의 설정 및 관리, 제4장은 해상교통 안전관리, 제5장은 선박 및 사업장의 안전관리, 제6장은 선박의 항법 등, 제7장은 보칙, 그리고 제8장은 벌칙으로 구성되어 있다.

이 법은 처음에는 국제규칙과 유사한 내용의 해상교통법만을 내용으로 하고 있었으나, 2013년 3월 국제안전관리규약(ISM Code)의 내용을 포함하는 등 우리나라 해상교통의 지리적 환경을 고려하여 국제규칙에서 볼 수 없는 독자적인 해상교통규정인 조성규범을 두고 있다. 또한 국제규칙은 집행상의 문제로 벌칙규정이 없으나 해사안전법에는 벌칙규정이 있다.

1.4.4 특 징

1. 정부기관의 관리자의무와 사용자의 질서의무를 규정

① 정부기관의 해상안전관리

(제6조) 국가해사안전기본계획

(제8조) 항행보조시설의 설치 및 관리

(제9조) 보호수역의 입역

(제10조) 특정해역의 설정과 관리(항로지정 및 해상교통관리)

(제36조) 항만국통제

(제55조) 사업장에 대한 지도감독

(제59조) 개선명령

② 선박소유자

(제5조) 선박. 해양시설 소유자의 책무

(제46조, 제47조) 선박의 안전관리체제수립 등

③ 선장의 의무

(제41조, 제41조의2) 주취 중 및 약물복용상태에서 조타기 조작금지, 조작지시금지

(제43조) 해양사고가 발생한 경우의 조치

2. 항법, 등화 및 형상물의 표시방법

항법, 등화 및 형상물의 표시방법은 국제해상충돌방지규칙과 거의 동일하며 선박의 입항 및 출항 등에 관한 법률이나 조약에 따른 규정이 있을 때는 그 규정의 정하는 바에 따른다.

3. 적용수역

① 모든 나라의 선박 : 한국의 영해와 내수(다만 해항선이 항해를 계속할 수 없는 하천 및 호소를 제외)

② 한국선박 : 공해를 포함한 해항선의 가항수역 전역에서 적용한다(영해 밖에서도 적용)

4. 국제규칙에 없는 정의 항목

① 항로지정방식이라 함은 선박이 통항하는 항로, 속력, 기타 선박 운항에 관한 사항을 지정하는 제도를 말한다. 특히 교통안전특정해역외의 해역으로서 그 해역의 지형·조류 기타 자연적인 조건 등으로 해양사고가 발생할 우려가 있다고 인정되는 경우에 그 해역의 범위, 선박의 항로 및 속력 등 선박의 항행안전에 필요한 사항을 정하고 있다.

② 거대선이라 함은 길이 200m 이상의 선박을 말한다. 국제규칙에서는 거대선의 개념을 규정하지 않고 대신 흘수가 깊어서 선박의 조종성에 제약을 받는 흘수제약선을 규정하고 있다.

③ 위험화물운반선이라 함은 선체구조의 일부로서 형성되는 화물창 또는 선체에 고정된 탱크 등에 해양수산부령이 정하는 위험물을 싣고 운반하는 선박을 말한다.

④ 고속여객선이라 함은 시속 15노트이상의 속력으로 항행하는 여객선을 말한다.

⑤ 해상교통관제방식이라 함은 선박의 안전하고 원활한 통항을 목적으로 해상교통관제시설을 설치하여 항행하는 선박에 대하여 적절한 항행정보를 제공하고 당해 선박이 적법하게 항행하는지의 여부를 감시·지도하는 제도를 말한다.

1.4.5 주요 내용

1. 선박출항통제

해상에 기상특보가 발표되거나 안개 등으로 인하여 시계(視界)가 제한되어 선박의 안전운항에 지장을 초래할 우려가 있다고 판단할 때에는 선박소유자 또는 선장에 대하여 선박의 출항을 통제할 수 있도록 함으로써 선박의 안전을 확보하고 있다.

2. 주취 중의 조타기 조작 등 금지

해상통항량의 증가로 인해 선박에서의 음주는 중대한 해양사고로 직결될 수 있다. 그러므로 술에 취한 상태에서 혹은 약물복용상태에서 운항을 위하여 조타기를 조작하거나 그 조작을 지시하지 못하도록 규정함으로써 음주로 인한 충돌과 좌초의 위험을 감소시키고 해상교통의 안전을 확보하기 위하여 이러한 규정을 두고 있다(법 제41조, 제41조의2).

3. 선박의 안전관리체계

해상안전관리에 관한 해상교통안전법의 내용 중 제10조부터 제10조의12까지는 국제해사기구(IMO)의 국제안전관리규약(ISM Code)을 수용하고 있다. 이는 해상에서의 인적과실로 인한 해양사고를 방지하기 위하여 지속적으로 선원의 자질향상을 위한 노력을 수행함과 동시에 해운기업의 육해상 모든 부서를 망라하는 안전관리시스템을 수립하여 시행하기 위한 규정이다. IMO에서는 SOLAS 협약의 제9장에 추가하여 강제화시키고 있다(제5장 선박 및 사업장의 안전관리, 제1절 선박의 안전관리체제).

4. 교통안전특정해역의 설정 및 관리

주요 개항 부근의 해상교통량이 밀집하는 지역에서는 그와 비례하여 충돌의 위험성이 증대될 수밖에 없다. 이러한 위험성을 제거하기 위하여 해상교통이 집중하는 연안해역을 해상교통의 특정해역으로 설정하여 교통위험을 인위적으로 관리하기 위한 제도가 교통안전특정해역제도이다.

특정해역이 설정된 해역은 인천, 부산, 울산, 포항 및 여수구역으로 총 5군데 해역이며, 이와 병행하여 항로지정방식을 사용하고 있다. 이는 교통안전특정해역을 항해하는 선박은 법에서 정한 지정항로를 따라 항행하도록 규정하고 있는 것으로 인천항 출입항로, 부산항 출입항로 및 광양만 출입항로에 대하여 분리선 및 분리대를 설치하여 출입항의 안전을 도모하고 있다.

또한 교통안전특정해역에서 어로에 종사하고 있는 선박으로 하여금 항로지정방식에 따라 항행하는 다른 선박의 통항에 지장을 주지 않도록 하며, 어망 기타 어구 등의 설치 및 양식어업을 금지시키고 있다(제2절 교통안전특정해역 등의 설정과 관리).

5. 유조선 통항금지해역 설정

유조선 통항금지해역이라 함은 선장 또는 항해당직을 수행하는 항해사가 유조선 등의 안전운항을 확보하고 해양사고로 인한 해양오염을 방지하기 위하여 유조선 등의 통항을 금지한 해역을 말한다. 이는 해양환경상의 대재앙이라고 할 수 있는 유류오염사고가 연안해역에서 발생하는 경우에 그 심각성이 더할 수 있기 때문에 이를 미연에 방지하고자 하는 목적이기도 하다.

즉 연안항해의 경우에 통항량의 증가 및 각종 항행상 장애물이 많을 수밖에 없으므로 연안으로부터 일정한 거리만큼 떨어져 항행하도록 함으로써 항행상의 안전을 확보할 뿐만 아니라 만일의 해양사고의 경우에도 해양생태계를 보호하고자 하는 목적이라 할 수 있다.

이러한 통항금지해역에 해당되는 선박으로는 경유 또는 중유 1천500킬로리터 이상을 화물로 싣고 운반하는 유조선 및 해양오염방지법상 유해액체물질로 규정된 물질을 1천500톤 이상 싣고 운반하는 케미컬선 등이 해당된다(제3절 유조선통항금지해역의 설정 및 관리).

6. 항로 등의 보전

선박이 출입하는 항로는 기본적인 교통로이므로 선박의 통항을 최우선적으로 보장하기 위하여 이를 제도화하고 있다. 즉, 항로를 보호하기 위하여 해양경찰서장을 감독권자로 하여 절대금지행위와 허가행위를 구분하여 법적으로 규제하고 있는 것이다.

절대금지행위로는 선박의 방치와 어망 등의 설치 및 투기가 있으며, 허가행위로는 항만해역의 해상교통의 질서를 보호하는 차원에서 해양스포츠를 허가사항으로 규제하고 있다. 즉, 개항 및 지정항, 항만수역 등에 있어서는 해상교통여건 등을 고려하여 스쿠버다이빙, 윈드서핑 및 수상스키 등과 같은 해양스포츠가 제한될 수 있는 것이다(제3장 수역안전관리).

1.5 선박의 입항 및 출항 등에 관한 법률

1.5.1 의 의

선박의 입항 및 출항 등에 관한 법률은 무역항의 수상구역 등에서 선박의 입항·출항에 대한 지원과 선박운항의 안전 및 질서유지에 필요한 사항을 규정함을 목적으로 한다. 동법은 무역항의 수상구역 등에서의 선박교통의 질서를 유지함을 목적으로 하는 법이므로 해사안전법에 대하여 특별법적인 지위에 있다. 그 입법목적을 구체적으로 분류하면 다음과 같다.

1. 해상교통의 안전유지

무역항의 항계안에는 출입항선박뿐만 아니라 우선피항선 등의 항내교통, 항내하역, 항내관광, 수로의 준설, 보수, 항로시설의 설치 및 정비로 인하여 선박교통이 복잡하여 위험이 발생하기 쉬우므로 이와 같은 위험을 방지하여 선박교통안전을 확보하기 위한 법이다.

2. 해상교통의 질서유지

무역항의 항계안에서 항로의 지정, 수리와 계선의 제한, 어로의 제한, 하역선석의 지정, 정박지의 지정 등으로 항만수역의 교통질서를 유지하여 해양사고를 예방하기 위한 법이다.

1.5.2 법적 성질

선박의 입항 및 출항 등에 관한 법률은 무역항의 항계내나 부근에서 선박교통의 안전과 질서유지의 목적을 위하여 해양수산부와 국민간의 관계에 대하여 규정한 행정법규이며, 내용의 지시, 명령, 제한, 금지, 허가 등을 준수하도록 하는 강행법규이며 또한 선박 및 관계자의 행동 또는 행위를 규제하는 교통규제법규이다.

또한 선박의 입항 및 출항 등에 관한 법률은 해사안전법에 대하여 특별법적 성격을 가지므로, 선박의 입항 및 출항 등에 관한 법률과 해사안전법에 상충하는 내용이 있으면 선박의 입항 및 출항 등에 관한 법률이 우선 적용되고 선박의 입항 및 출항 등에 관한 법률에 없는 규정은 해사안전법의 규정이 보충적으로 적용된다(법률 제3조 무역항의 수상구역 등에서의 선박 입항 · 출항에 관하여는 다른 법률에 특별한 규정이 있는 경우를 제외하고는 이 법에 따른다).

1.5.3 적용범위

1. 적용수역

선박의 입항 및 출항 등에 관한 법률 제1조는 "무역항의 수상구역 등에서 선박의 입항 · 출항에 대한 지원과 선박운항의 안전 및 질서유지에 필요한 사항을 규정함을 목적으로 한다"고 규정하고 있다. 그러므로 무역항의 수상구역 등의 항계밖에서는 선박의 입항 및 출항 등에 관한 법률이 적용되지 않고 해상교통안전법이나 국제규칙이 적용된다. 우리나라의 무역항은 부산, 인천, 목포, 포항, 태안항 등을 포함한 28개항이다.

그러나 유해물의 투기금지구역에는 무역항의 수상구역의 항계내 및 항계밖 10킬로미터 이내까지 준용하고 있는 등 "무역항의 수상구역의 항계내 또는 항계 부근"까지 적용하고 있다. 또한 위험물의 하역, 운반조항에는 항계밖의 일정한 장소를 지정하여 하역을 할 수 있게 하고, 이 경우 항계밖의 일정한 장소는 항계내에 있는 것으로 간주하고 있다.

2. 적용대상

무역항의 수상구역의 항계내에 있는 내외국적의 모든 선박이 그 대상이 되며, 인적 범위로써는 선장에게 적용되는 것이 원칙이나 선장에 관한 규정은 선장에 갈음하여 그 직무를 행하는 자에게도 이를 적용한다. 즉 선박의 운항지휘자로서 선장뿐만 아니라 선장의 지휘, 감독 아래 해상교통의 안전의무를 지고 있는 항해사와 도선사를 포함한다.

1.5.4 구 성

선박의 입항 및 출항 등에 관한 법률은 제1장 총칙, 제2장 입출항 및 정박, 제3장 항로 및 항법, 제4장 선박교통관제, 제5장 예선, 제6장 위험물의 관리 등, 제7장 수로의 보전, 제8장 등화 및 신호, 제9장 보칙 그리고 제10장 벌칙으로 구성되어 있다.

1.5.5 주요 내용

1. 입항·출항 및 정박

① 무역항의 항계내에 입항 또는 출항하는 선박은 규정에 따라 관할 지방해양수산청장에게 신고하여야 한다.

② 무역항의 항계내에 정박하는 선박(우선피항선은 제외)은 선박의 종류·톤수·흘수 또는 적재물의 종류에 따라 대통령령이 정하는 바에 의하여 지정된 정박구역 또는 정박지에 정박하여야 하며, 이동이 있을 경우에 관할 지방해양수산청장에게 신고하여야 한다.

③ 지방해양수산청장은 개항의 효율적인 운영을 위하여 필요하다고 인정하는 경우에

는 무역항의 항계내에 있는 선박에 대하여 항계내이나 항계밖의 다른 장소로의 이동을 명하거나 전시 · 사변 또는 이에 준하는 국가비상사태하나 국가안전보장상 필요한 경우에는 다른 무역항으로의 이동을 명할 수 있다.

2. 항로와 항법

무역항의 항계내에서 항행하는 선박은 다음의 항법원칙을 준수함은 물론 다른 선박에 위험을 주지 아니할 정도의 속력으로 항행하여야 한다.

1) 기본항법

① 항로우선원칙 : 지정항로내에서 항행하는 선박은 우선적 항행권을 가진다.
② 오른쪽 항행원칙 : 지정항로내에서 다른 선박과 마주칠 경우에는 각 선박은 항로의 오른쪽으로 항행하여야 한다.
③ 나란한 항행의 금지 : 항로의 폭이 좁으므로 2척 이상의 선박이 나란히 항행할 수 없다.
④ 추월금지 : 항로내에서는 앞에 있는 선박을 추월할 수 없다.
⑤ 흘수제한선, 위험물적재선 우선원칙이다.

2) 방파제 부근의 항법

방파제 부근에서 입항하는 선박과 출항하는 선박이 마주치는 경우에는 출항선이 방파제를 완전히 통과할 때까지 입항선은 방파제 밖에서 기다려야 한다. 이를 출항선 우선원칙이라 한다. 그리고 동시에 2척의 선박이 입항하거나 출항할 경우에는 이곳은 인위적인 좁은 수로이기 때문에 좁은 수도에서의 항법을 적용하여야 한다.

3) 기타의 항법

① 속력의 제한 : 무역항의 항계내와 항계 부근에 있는 선박은 다른 선박에 위험을 미치지 아니할 정도의 속력으로 항행하여야 한다.
② 항내 시설물 부근의 항법 : 방파제, 부두, 해안으로 길게 뻗은 육지, 잔교 등의 시설물 또는 정박선의 부근을 항행할 때는 오른쪽 항행원칙을 적용하여 이들을 오른쪽으로 보고 항행할 때는 이에 접근하여 항행하고, 이를 왼쪽으로 두고서 항행할 때는

가능한 멀리 떨어져서 항행하여야 한다.

③ 우선피항선의 피항의무 : 우선피항선은 항계내에서 동력선과 범선의 진로를 피해야 한다.

④ 예항의 제한 : 무역항의 항계내에서 4척 이상의 단정, 3척 이상의 부선 및 소선 또는 2척 이상의 항양선을 예항하고자 할 때에는 지방청장의 허가를 받아야 한다.

3. 위험물 반입 · 하역

① 위험물을 무역항의 항계내로 반입하고자 하는 자는 입항 전에 해양수산부령이 정하는 바에 따라 지방해양수산청장에게 신고하여야 한다.

② 선박이 무역항의 항계내에서 위험물을 하역할 때에는 지방해양수산청장의 허가를 받아야 한다.

4. 수로의 보전

① 누구든지 무역항의 항계내 및 항계밖 10킬로미터 이내의 수면에 선박의 안전운항을 해할 우려가 있는 폐기물을 버려서는 아니 된다.

② 지방해양수산청장은 무역항의 항계내 또는 항계의 부근에서 선박의 항행을 방해하거나 항행에 위험을 미칠 우려가 있는 표류물 · 침몰물 등의 물건을 발견한 때에는 그 물건의 소유자 또는 점유자에 대하여 그 제거를 명할 수 있다.

1.5.6 1972년 국제해상충돌방지규칙 및 해사안전법과의 차이

선박의 입항 및 출항 등에 관한 법률은 국제규칙 및 해사안전법에 대하여 특별법적인 성격을 가지므로 다음과 같은 차이점을 가지고 있다.

① 국제규칙과 해사안전법에 따르면 상호시계내에서는 자신의 우현측에 접근하는 선박을 두고 있는 선박이 피항선이 된다(국제규칙 제15조 및 해사안전법 제73조). 그런데 선박의 입항 및 출항 등에 관한 법률에서는 우선피항선은 항상 동력선과 범선의 진로를 피하여야 하며, 항로밖에서 항로에 들어오거나 항로에서 항로밖으로 나가는 선박은 항로를 항행하는 다른 선박의 진로를 피하도록 규정되어 있다.

방파제부근에서도 입항하는 선박은 방파제 밖에서 출항하는 선박의 진로를 피하도록 되어 있다. 이와 같이 횡단관계가 형성된다고 하여도 선박의 입항 및 출항 등에 관한 법률의 각 특칙이 우선 적용되어 횡단규정의 적용이 배제된다.

② 국제규칙과 해사안전법에 따르면 협수로에서 길이 20미터 미만의 선박이나 범선 그리고 어로작업에 종사하는 선박은 '항로내에서' 동력선이나 모든 선박에 대하여 통항불방해의무를 부담하나(동규칙 제9조 및 동법 제67조), 선박의 입항 및 출항 등에 관한 법률에서는 항로내에서 추월자체가 금지되어 있다. 또한 항로내에서의 병렬항행도 금지된다.

③ 국제규칙과 해사안전법에 따르면 협수로에서 길이 20미터 미만의 선박이나 범선 그리고 어로작업에 종사하는 선박은 '항로내에서' 동력선이나 모든 선박에 대하여 통항불방해의무를 부담하나(동규칙 제9조 및 동법 제67조), 선박의 입항 및 출항 등에 관한 법률에서는 잡종선, 예인선 및 통선 등은 항계내에서 다른 선박의 진로를 방해하여서는 아니된다.

④ 국제규칙이나 해사안전법에서는 협수로에서 어선은 통항불방해의무를 부담할 뿐이나(국제규칙 제9조, 해사안전법 제67조), 선박의 입항 및 출항 등에 관한 법률에서는 항로내에서는 원칙적으로 어로작업이 금지된다.

1.6 벨 북(Bell Book)

어원은 통상 선박이 엔진을 사용하기 위해서는 선교에 있는 텔레그래프(telegraph)를 움직여 원하는 위치로 이동하는데 이때에 텔레그래프가 움직이면 종(bell)소리와 비슷한 소리가 울리는 데에서 유래가 있다고 한다.

Stand by Book이라고도 하며 본선이 출입항(departure & arrival), 투묘(anchoring) 혹은 양묘(heaving anchor) 또는 전선(shifting) 등 본선 움직임이 변화하면 이 벨 북을 사용하여 간단하게 메모(memo)식으로 기록하는 소책자로써 기입해야 할 중요사항은 출입항 준비시간, 묘 및 계류삭(hawser)의 사용현황, 출입항시각, 도선사 승하선시각 및 성명, 예선(tug boat)의 성명, 사용시각 등의 중요자료를 기록한다. 작업이 종료되면 벨 북(Bell Book)에 기재된 중요사항 등은 나중에 로그북에 옮겨 적는다.

1.6.1 Bell Book 기재의 예문

0700 Arrival Oryuk Do, 010°, 0.8'(mile) off. 0715 Let go port anchor, 4shackles in water. 1020 Pilot on board, capt. W.H Kim. 1023 commence heave in anchor. 1030 S.B.E(stand by engine). 1032 Anchor aweigh(Up and down anchor). 1050 Tugs Hwang Ryung #1, 2 alongside stbd. 1054 Breakwater abeam. 1058 Danger signal. 1100 Taking lines to tugs. 1101 Let go stbd anchor in water. 1120 First line to shore. 1130 All lines made fast. 1132 In potion fore & aft. 1137 F.W.E(finished with engine).

1.6.2 사용 중인 Bell Book의 실례

Arr./Dep. From : To : Date : 20 . . .

AHEAD				STOP	ASTERN			
FULL	HALF	SLOW	DEAD SLOW		DEAD SLOW	SLOW	HALF	FULL

Pilot

No	Name	Embark	Disembark
1			
2			
3			

Tug Boat

No	Name	Alongside	Away
1			
2			
3			
4			
5			

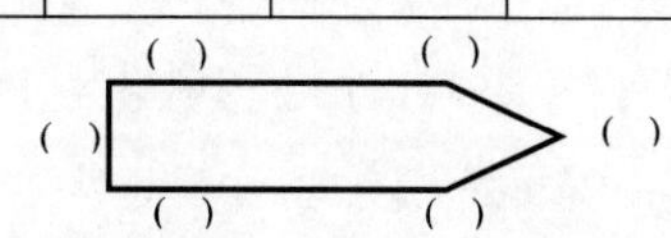

Berth. Name :

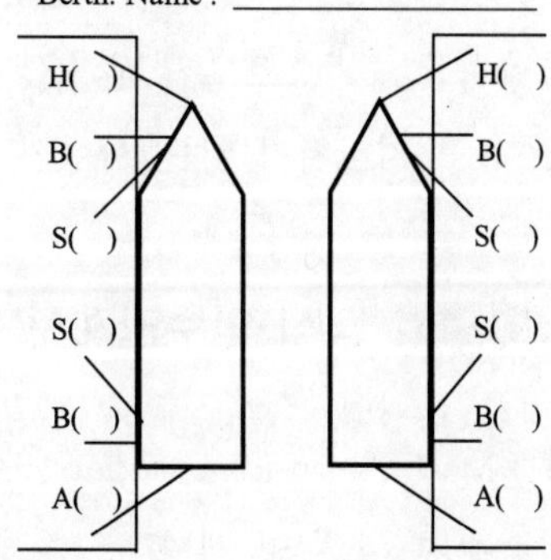

Remarks

[그림 1-33]

1.7 당직 인수인계(Bridge Hand-over)

항해당직사관이 4시간의 긴 항해당직을 수행하다 보면 당직종료 시점에 가서는 피곤이 엄습하며 집중력이 떨어지고 판단이 흐려지게 되며 후임 당직사관이 오기만을 기다리게 된다. 마침내 교대사관이 선교에 올라오면 안도감과 함께 신속하게 침실로 내려가 쉬고 싶은 생각에 후임자에게 당직 중 주의해야 하거나, 현재의 위험상태에 대하여 피상적이고 형식적인 인수인계(hand-over)를 실시하고 당직교대를 마치게 된다.

이러한 관행이 상선에서 특히 우리나라 항해사에게 많은 관습이 되어 왔는데 대단히 위험한 상황인 것이다. 다행히 예전에는 선박의 속력도 그리 빠르지도 또한 선박통항량도 많지 않은 시대여서 사고가 빈번하게 발생하지는 아니하였으나, 최근의 상황은 선복량(船腹量)의 폭발적인 증가와 더불어 선박의 대형화, 고속화의 진전에 따라 속력 25노트(knots), 선장이 300미터가 넘는 20,000 TEU급 초대형 Full Container 선박의 출현이 임박한 시점에서 우리 항해당직사관이 항해당직을 수행하는데 더욱더 주의해야 하고 당직 인수인계를 철저히 수행해야 할 필요가 있다고 하겠다.

특히 충돌사고의 발생시점을 살펴보면 항해당직 교대시점 전후가 많은 비중을 차지하고 있어 실증적으로도 당직의 정확하고 철저한 인수인계는 정말 중요하고 또한 아무리 강조해도 지나치지 않는 사항인 것이다.

1.7.1 당직 인수인계시 인계되어야 할 주요 사항

① Master's standing orders, Bridge order & Supplementary master's instructions의 숙지 여부 및 서명
② 해역주위에 발령된 혹은 예상되는 항행경보(navigational warnings)의 숙지
③ 본선의 현위치(position), 침로(course), 속력(speed) 및 흘수(draft)의 숙지
④ 현재의 조류(current), 조석(tide), 천후(weather), 시정(visibility)의 숙지
⑤ 당직 수행기간동안의 항해 및 안전장치의 상태
⑥ 자이로(gyro) / 마그네틱(magnetic) 컴퍼스(compass)의 에러(errors)
⑦ 본선 주위의 부이(buoys) 및 육지의 등화(shore lights)

⑧ 당직수행 중 조우가 예상되는 위험 및 상태에 대한 숙지 여부

1.7.2 해운선사에서 사용 중인 당직 인수인계 절차서 실례

CHANGING OVER THE WATCH	SHIP: DATE: 20 . . .					
(N.B. Changeover should be postponed when the ship is, about to be, engaged in a collision avoidance maneuver or navigational alteration of course.)	04	08	12	16	20	24
1. Has the relieving officer of the watch read the following directives?						
- Standing orders	☐	☐	☐	☐	☐	☐
- Bridge order	☐	☐	☐	☐	☐	☐
- Supplementary master's instructions	☐	☐	☐	☐	☐	☐
- Navigational warnings	☐	☐	☐	☐	☐	☐
- Are all members of the relieving watch capable of carrying out their duties?	☐	☐	☐	☐	☐	☐
2. Has the relieving officer of the watch been acquainted with the following information?	☐	☐	☐	☐	☐	☐
- Position, course, speed and draught of ship	☐	☐	☐	☐	☐	☐
- Course plotted on chart	☐	☐	☐	☐	☐	☐
- Prevailing/predicted tides, currents, weather and visibility and the effect of these factors upon course and speed	☐	☐	☐	☐	☐	☐
- Procedures for the use of main engines to maneuver when the main engines are on bridge control	☐	☐	☐	☐	☐	☐
- Operational condition of all navigational and safety equipment being used or likely to be used during the watch	☐	☐	☐	☐	☐	☐
- Gyro/magnetic compass errors	☐	☐	☐	☐	☐	☐
- Presence and movement of ships in sight or known to be in vicinity/effect on own ship	☐	☐	☐	☐	☐	☐
- Identification of shore lights, buoys etc.	☐	☐	☐	☐	☐	☐
- Conditions/hazards likely to be encountered during the watch	☐	☐	☐	☐	☐	☐
- Possible effects of any heel, trim, water density, 'squat' under keel clearance	☐	☐	☐	☐	☐	☐
3. Is the vision of all relieving member adjusted to the prevailing conditions?	☐	☐	☐	☐	☐	☐

RELIEVED:2/O	RELIEVING:C/O	RELIEVED:C/O	RELIEVING:3/O	RELIEVED:3/O	RELIEVING:2/O
0400 LT		0800 LT		1200 LT	

RELIEVED:2/O	RELIEVING:C/O	RELIEVED:C/O	RELIEVING:3/O	RELIEVED:3/O	RELIEVING:2/O
1600 LT		2000 LT		2400 LT	

MASTER : ____________________

[그림 1-34]

1.8 선교 체크리스트(Bridge Check List)

사람의 기억에는 한계가 있기 때문에 인류는 오래전부터 문자를 만들어 사용해 왔으며 중요한 일에 대해서는 기록을 남긴다. 위험이 많은 선박이라는 운항하는데 있어서 본선이 움직이거나 혹은 정지 등의 경우에는 선박에 탑재된 장치, 선박승무원 또는 주위의 장비의 이상 여부를 확인해야 할 필요성이 발생한다. 이전에는 이러한 위험성 여부의 판단 혹은 확인을 개인의 눈으로 혹은 경험만으로 실시하여 필수 확인사항을 누락하거나 간과하여 결과적으로 인명이 손상되거나 혹은 기기의 손상으로 이어져왔다. 즉 기기 자체의 결함에 의한 사고보다는 휴먼에러(Human error)에 의한 사고가 월등히 많은 비율을 차지한다. 체크리스트의 사용목적이 바로 이 휴먼에러를 줄여 인간의 경험에 의한 위험을 사전에 예방하고 기기의 결함을 파악하여 선박의 안전운항을 성취하고자 함이다.

1.8.1 체크리스트의 종류

현재 상선에서 사용하고 있는 체크리스트의 종류는 선박의 종류에 따라 선사(船社)에 따라서 조금씩 차이가 있으며, 여기서는 대부분의 선박에서 사용하고 있는 일반적인 체크리스트에 대하여 알아보도록 한다.

1.8.2 출항 전(Before Sailing) 체크리스트

출항 스텐바이 전, 이안작업 전 및 출항 전으로 세분하여 확인하고 체크한다.

① Pilot card 작성하여 도선사에게 전달
② 기상예보 및 항행경보 수신
③ Telegraph 및 Bow thruster 작동상태 확인
④ 화물고박 상태, 수밀문, 해치커버 및 램프 등의 고박상태 점검
⑤ 전승무원 재선 여부 등을 확인
⑥ 비상투묘 준비 및 계류설비 작동시험
⑦ 도선사 음주 / Drug 확인

⑧ Tug의 마력 및 척수, 작업방법 확인

⑨ 항로상의 교통량, 풍향/풍속, 해상상태, 조석/조류 및 시정상태 확인

⑩ 수심/Air draft의 여유 확인

CHECK LIST FOR BEFORE SAILING

<table>
<tr><td colspan="3">CHECK LIST N1</td><td colspan="3">BEFORE SAILING</td></tr>
<tr><td colspan="3">SHIP'S NAME :</td><td colspan="3">PORT :</td></tr>
<tr><td rowspan="2">TIME</td><td rowspan="2">NO</td><td rowspan="2">CHECK ITEMS</td><td colspan="2">VERIFY</td><td rowspan="2">REMARKS</td></tr>
<tr><td>Y</td><td>N</td></tr>
<tr><td rowspan="12">출항
S/BY 전</td><td>1</td><td>PILOT CARD(N4) 작성 (PILOT 에게 전달)</td><td></td><td></td><td></td></tr>
<tr><td>2</td><td>MASTER'S PASSAGE PLAN 점검
- 기항지 또는 항로의 변경 여부
- 계획 항로의 해도/수로지 소개정 여부</td><td></td><td></td><td></td></tr>
<tr><td>3</td><td>기상예보 및 항행경보 수신</td><td></td><td></td><td></td></tr>
<tr><td>4</td><td>기관실 제어 상태로 TELEGRAPH 테스트 및
CHECK LIST E1 에 의한 기관실 준비상태 확인</td><td></td><td></td><td>UMA 선박:시험 후 W/H 로 전환</td></tr>
<tr><td>5</td><td>BOW THRUSTER
- 사용/작동 시험(설치된 경우)
- WING BRIDGE, CONTROL KNOB 중립(설치된 경우)</td><td></td><td></td><td></td></tr>
<tr><td>6</td><td>화물고박 □, 선상 이동물 고박 □ 점검</td><td>X</td><td>X</td><td></td></tr>
<tr><td>7</td><td>HATCH COVER 또는 외부 RAMP 고박</td><td></td><td></td><td></td></tr>
<tr><td>8</td><td>수밀문/풍우밀문/방화문 폐쇄</td><td></td><td></td><td></td></tr>
<tr><td>9</td><td>TRIM & STABILITY 계산 및 점검</td><td></td><td></td><td>흘수, GM, BM, SF 등</td></tr>
<tr><td>10</td><td>전승무원 재선 □ 밀항자 및 마약 검색 □</td><td></td><td></td><td></td></tr>
<tr><td>11</td><td>서류 및 증서 확인</td><td></td><td></td><td>(R/O)</td></tr>
<tr><td>12</td><td>PILOT 승하선 준비
(PILOT LADDER, 승하선 장소, 통로, 조명 등)</td><td></td><td></td><td></td></tr>
<tr><td rowspan="4">이안작업전</td><td>13</td><td>선수미/BRIDGE 간 통신장비의 준비 및 점검</td><td></td><td></td><td>트랜시버, Mic 등</td></tr>
<tr><td>14</td><td>비상투묘 준비 및 계선 설비(MOORING EQUIP.)
작동 시험</td><td></td><td></td><td></td></tr>
<tr><td>15</td><td>작업안전수칙 환기
- PROPELLER 및 BOW THRUSTER 의 장애물 CLEAR 확인
- MOORING ROPE/TUG LINE 의 과도한 장력으로 인한 작업자 안전 확보
- 작업신호의 확인</td><td></td><td></td><td></td></tr>
<tr><td>16</td><td>육상측과 연결된 설비의 제거 확인</td><td></td><td></td><td></td></tr>
<tr><td rowspan="4">출항전
PILOT 와
협의</td><td>17</td><td>PILOT 음주/DRUG 확인</td><td></td><td></td><td></td></tr>
<tr><td>18</td><td>항로상의 교통량, 풍향/풍속, 해상, 조석/조류 및 시정상태</td><td></td><td></td><td>풍향: ,풍속:</td></tr>
<tr><td>19</td><td>수심/AIR DRAFT 의 여유</td><td></td><td></td><td>수심:</td></tr>
<tr><td>20</td><td>TUG 의 마력 및 척수, 작업방법</td><td></td><td></td><td>마력: ,척수:</td></tr>
<tr><td colspan="6">※ 묘박시는 CHECK LIST N3 ANCHORING 을 동시 사용할 것.</td></tr>
<tr><td colspan="3">ALL OF ABOVE WERE CHECKED BY

3RD OFFICER   DATE & TIME</td><td colspan="3">ALL OF ABOVE WERE VERIFIED BY

MASTER   DATE & TIME</td></tr>
</table>

[그림 1-35]

1.8.3 입항 전(Preparation for Arrival in Port) 체크리스트

입항준비 단계, 도선사 승선 전 단계, 도선사 승선 후 및 접안작업 전 체크 단계로 나누어 실시한다.

① 최근의 항행경보
② 입항서류(Customs Immigration & Quarantine, CIQ) 이상 유무 확인
③ Pilot card 작성
④ 비상투묘 준비 및 계류설비 작동시험
⑤ 엔진(main engine) 전 후진시험 및 Bow thruster 작동상태 확인
⑥ 통신장비(walki-talki) 점검
⑦ 도선사 승하선 준비
⑧ Tug boat의 마력 및 척수, 작업방법 확인
⑨ 항로상의 교통량, 풍향 / 풍속, 해상상태, 조석 / 조류 및 시정상태 확인
⑩ 작업안전수칙 환기(propeller, bow thruster의 장애물, 계류삭/tug line의 과도한 장력 지양, 안전한 장소의 확보 및 작업신호의 확인)

CHECK LIST FOR PREPARATION FOR ARRIVAL IN PORT

CHECK LIST N2	PREPARATION FOR ARRIVAL IN PORT
SHIP'S NAME :	PORT :

TIME	NO	CHECK ITEMS	VERIFY Y	VERIFY N	REMARKS
입항준비	1	입항전 점검항목	X	X	
		최근 항행경보 □ 조석/조류(TIDE TABLE 작성)□ UKC(M) 묘박□ or 접안□ PILOT 승선시간 (일시) 접안시간 (일시)			항만정보는 MASTER'S PASSAGE PLAN 을 참고
	2	서류 및 증서 준비(CIQ)			R/O
	3	PILOT CARD 작성 (PILOT 에게 전달)			- S/G TEST 포함
	4	계류설비 워밍 및 작동 시험			
PILOT 승선전	5	감속 장소 결정/ENG. S/BY 시각 기관실 통보.			
	6	ENG. 전/후진 시험 CHECK LIST E2 에 따른 기관실 준비상태 확인			
	7	BOW THRUSTER - 사용/작동 시험 - WING BRIDGE, CONTROL KNOB 중립확인			환기 FAN 운전 확인
	8	통신장비 준비 (선수미 부서/선교)			트랜시버, Mic 등
	9	PILOT 승/하선 준비 (PILOT LADDER, 승하선 장소, 통로, 조명 등)			
PILOT 와 협의	10	음주/DRUG 확인			
	11	사용 BERTH 길이 및 높이 확인			길이:
	12	흘수/AIR DRAFT, 수심의 확인			수심:
	13	항로상의 교통량, 풍향풍속, 해상, 조석조류 및 시정상태			풍향: ,풍속:
	14	TUG 의 마력 및 척수, 작업방법			마력: ,척수:
※ 묘박시에는 CHECK LIST N3 ANCHORING 을 동시 사용할 것					
접안작업전	15	비상투묘 준비, 계류색, HEAVING LINE, MESSENGER LINE, STOPPER, WIRE 등 준비 및 점검			
	16	작업안전수칙 환기 - PROPELLER 또는 BOW THRUSTER 의 장애물 유무 - 계선/TUG LINE 의 과도한 장력 회피 - 안전한 장소의 확보 및 작업신호의 확인			

ALL OF ABOVE WERE CHECKED BY		ALL OF ABOVE WERE VERIFIED BY	
WATCH OFFICER	DATE & TIME	MASTER	DATE & TIME

[그림 1-36]

1.8.4 투묘 전(Anchoring) 체크리스트

투묘작업 전, 작업 중, 작업완료 후로 각각 나누어 확인한다.

① 투묘위치(anchor position) 및 시간 확인

② 투묘예정시간 관련 부서에 통보(기관실, 작업자 등)

③ 투묘계획 수립(정박지 수심 및 해저저질, 접근침로의 결정, 조류/바람의 영향력 확인)

④ 투묘설비의 점검

⑤ 정박지 여유수역 및 타 선박동정 확인

⑥ 투묘(양묘) 위치/시간 통보(VTS, pilot, 항만당국 등)

⑦ 형상물 게시(강하) 또는 등화 점등(소등)

⑧ 악천후에 대한 대책(농무, 강풍 등)

⑨ Anchor의 고박상태 확인(windlass brake, brake handle, compressor bar, anchor stopper)

CHECK LIST FOR ANCHORING

CHECK LIST N3			ANCHORING				
SHIP'S NAME :			PORT / POSITION :				
TIME	NO	CHECK ITEMS	투묘		양묘		REMARKS
			Y	N	Y	N	
작업 시작전	1	ANCHOR POSITION 및 시간 확인 (항만 , VTS , PILOT)					
	2	투묘[양묘] 예정시간 통보(선장, 기관실, 작업자)					
	3	투묘[양묘] 계획 수립	X	X	X	X	
		- 묘박지 수심 및 해저 저질 확인			X	X	심해투묘 등
		- 접근[출발]침로의 계획 확정					
		- 조류/바람의 영향력 점검					
	4	설비/기구의 점검 및 준비 WINDLASS□ANCHOR□등화/형상물□음향신호□					
	5	묘박지 여유수역 및 타 선박 동정 점검					
	6	위치측정을 위한 육상물표 선택			X	X	
	7	ENG.ROOM 준비상태 확인 (CHECK LIST E1 또는 E2 에 따름)					
	8	선수하부 소형선박 또는 장애물 유무					
	9	WINDLASS 작동자 보안경 및 헬멧 착용			X	X	
	10	작업자 전원 보호된 장소에 위치					
작업중	11	ANCHOR CHAIN 에 과도한 장력발생시 원인파악					
	12	BROUGHT UP[WEIGHED] ANCHOR 확인					
작업 완료후	13	투묘[양묘] 위치/시간 통보(VTS, PILOT, 항만당국)					
	14	투묘위치 확인 및 위치측정용 육상물표의 해도상 표시			X	X	
	15	형상물 게시[강하] 또는 등화 점등[소등]					
	16	악천후(농무, 강풍 등)에 대한 대책			X	X	
	17	ANCHOR 고박상태 확인 - Windlass Brake □ - Brake Handle □ - Compressor Bar □ - Anchor Stopper □	X	X			

※ 양묘시 이 점검표 작성전 점검표 N1(출항시) 또는 N2(입항시)를 작성해야 한다.

투묘	양묘	투묘	양묘
OOW: DATE & TIME	OOW: DATE & TIME	MASTER: DATE & TIME	MASTER: DATE & TIME

[그림 1-37]

1.8.5 Pilot Card 체크리스트

도선사 승선시 도선사에게 제공되는 본선의 정보를 한눈에 파악할 수 있도록 만든 카드로 여기에는 본선제원(ship's particular), 기관 및 Bow thruster 마력, Main engine의 각 단계별 속력 등을 표시하였고, 뒷면에는 항해사가 작성하는 각종 항해장비의 체크리스트가 있어 도선사 승선 전 항해사는 이를 확인하고 이상 유무를 표시한다.

PILOT CARD

PILOT CARD N4			M /
			Voy. No. :
PORT :	A	D	DATE : 2001 . . .

Call sign : 3FJT8 Flag : PANAMA Year Built : 1998

Gross Ton : 56,955 Net Ton : 29,584 Deadweight : 99,991.6 M/T

Draft	Aft : m(ft) cm(in)	Fwd : m(ft) cm(in)

Summer draft : m

Ship's particulars

Length overall	:	243.85 m	Anchor chain	: Port	13	shackles
Breadth	:	42.0 m		Stbd	13	shackles
Depth	:	21.0 m		(1 shackle : 27.5 m)		

Type of Engine : Max. Power : PS

Engine Order	RPM	Speed (knots) BALLAST	LADEN
Nav. Full ahead	98	15.0 KTS	14.5 KTS
Full ahead	83	13.2 KTS	12.7 KTS
Half ahead	65	10.4 KTS	9.8 KTS
Slow ahead	45	7.2 KTS	6.8 KTS
Dead slow ahead	32	5.1 KTS	4.8 KTS
Dead slow astern	32	Time limit astern : N/A min	
Slow astern	45	Full ahead to full astern : sec	
Half astern	65	Max. No. of consecutive starts : 19 TIMES	
Full astern	83	Min. RPM : 21.7 , Min, Speed : 3.4 KTS Astern power : 50 % ahead	

Bow Thruster Power : N/A PS

* The Rear Side : Check List For Navigational Equipment.

[그림 1-38] PILOT CARD (1/2)

CHECK LIST FOR NAVIGATIONAL EQUIPMENT

DATE & TIME : ____________________

I. NAVIGATION AIDS

1 MAGNETIC COMPASS ☐

2 GYRO COMPASS ☐
GYRO REPEATER ☐
GYRO ERROR _ _ _ _ _ ° E/W

3 AUTO PILOT ☐

4 CLEAR VIEW SCREEN ☐
WINDOW WIPER ☐

5 RPM INDICATOR ☐

6 RUDDER ANGLE INDICATOR ☐

7 STEERING GEAR ☐
NUMBER OF POWER UNITS OPERATIONS(2) ☐ or ☐
HYDRAULIC UNIT ☐

II. SEARCHING EQUIPMENT

1 ECHO SOUNDER ☐

2 SPEED LOG ☐

3 RADAR ☐ 3cm ☐ 10cm
ARPA ☐

4 ANEMOMETER ☐

5 ELECTRIC CLOCK ☐

III. NAVIGATOR

1 GPS/DGPS NAVIGATOR ☐

2 NAVTEX RECEIVER ☐

3 TOTAL NAVIGATION SYSTEM ☐

IV. LIGHTING

1 NAVIGATION LIGHT ☐

2 SIGNAL LIGHT ☐

3 ANCHOR LIGHT ☐

4 OUTSIDE LIGHT ☐

5 FLOOD LIGHT ☐

V. FLAGS & SHAPES ☐

OTHER INFORMATION :

Function	Duty officer	Master	Pilot
Name/Signature			
Date	/	/	/

[그림 1-38] PILOT CARD (2/2)

1.8.6 Helicopter / Ship Operation 체크리스트

항해 중 본선 승무원의 위급한 질병발생, 사고로 인한 긴급 후송환자 등이 발생하여 본선이 병원이 있는 항구에 도착할 시간적 여유가 없을 때 혹은 중동지역의 원유 선적항(보통은 육지에서 멀리 떨어진 sea island에 터미널이 있다)에서는 도선사가 헬리콥터를 이용하여 승하선을 한다. 헬리콥터를 이용하여 환자를 수송하거나 도선사 승하선시는 작업의 성질상 특별한 주의가 필요하다.

① 착륙위치표시 점검 및 주변장애물 고박 / 정리

② Wind-sock 또는 Pennant 게양

③ 선교와 현장간의 통신수단 / 방법 확인

④ Deck 소화라인 압력유지 및 Fire hose 사용 준비

⑤ 휴대용 소화기 사용 준비

⑥ 현장인원 배치(헬멧, 보안경 및 작업복 착용)

⑦ 구조정 강하 준비

⑧ Fire axe, Wire cutters, 홍색비상전등 및 구급상자 사용 준비

⑨ 화물탱크(cargo tank) 내 압력을 정압으로 유지(유조선의 경우)

CHECK LIST FOR HELICOPTER / SHIP OPERATION

CHECK LIST N5		HELICOPTER/SHIP OPERATION			
SHIP'S NAME :		PORT :			
NO		CHECK ITEMS	VERIFY Y	VERIFY N	REMARKS
A	착륙(LANDING ON) 또는 WINCHING 점검		■	■	
	1	착륙위치표시 점검 및 주변장애물 고박/정리			
	2	WINDSOCK 또는 PENNANT 게양			
	3	BRIDGE 와 현장간의 통신수단/방법 확인			
	4	DECK 소화 LINE 압력유지 및 FIRE HOSE 사용준비			
	5	휴대용 소화기 사용준비			
	6	현장인원 배치(헬멧, 보안경 및 작업복 착용)			
	7	구조정 강하준비			
	8	FIRE AXE, CROWBARS, WIRE CUTTERS, 홍색비상휴대전등 및 구급상자 사용준비			
	9	CARGO TANK 내 압력을 정압으로 유지 (TANKER)			
	10	CARGO 의 VAPOUR GAS 방출금지(GAS CARRIER)			GAS CARRIER ONLY
B	착륙(LANDING ON)점검 [WINCHING 시 작성불요]		■	■	
	1	현장인원 착륙사실 인지			
	2	착륙장소에 해수 유입 금지			
	3	착륙시 방해되는 장애물 제거			
	4	HELICOPTER 를 고박할 수 있는 ROPE 의 준비			
	5	HELICOPTER 의 ROTOR 회전 및 배기가스의 위험성 인지			
비고	착륙시는 A+B 모두 점검, WINCHING 시는 A 만 점검한다.				

ALL OF ABOVE WERE CHECKED BY		ALL OF ABOVE WERE VERIFIED BY	
WATCH OFFICER	DATE & TIME	MASTER	DATE & TIME

[그림 1-39]

1.8.7 제한수역(선박폭주 해역, 운하, 해협, TSS 해역) 항해시 체크리스트

선박의 통항이 빈번하거나 운하, 말라카 해협 등 특별한 주의와 사전 준비를 요하는 해역통과가 예상될 때 사용한다.

① 해당 수역통과 예상시간

② VTS 참가 여부

③ 해당 수역항행경보 / 기상예보 점검 확인

④ 해당 수역해도 / 항행규칙/수로서지의 확인

⑤ 계획항로의 안전 확인/피험선 설정

⑥ 뚜렷한 육상물표의 표시 / 해안선의 식별표시

⑦ 제한수역 통과 전 점검 항목

- 예상되는 교통량
- 흘수 / 최저수심
- 조석 / 조류 영향
- 항해 / 조타설비의 작동상태
- Echo sounder 작동

⑧ 안전속력 설정

⑨ 해적피습 방지대책 수립

CHECK LIST FOR RESTRICTED AREAS

<table>
<tr><td colspan="3">CHECK LIST N6</td><td colspan="3">PASSAGE PLANNING IN RESTRICTED AREA (COASTAL, TSS, STRAIT, CANAL, ETC)</td></tr>
<tr><td colspan="3">SHIP'S NAME :</td><td colspan="3">PORT :</td></tr>
<tr><td rowspan="2">TIME</td><td rowspan="2">NO</td><td rowspan="2">CHECK ITEMS</td><td colspan="2">VERIFY</td><td rowspan="2">REMARKS</td></tr>
<tr><td>Y</td><td>N</td></tr>
<tr><td rowspan="23">해당수역진입전/통과중</td><td>1</td><td>해당수역의 통항예상 시간 (FM : ~ TO :)</td><td></td><td></td><td></td></tr>
<tr><td>2</td><td>VTS 참가여부(VHF CH. : VTS 명칭)</td><td></td><td></td><td>보고지점 확인</td></tr>
<tr><td>3</td><td>해당수역 항행경보/기상예보 점검</td><td></td><td></td><td></td></tr>
<tr><td>4</td><td>해당수역의 해도/규칙/서지의 확인</td><td></td><td></td><td>최신판, 소개정</td></tr>
<tr><td>5</td><td>계획항로의 안전 확인/피험선 설정</td><td></td><td></td><td></td></tr>
<tr><td>6</td><td>계획항로 변경대비 2차 계획항로 설정(적용가능시)</td><td></td><td></td><td>수로, 해협, 협수도 등</td></tr>
<tr><td>7</td><td>뚜렷한 육상물표의 표시, 해안선의 식별표시</td><td></td><td></td><td></td></tr>
<tr><td rowspan="5">8</td><td>제한수역 통과전 점검항목</td><td rowspan="5">X</td><td rowspan="5">X</td><td></td></tr>
<tr><td>예상되는 교통량□, 흘수/최저수심 (M/ M)</td><td></td></tr>
<tr><td>SQUAT (M) , 조석/조류영향 □</td><td>조석표, 기상도 등</td></tr>
<tr><td>항해/조타설비의 작동상태 □</td><td></td></tr>
<tr><td>ECHO SOUNDER 작동 □</td><td></td></tr>
<tr><td>9</td><td>BRIDGE/ENG. ROOM 당직자 배치</td><td></td><td></td><td></td></tr>
<tr><td>10</td><td>선장에게 통항시간 보고 (선장직접지휘) (HRS)</td><td></td><td></td><td>"직접지휘" 규정참조</td></tr>
<tr><td>11</td><td>안전속력 설정(KTS or RPM)</td><td></td><td></td><td>통항량,시정,기상,
통항규칙 고려</td></tr>
<tr><td>12</td><td>통항중 선위 측정 주기설정(MIN)</td><td></td><td></td><td></td></tr>
<tr><td>13</td><td>AUTO PILOT 자동/수동 전환시험</td><td></td><td></td><td></td></tr>
<tr><td>14</td><td>조타기 POWER UNIT NO.1/NO.2 동시 작동여부</td><td></td><td></td><td></td></tr>
<tr><td>15</td><td>해적피습 방지 대책수립</td><td></td><td></td><td>해당구역 통항시</td></tr>
<tr><td></td><td></td><td></td><td></td><td></td></tr>
<tr><td></td><td></td><td></td><td></td><td></td></tr>
<tr><td></td><td></td><td></td><td></td><td></td></tr>
</table>

<table>
<tr><td>진입전</td><td colspan="3">당직교대시 확인</td><td rowspan="2">ALL OF ABOVE WERE VERIFIED BY

MASTER   DATE & TIME</td></tr>
<tr><td>(OOW)
/</td><td>3/0
/</td><td>2/0:
/</td><td>C/0:
/</td></tr>
</table>

[그림 1-40]

1.8.8 제한시정 해역 항해시 체크리스트

① 선장에게 보고, 필요시 선장직접지휘

② 예상 시정거리 확인

③ E/R 유인당직 배치(UMA 선박)

④ 레이더 2대 작동 및 움직이는 물표 Plotting 실시

⑤ 무중신호 취명

⑥ 전속력 결정/시행

⑦ 추가 견시인원 배치(필요시)

CHECK LIST FOR RESTRICTED VISIBILITY

CHECK LIST N7	NAVIGATION UNDER RESTRICTED VISIBILITY
SHIP'S NAME :	AREA : Arr/Dep.□ Coastal□ Ocean□

TIME	NO	CHECK ITEMS	VERIFY		REMARKS
			Y	N	
제한시정상태 (3 마일 이하)	1	선장에게 보고, 필요시 선장직접지휘			(통보시간:)
	2	예상 시정거리 (miles)			
	3	BRIDGE/ENG. ROOM 당직배치 확인			
	4	E/R 유인당직배치(제한시점 1 마일 이하)			출입항
	5	항해등 점등(주간)			
	6	RADAR 2 대 작동 및 움직이는 물표 PLOTTING 실시			ARPA 작동
	7	무중신호 취명			
	8	안전속력의 결정(KTS or RPM)			
	9	VHF 지정채널 청취(CH.16 or 지정 CH.)			
	10	추가 견시요원 배치 (필요시)			
	11	근접선박이 있는 경우 VHF 로 상대방 의도 확인			가능한 경우
	12	예기치 않는 상황 대비 투묘 또는 대기 가능성 검토			입출항시 연안항해시

* 제한시계하에서의 운항안전 지침

시 계	해 역 구 분		
	출 입 항	연 안	대 양
0.5 마일 이내	1. 출입항 보류 및 관련부서 통보	1. 안전속력으로 감속 2. 필요시 투묘 대기 3. 견시원 추가 배치	1. 안전속력으로 항행 2. RADAR 2 대 운용 3. 선교 2 인 당직체계 유지
1 마일 이내	1. 출입항 안전성 검토 – 도선사 및 항만당국 정보수집 2. TUG ESCORT 요청 3. 필요시 투묘 대기 4. 관련부서 통보	1. 안전속력으로 감속 2. 견시원 추가 배치	
3 마일 이내	1. 보침기능 최저 속력 유지 2. 출입항 부서배치	1. 안전속력으로 항행 2. 선장 선교 직접지휘	

시계제한 발생시	통과중 교대시			ALL OF ABOVE WERE VERIFIED BY	
(OOW) /	3 / 0: /	2 / 0: /	C / 0: /	MASTER	DATE & TIME

[그림 1-41]

1.8.9 악천후 항해시 체크리스트

악천후 예상 / 조우 / 통과시로 구분하여 점검한다.

① 기상정보 수신주기 증가

② 화물고박 사전점검 및 필요시 재고박 실시

③ 이동물의 고박상태 및 수밀문 / 풍우밀문 폐쇄상태 재확인

④ 필요장소에 Safety / Man rope 설치

⑤ 당직기관사에 예상되는 악천후 통보

⑥ 전 승무원에게 예상되는 악천후 공지 및 외부출입통제

⑦ 침로변경 고려 및 필요시 피항지 확인

⑧ 필요시 선장 직접지휘

⑨ 수동조타 시행

⑩ 주기적 거주구역 내부 순찰실시

⑪ Cargo hold, Bilge 주기적 배출(container ship)

⑫ 선체 손상 유무 점검

⑬ 선체의장품 손상 유무 점검

⑭ Ballast tank 및 Hold bilge 측심(sounding) 시행

⑮ 화물고박(securing)상태 재점검

⑯ 구명 / 소화설비 상태 점검

CHECK LIST FOR HEAVY WEATHER

CHECK LIST N8	NAVIGATION UNDER HEAVY WEATHER
SHIP'S NAME :	PORT :

TIME	NO	CHECK ITEMS	VERIFY		REMARKS
악천후 예상시 (BF7 이상)	1	기상정보 수신 주기 증가 및 O.S.R.S 권고 등의 분석			필요시 담당부서에 기상정보 요청
	2	화물고박 사전점검 및 필요시 재고박시행			화물취급설비 등
	3	이동물의 고박상태 및 수밀문/풍우밀문 폐쇄 상태 재확인			
	4	필요장소에 SAFETY/MAN ROPE 설치			
	5	당직기관사에게 예상되는 악천후 통보			
	6	전 승무원에게 예상되는 악천후 공지 및 외부 출입 통제			
	7	침로변경 고려 및 필요시 피항지 확인			
	※ 열대성저기압 조우 예상시는 "비상대응 절차 참조				
악천후 조우시	1	필요시 선장직접지휘			
	2	침로 조정 또는 선속 감속 (DEG, KTS.)			선체 및 화물에 충격 억제
	3	수동조타 시행			
	4	주기적 거주구역 내부 순찰실시 (주기 : HRS)			
	5	CARGO HOLD, BILGE 주기적 배출 (CONTAINER)			
악천후 통과시	1	선체 손상유무 점검			
	2	선체의장품 손상유무 점검			
	3	화물 SECURING 상태 재점검			
	4	BALLAST TANK 및 HOLD BILGE 측심시행			
	5	구명/소화 설비상태 점검			

ALL OF ABOVE WERE CHECKED BY					ALL OF ABOVE WERE VERIFIED BY MASTER		
악천후 예상시	조우시			통과후	예상시	조우시	통과후
C / O	3 / O	2 / O	C / O	C / O			
/	/	/	/	/	/	/	/

[그림 1-42]

1.8.10 유빙해역 항해시 체크리스트

유빙해역 및 결빙항구 통과시로 구분하여 점검한다.

① 유빙정보 확보

② 선장에게 유빙정보 통보(육상 혹은 기상정보 수신시)

③ 유빙해역 회피항로 설정

④ 수밀문 폐쇄 재확인

⑤ 선장직접지휘

⑥ 노출된 파이의 결빙방지를 위한 Drain 실시

⑦ 선속 및 침로조정

⑧ 선수에 경계원 배치(필요시)

⑨ Sea chest High에서 Low로 전환

⑩ 흘수조정

CHECK LIST FOR NAVIGATION IN ICE

CHECK LIST N9			NAVIGATION IN ICE(유빙해역 또는 항구)			
선 명:			위치:	DATE/TIME: . . . / : ( )		
TIME		No	CHECK ITEMS	VERIFY Y	VERIFY N	REMARKS
유빙해역	접근시	1	유빙정보 확보 및 MONITORING (FAX, 기상도, NAVTEX 등 유빙경보)			
		2	선장, 항해사, 기관사 및 해당승무원에게 유빙 정보 통보			
		3	유빙해역 회피토록 항로 조정			
		4	수밀문 폐쇄 재확인			
		5	RADAR 작동, 정체불명 물표 발견시 충분한 거리 유지 통과			
		6	노출된 PIPE 의 결빙 방지를 위한 DRAIN 실시			
	통과시	7	선장직접지휘			
		8	선속 및 침로 조정			
		9	선수(FORECASTLE)에 견시원 배치(필요시)			
		10	SOLAS CH.V 에 따라 MESSAGE 발송			
		11	유빙에 의한 SEA CHEST 폐쇄 방지를 위해 SEA CHEST 를 HIGH 에서 LOW 로 변환			
결빙항구	입항전	1	동계 및 고위도 항구 입항전 ICE 정보 유무 확인 및 입항 가능 여부 확인			대리점에 문의 및 기상정보 참조
		2	결빙정보가 있는 경우, '특정사안 협의' 발의			
		3	SEA CHEST 는 HIGH 에서 LOW 로 변경			
		4	BOW THRUSTER 및 PROPELLER IMMERSION 을 위해 DRAFT 는 가능한 한 깊게 유지			수로 또는 부두수심의 허용범위 내에서 조정
		5	선체/설비/기기/PIPE LINE 등 동파 방지조치			
	입항시	6	강의 상류지역으로부터 또는 항내의 결빙상태 및 유빙 부유여부 확인			얼음 상태(두께) 확인
		7	유빙 존재시 안전속력 유지 및 PILOT 와 ASTERN ENG. 사용 위험성 및 안전한 조선방법 사전협의(필요시, TUG BOAT 추가 수배)			선박상세 및 'NON-ICE STRENGHENED SHIP' 등을 PILOT 에게 제공/쇄빙작업 등 확인 ■ PILOT 감시감독
	접안중	8	선체/설비/기기/PIPE LINES/PROPELLER 및 기타에 대한 동파 점검 시행			
		9	MAXIMUM SAILING DRAFT 로 조정			
	출항시	10	기상정보 수집 및 대리점/PILOT 와 협의 (출항여부 결정 및 TUG BOAT 추가 수배 등)			선박상세 및 'NON-ICE STRENGHENED SHIP' 등을 PILOT 에게 제공/쇄빙작업 등 확인 ■ PILOT 감시감독
		11	안전속력 유지 및 ASTERN ENG. 사용 제한			
		12	선수미 견시원 배치(필요시)			

ALL OF ABOVE WERE CHECKED BY				ALL OF ABOVE WERE VERIFIED BY MASTER			
접근시/입출항시	통과시/입출항시			접근시	통과시	입항시	출항시
OOW	3 / O	2 / O	C / O				
/	/	/	/	/	/	/	/

[그림 1-43]

1.9 Condition Report 작성

선박이 항구에 입항/출항하거나 혹은 투묘/양묘시 그 당시 본선 상태(condition)를 양식에 따라 기록하게 되는데 이것이 컨디션 리포트이다.

여기에는 당시의 연료유, 윤활유, 청수밸러스트 보유량(수급량) 등이 기록되며 또한 도선사 승선/하선 시간, 접안/이안 시간, 투/양묘 시간, 흘수, 화물의 선적/양하량 등이 기재되어 본선 입출항 수속시 필요하다.

CONDITION REPORT

ARR/DEP. CONDITION REPORT

(for CNTR)

M.V.				VOY. NO.		
PORT				DATE	20 . . .	
DRAFT	FWD:		AFT:		MID:	
BUNKER	H.F.O (M/T)	I.F.O (M/T)	M.D.O (M/T)	M/E CYL (LTR)	M/E SYS (LTR)	D/G SYS (LTR)
SUP'D						
R.O.B						
F.W(M/T)	SUP'D		R.O.B.		BALLAST	
CNTR	FULL		EMPTY		TTL	
	20FT	40FT	20FT	40FT	20FT	40FT
LOAD.						
DISCH.						
R.O.B						
REMARKS						

[그림 1-44]

1.10 Noon Report 작성

상선의 모든 데이터의 기준은 정오를 기본으로 삼는다. 범선시대에는 태양이 정중하는 자오선정중시(子午線正中時)가 그날 선박의 기준이 되었다. 현재에도 위치보고시나 일일 항해거리, 항해한 항정 및 남은 항정 등이 정오위치를 기준으로 삼는다. Noon Report는 전일 정오부터 당일 정오까지의 항정(distance), 평균속력, 평균 RPM, 슬립, Distance to go(miles), 조수량(造水量), 청수소모량, 연료유 및 윤활유 소모량 및 잔량, 기상상태(천후, 바람의 방향/세기, 해상상태 등), 전 항구로부터 총 항정(total distance), 평균속력, 평균 RPM, 평균 슬립(slip), 다음 항구의 ETA(Estimated Time of Arrival) 및 Target speed 등을 2등항해사가 3등기관사의 협조로 5부(선교, 기관실, 통신실, 선장, 기관장)를 작성 배포한다.

NOON REPORT

NOON REPORT

Voy. No ________ Date 20 . . .

Position Lat. ________ Long. ________ T co. ________

	Running Hours	Distance By Obs.	Average Speed	Average R.P.M	Slip(%)
Noon to Noon					
Total					
Distances to Go	Weather	Wind Dir/For.	Sea State	Evapor. (Day)	F.W Cons/ROB
	H.F.O (M/T)	I.F.O (M/T)	M.D.O (M/T)	Cyl. oil (LTR)	Sys. oil (LTR)
Consump/day					
R.O.B					

E.T.A :	Remarks:

Noon Report:

C/Eng.		Master	

[그림 1-45]

1.11 AB-Log(Abstract Log) 작성

1.11.1 정 의

선박이 1항차(one voyage)를 종료하면 당해 항차의 실적을 한눈에 볼 필요가 있다. 특히 선박이 용선(傭船, chartered ship)상태로 운항이 되는 경우 선박의 운항실적은 선주(船主, owner)와 용선주 양측의 이해관계가 첨예하게 대립하여 차기(次期)의 용선가격을 결정한다. 즉 AB-Log는 당해 선박의 당해 항차의 운항실적을 한눈에 파악할 수 있는 '운항자료'라고 정의된다.

1.11.2 작 성

통상 2등항해사가 3등기관사의 도움으로 작성하며 항차 종료 후 본사의 영업부 및 운항 관련 부서로 송부하며 선박이 용선되었을 경우에는 용선주(charterer)에게도 1부 송부한다.

1.11.3 주요 내용

① 항차시작 / 종료시점

② 당해 항차의 총 운항시간

③ 당해 항차의 기항항구의 정박시간 / 하버(harbour)시간 / 대양 항해시간

④ 당해 항차의 기항항구의 하버(harbour)항해거리 / 대양 항해거리

⑤ 당해 항차의 기항항구의 평균속력 / 평균 RPM / F.O, D.O 소모량

⑥ 당해 항차의 총 정박시간 / 하버(harbour)시간 / 대양 항해시간 등

1.11.4 작성시 주의사항

1. AB-Log에 기록되는 모든 시간은 6분단위로 기록한다.

㉮ 접안완료 시간이 12시 13분이라면 6분단위로 12시 12분 혹은 12시 18분으로 기록해야 한다(왜냐하면 시간은 소수 첫째자리까지만 기록해야 하므로 위의 예처럼 12시 12분이면 12.2로, 12시 18분이면 12.3으로 표시할 수 있다).

2. 용어의 정의

① 입항시간(arrival time) : 당해 항구에 입항하기 위하여 투묘한 시간(let go anchor), 첫 줄을 부두에 잡은 시간(first line to pier) 및 첫 줄을 부이에 잡은 시간(first line to buoy) 중 최초로 발생한 시각(first motion time)이다.

② 출항시간(departure time) : 당해 항구의 출항을 위하여 마지막 줄이 부두에서 떨어진 시간(cast off last line) 또는 Up & down anchor 한 시각 중 최후 발생한 시각(last motion time)이다.

③ 정박시간(hours in port) : 당해 항구의 입항시부터 출항시까지의 시간을 말하며 이 정박시간에는 항내의 전선시간(shifting hour), 순수 급유시간(bunkering hour), 순수 수리시간(repairing hour) 등 모든 지연시간(detention hour)을 포함한다.

④ 항해시간(hours under way) : 한 항구의 출항시로부터 타 항구의 입항시까지의 시간을 말하며 Harbour voyage hour와 Sea voyage hour로 구분한다.

㉠ Harbour voyage hours

- 입항시 : S/B ENG.부터 입항시(arrival time)까지 시간
- 출항시 : 출항시(departure time)부터 R/UP ENG.까지 시간

⑤ 지연시간(detention hour) : 사고로 인한 정선, 감속운전, 이로(deviation), 하역지연, 입출항지연 및 기타 Delay 시간을 말하며 다음 사항을 포함한다.

㉠ 항해 중

- 사고(fire, collision, grounding), 기관고장 등으로 인한 정선
- 사고, 기관고장 등으로 인한 감속운전
- 태풍 등을 피항하기 위한 이로
- 인명손상, 환자발생 등으로 인한 회항

- 회사의 지시에 의한 Waiting

㉡ 정박 중

- 하역기기나 기관 고장 등으로 인한 지연
- 인명손상이나 환자발생, 미귀선자로 인한 지연
- 입출항을 위해 Pilot나 Tug boat 등을 기다리는 시간
- 선박 자체의 원인이 아닌 사유 또는 사고로 인해 선박의 정상운항이 정지된 시간

⑥ 입거항차(docking voyage) : 수리 및 검사를 위하여 입거하는 경우에는 그 전 항차의 항차 종료시부터 수리를 완료하고 Dock 출항하는 시점(출항시 : departure time)까지를 Docking voyage라 한다.

1.11.5 사용 중인 AB-Log의 예

ABSTRACT LOG

Ship's Name		Voy. No.		Date	20			Master	
Code		Voy. Com'ced	20	Voy. Com'ted	20	Voy. Total Hours	Day / Hour / Min.	C/Eng.	

Port(ZD)		Arrival / Departure	Hours: In Port	Hours: Harbour Voyage	Hours: Sea Voyage	Distances: Har-Bour Voyage	Distances: Sea Voyage	Speed & RPM: Sea Voyage Ave. Sp'd	Ave. RPM	Slip (%)	F.O Consumption (M/T): Sea Voyage M/E D/G | BLR	F.O Harbour Voyage M/E	D/G	BLR	F.O In Port ME/DG/BLR	D.O Consumption (M/T): Sea Voyage M/E D/G | BLR	D.O Harbour Voyage M/E	D/G	BLR	D.O In Port ME/DG/BLR
		M D H M	D H M	D H M	D H M															
	C																			
	D																			
	A																			
	D																			
	A																			
	D																			
	A																			
	D																			
	A																			
	D																			
	A																			
	D																			
	A																			
	F			X	X		X		X			X					X			
Total or Average																				

Oil and F.W Statements

	DATE M D	Port	F.O (M/T)	D.O (M/T)	M/E Cyl. (Ltr)	M/E Sys. (Ltr)	D/G Sys. (Ltr)	F.W (W/T)
Commenced								
Received								
Received								
Received								
Total	X	X						
Balance								
Voyage Total Consumption								

M/E	Gr/BHP.Hr
Cyl.oil	Ltr/Day
Sys.oil	Ltr/Day
D/G Sys.oil	Ltr/Day
Boiler Water	M.T/Day

Last Dry Docking	20		
Last Bottom Cleaning	20		

DETENTIONS (STOPPAGE, DELAYS, DEVIATION, ETC.)

From	To	Remarks

[그림 1-46]

1.12 VTS 시스템(Vessel Traffic Service System)

1.12.1 정 의

VTS는 항만통제(port control)를 의미하는 것이 아니라 항만해역을 통항 또는 이용하는 선박의 교통을 관리하는 제도이다. 즉 좁은 뜻의 해상교통서비스시스템(Vessel Traffic Service System)을 의미하는 것이다.

1.12.2 기 능

① 항만, 접근해역 및 정박지의 일반적 감시

② 접근하는 선박, 항내에서 이동하는 선박과의 효과적인 통신

③ 선장과 도선사에게 항해정보 제공

④ 항로감시와 항로상 잡종선 출입규제

⑤ 항만/항로 주변해역에서의 해난수색 및 구조의 협력

1.12.3 우리나라 선박의 입항 및 출항 등에 관한 법률의 관련 조항

① 선박이 선박교통관제구역을 출입·통과하거나 선박교통관제구역에서 이동·정박·계류할 때에는 선박교통관제에 따라야 한다(제20조 1항).

② 선박교통관제 적용대상 선박이 선박교통관제구역에서 운항하는 경우에는 해양수산부장관의 의견을 들어 총리령으로 정하는 무선설비를 갖추고 해양수산부장관의 의견을 들어 총리령으로 정하는 호출응답용 관제통신을 항상 청취·응답하여야 한다(제22조).

③ 무역항의 수상구역 등에 출입하려는 선박의 선장(이하 이 조에서 "선장"이라 한다)은 대통령령으로 정하는 바에 따라 해양수산부장관에게 신고하여야 한다(제10조 1항).

④ 무역항의 수상구역 등에 정박하려는 선박(우선 피항선은 제외한다)은 제1항에 따른 정박구역 또는 정박지에 정박하여야 하며, 정박구역 또는 정박지가 아닌 곳에 정박한 선박의 선장은 즉시 그 사실을 해양수산부장관에게 신고하여야 한다(제5조).

1.12.4 해상교통서비스시스템의 구조

1. 구조의 변화

국제적으로 채택된 해상교통서비스시스템은 다음의 단계를 거쳐 초기에는 대형 탱커와 고속 컨테이너선 등의 입출항을 관제하기 위하여 항공관제시스템을 기초로 하여 도입하였다.

① Vessel Traffic Control system(VTC) : 1960년대

② Vessel Traffic Management system(VTM) : 1970~80년대

③ Vessel Traffic Service system(VTS) : 1980년대 이후

그러나 항만당국이 입출항 선박을 직접관리(management) 또는 통제(control)함으로써 발생하는 해난사고에 대하여 민사상 / 형사상 책임이 항만당국에 귀속하는 법률문제 때문에 근래에 와서는 해상교통정보를 제공하는 소극적 시스템으로 바꾸어졌다. 즉 해상교통정보센터(vessel traffic service center)로 그 기능을 일원화시키고 있다. 물론 이 기구에는 해상교통법 위반선박을 감시하는 기능을 엄격하게 시행하고 있는 것은 물론이다.

2. 구조 및 요소

1) 적극적 VTS 시스템(active VTS system)

이 시스템은 특정해역 안에서 해상교통의 이동을 직접 관리하는 것으로 항만당국의 권한에 의하여 실시된다. 이 시스템은 선박의 크기, 종류, 운송화물, 선석의 이용성, 도선사의 지원요청 및 다른 선박의 동정을 기본요소로 하여 계획된 교통 프로그램에 따라 선박교통을 직접 규제한다. 유럽 및 미주 항만의 경우 이 방식을 채택하고 있는 항만당국이 가끔 있으며 미국 롱비치(Long beach)항이 대표적이다.

2) 소극적 VTS 시스템(passive VTS system)

이 시스템은 항로 분리해역에 있어서 선박의 항행을 지원하기 위하여 계획된 시설이다. 이 시스템은 현재 선박교통의 정보 및 비정상적인 교통환경에 대한 정보를 제공하는 것으로 그 기능이 제한적이다. 현재 대부분의 항만당국이 채택한 방식이며, 대표적으로 싱가포르(Singapore) VTS가 대표적이다.

1.13 Ship's Reporting System의 개요

1.13.1 목 적

선위통보시스템(ship's reporting system)으로 번역되며 선박이 해상을 항해할 때 맞이하는 해상고유의 위험(충돌, 침몰 좌초, 화재 및 긴급후송 환자발생 등)이 발생하였을 경우에 선박 상호간에 도움을 주거나 혹은 수색과 구조에 협조할 수 있도록 하는 제도로써 미국의 AMVER(Automated Mutual - assistance Vessel Rescue System)가 대표적인 선위통보시스템이다.

수년전 우리나라 국적 풀 컨테이너선박이 미국연안에서 표류하던 요트를 AMVER의 요청으로 구해준 일이 있어 그 당시 미국에서 언론에 크게 보도되었고, 해당 선박은 미연안경비국(United States Coast Guard, USCG)으로부터 감사패를 받았고 또한 그 회사와 우리나라에게 많은 홍보효과를 가져온 사실이 있다.

1.13.2 종 류

이러한 선위통보제도는 전 세계가 통일규정에 의해 실시되는 것이 아니라 각국의 필요에 의해서 실시되므로 인해 지역마다 관장하는 국가가 다르고 또한 종류도 많은 것이 사실이다(그림 1-37 참조).

① AMVER(The Automated Mutual-assistance Vessel Rescue System) : 미국의 연안경비국(USCG)이 관장하는 시스템으로 선위통보제도의 효시이며 특정지역을 제외한 전 세계를 커버하고 있는 선위통보시스템의 근간이다.

② AUSREP(Australian Ship Reporting System) : 호주연안 및 남극부근을 항행하는 선박들이 해당된다.

③ CHILREP(Chilean Ship Reporting System)

④ SECOSENA(Argentina Safety of Navigation Communication Service)

⑤ SISTRAM(Brazil Maritime Traffic Information System)
⑥ SAFREP(South African Ship Reporting System)
⑦ JASREP(Japanese Ship Reporting System)
⑧ KOSREP(Korea Ship Reporting System) 등이 있다.

1.13.3 통보양식

① Sailing plan report : 출항 전 혹은 AMVER 보고라인에 진입시 보고양식
② Position report : 구역 진입 후 매일 보고하는 양식
③ Arrival report : 도착 후 보고양식
④ Deviation report : 이로 발생시 보고양식
⑤ AMVER의 통보양식의 예 : Form of messages

A/ Vessel's name/International Radio Call Sign//
B/ Date and time(UTC)//
C/ Latitude/Longitude//
E/ Current course//(3 digital group)
F/ Estimated average speed//(3 digital group in knots and tenths of knots without decimal point)
G/ Port of departure/Latitude/Longitude//
I/ Port of destination/Latitude/Longitude/Estimated time of arrival//
K/ Port name/Latitude/Longitude/Time of arrival//
L/ Route information//
M/ Current coast radio station or satellite number/next station, if any//
V/ On board medical resources//(one or more of the following MD for physician /PA for physicians assistant/Nurse/or None//
X/ Up to 65 characters of amplifying comments//
Y/ For forwarding of message to JASREP and MAREP(on request)//
Z/ EOR(End of report)—For computer processing of message//

1.13.4 전 세계 선위통보시스템 개요

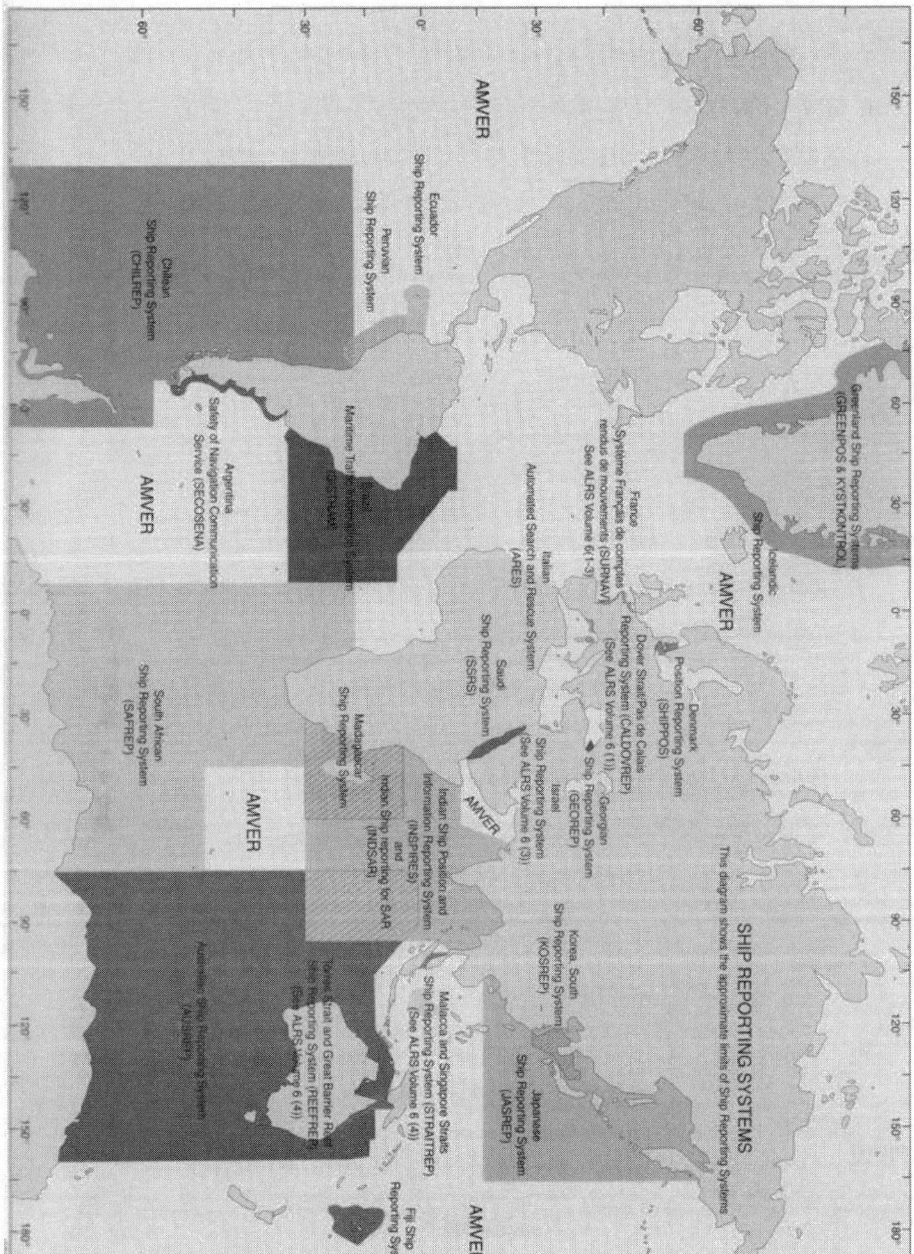

[그림 1-47]

실습과제

〈1.2〉 국제해상충돌방지규칙의 내용

1. 국제해상충돌방지규칙 용어의 개념을 숙지하시오.
2. 통항분리방식을 이용하는 선박의 항행상의 의무에 대하여 설명하시오.
3. 승선 선박이 피항선이 되었을 때 취해야 할 의무에 대하여 설명하시오.
4. 두 선박이 충돌의 위험을 안고 서로 접근할 때의 4가지 단계의 행동에 대하여 설명하시오.
5. 항해 중 조우하는 타 선박의 등화와 형상물을 확인하시오.
6. 승선 선박에 설치된 등화 및 형상물의 종류, 위치, 사용법 등을 확인하시오.
7. 제한된 시정에서 사용되는 기기의 종류 및 사용법을 확인하시오.

〈1.3〉 해사안전법

1. 1972년 국제해상충돌방지규칙과 해사안전법과의 차이를 설명하시오.
2. 항로지정방식과 교통안전특정해역이 설정되어 있는 곳을 조사하여 내용을 파악하시오.
3. 유조선 통항금지해역에 대하여 설명하시오.

〈1.4〉 선박의 입항 및 출항 등에 관한 법률

1. 선박의 입항 및 출항 등에 관한 법률과 해사안전법과의 관계에 대하여 설명하시오.
2. 선박의 입항 및 출항 등에 관한 법률상 항로와 항법의 특성에 대하여 설명하시오.

제2장

항해학(지문항해, 천문항해)

2.1 항해의 기초

항해학은 항해자에게 선박의 위치를 결정하게 하고, 지구 위의 한 지점에서 다른 지점으로 선박을 안전하고 효율적으로 이동시키는 과학 또는 기술이다.

2.1.1 항법의 분류

항법은 선박이 두 지점 사이를 해상을 통하여 가장 안전하고 정확하게 항행(航行)하도록 하는 기술로서 그 선위(船位)를 결정함에 있어서 이용하게 되는 관측의 대상이나 수단에 따라서 다음과 같이 구분할 수 있다.

1. 지문항법(Geo Navigation)

① 연안항법(沿岸航法) : 지상에 있는 육표, 항로표지(航路標識) 등의 방위, 거리 및 고도 등을 관측하거나 수심 등을 측정하여 선위를 결정하면서 연안을 따라서 항해하는 항법이다.

② 추측항법(推測航法) : 가장 최근에 구한 정확한 선위를 기초로 선박의 침로와 항정에 의하여 현재의 선위를 추측 및 추정하거나 또는 출발지와 목적지의 경도·위도를 알고 앞으로 취해야 할 침로 및 항정을 구하는 항법이다.

2. 천문항법(Celestial Navigation)

대양에서 태양, 달, 별 등과 같은 천체의 고도와 방위를 관측하여 선박의 위치 및 컴퍼스의 오차를 측정하는 항법을 천문항법(天文航法)이라 한다. 최근에 들어서 이 항법은 전파항법의 개발로 선위를 결정하는 효용성은 점점 감소되고 있으나 시의 개념을 이해하고 대양에서 컴퍼스의 오차를 측정하는 데는 여전히 이용되고 있다.

3. 전파항법(Radio Navigation)

육상의 물표나 천체를 관측하지 않고도 전파가 가지는 세 가지 특성, 즉 직진성, 반사성 및 등속성을 이용한 전파표지 및 항법장치를 이용하여 선위를 결정하면서 항해하는 항법을 전파항법(電波航法)이라 한다. GNSS 및 레이더 등이 대표적인 전파항법 장치이다.

2.1.2 거리와 속력

1. 해리(Nautical mile, Sea mile, M)

지구의 위도 45°에 있어서의 지리 위도 1'의 길이인 약 1,852m를 1 마일 또는 1 해리(海里)라고 하고, 해상에서 거리의 단위로 사용한다.

한편, 구미지역에서 많이 사용하는 마일을 육리(陸里, land mile 또는 statute mile)는 약 1,609.3m로 이것은 해리와는 전혀 다른 것으로 해상에서는 쓰지 않는다. 또한 해상에서 관행적으로 사용하는 거리단위로서 케이블(cable)이 있는데 이는 1해리의 10분의 1이다.

2. 노트(Knot, kn 또는 kt)

선박의 속력을 나타내는 단위로 1시간에 1해리를 항주하는 속력을 말한다. 선박은 물위에 떠서 항행하므로, 선박의 속력은 물에 대한 속력과 지면에 대한 속력이 다를 수 있으며, 이들을 대수(對水)속력(speed through water)과 대지(對地)속력(speed over the ground)으로 구별한다.

만일, 기관속력 12노트로 항행 중인 선박이 2노트 크기의 해류를 선박의 진행방향 뒷쪽에서 앞쪽으로 받으면서(순조) 항행을 하면 선박의 대수속력은 12노트이지만 대지속력은 14노트가 되고, 선박의 진행방향과 반대방향으로 이 해류를 받게 되면(역조) 이 선박의 대수속력은 12노트이지만 대지속력은 10노트가 된다.

[예제] 현재 선속이 16kts일 때 변침점까지의 거리가 4마일이다. 변침점까지의 도달시간은?

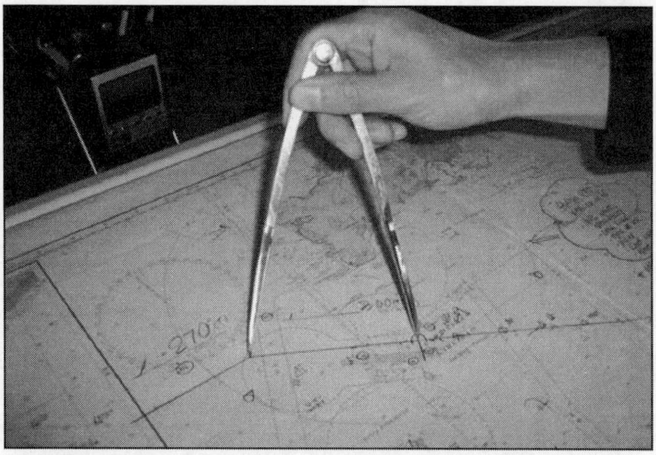

[그림 2-1] 거리와 속력계산(1)

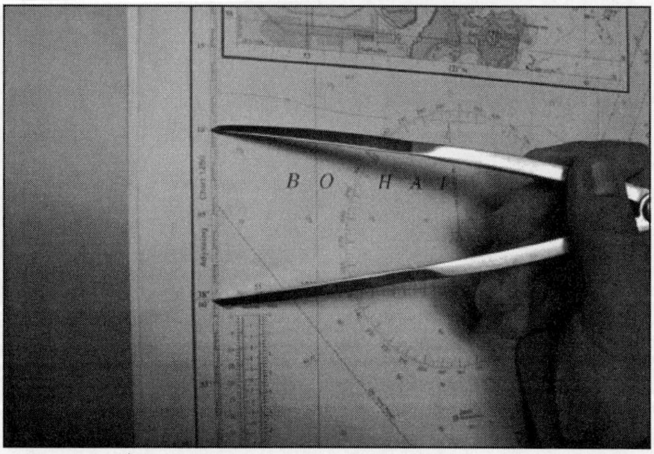

[그림 2-2] 거리와 속력계산(2)

2.1.3 지구상의 위치요소

지구는 완전한 구형이 아니고 적도반지름(약 6,378km)이 극반지름(6,357km)보다 약간 긴 회전타원체이다. 그러나 이는 지구의 평균반지름에 비하여 대단히 작은 값이므로, 특별한 경우가 아니면 이론과 계산을 간단하게 하기 위하여 지구를 평균반지름을 반지름으로 하는 구로 가정하여 항해에 관한 문제를 처리하고 있다.

1. 대권과 소권

① 대권(大圈, Great circle) : 지구의 중심을 지나는 평면으로 구를 자른다고 가정할 때 지구 표면에 생기게 되는 원이다.

② 소권(小圈, Small circle) : 지구의 중심을 지나지 않는 평면으로 자를 때 생기는 원이다.

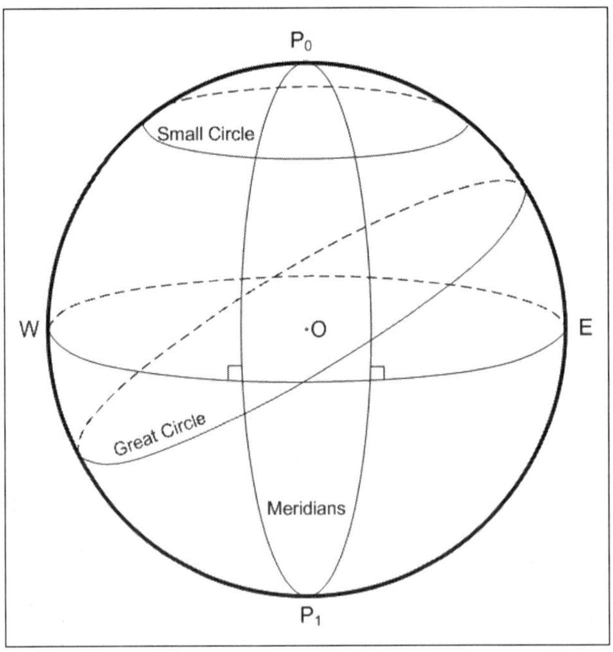

[그림 2-3] 지구상의 위치요소

2. 적도와 거등권

지축과 직교하는 대권을 적도(Equator)라고 하고, 적도에 평행한 소권을 거등권(Parallel of latitude) 또는 평행권이라 한다.

3. 자오선

지구의 양 극(極)을 지나는 대권을 자오선(Meridian)이라 하며 적도와는 직교한다. 많은 자오선 가운데 영국의 그리니치(Greenwich) 천문대를 지나는 자오선을 000°로 하여 본초 자오선(Prime meridian)이라 하고, 이를 기준으로 경도를 측정한다.

4. 위도와 경도

① 위도(Latitude, L) : 어느 지점에 있어서의 거등권과 적도 사이의 자오선상의 호(弧)의 길이 또는 이 호가 지구 중심에서 이루는 각(角)을 그 지점의 위도라 한다. 위도는 적도를 0°로 하여 남북으로 각각 90°까지 재는데, 재는 방향이 북쪽이면 북위라 하고 부호 N을 붙여 표시하며, 남쪽이면 남위라 하고 부호 S를 붙여 표시한다.

② 경도(Longitude, λ) : 어느 지점의 자오선과 본초 자오선이 이루는 적도의 호 또는 그 호가 지구 중심에서 이루는 각을 그 지점의 경도라 하며, 본초 자오선을 000°로 하여 동쪽과 서쪽으로 각각 180°까지 잰다. 동쪽으로 잰 경우에는 동경(East longitude)이라 하고 E를 붙여 표시하며, 서쪽으로 잰 경우에는 서경(West longitude)이라 하고 W를 붙여 표시한다.

5. 항정선

지구상의 모든 자오선과 같은 각도로 만나는 곡선으로 선박이 일정한 침로를 유지하면서 항해할 때 지구상에 그리는 항적을 말한다. 적도, 거등권 및 자오선도 항정선(Rhumb line)이다. 선박이 동서남북이 아닌 방향으로 항해하며 자오선과 같은 각을 유지하면 나선형의 곡선을 이루면서 극에 가까워지는데, 이는 자오선이 평행이 아니고 극에 가까워질수록 간격이 좁아지기 때문이다.

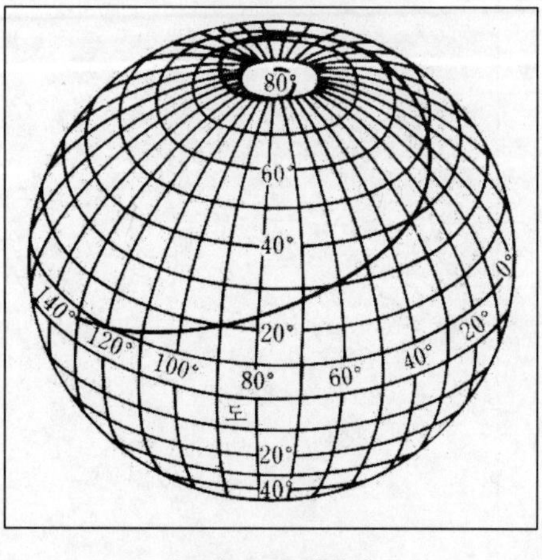

[그림 2-4] 항정선

2.1.4 방위와 침로

자이로 컴퍼스는 고속으로 돌고 있는 자이로스코프를 이용하여 지구상의 진북을 가리키는 장치로, 마그네틱 컴퍼스에서 나타나는 자차 및 편차가 없고 지북력도 강하다. 하지만 여러 가지 이유로 자이로 컴퍼스는 항상 진북을 정확히 가리키는 것은 아니며, 또한 전기로 구동되기 때문에 선박에서 전원이 상실되면 더 이상 사용이 불가능하게 된다. 따라서 자이로 컴퍼스를 대신하여 사용할 수 있는 자기 컴퍼스를 선박에서는 추가로 설치하고 있다.

지구는 하나의 거대한 자성체이기 때문에 자석을 회전할 수 있도록 설치하면 이 자석은 지구의 자력선의 방향과 일치한 채 정지하게 된다. 그러나 이러한 자력선의 방향은 진남북 방향과 약간의 오차를 가지고 있다(편차). 또한 이런 자석으로 만들어져 선박에 설치된 자기 컴퍼스는 자력선의 방향도 가리키지 못하고 또 오차가 생기게 된다(자차). 이러한 원인은 편차와 자차에 기인한 것으로 자기 컴퍼스를 이용하기 위해서는 이러한 편차 및 자차값을 확인하여 정확한 진방위로 수정할 필요가 있다.

1. 편 차

어느 지점을 지나는 진자오선과 자기 자오선이 이루는 교각을 그 지점의 편차(偏差, Variation, Var.)라 한다.

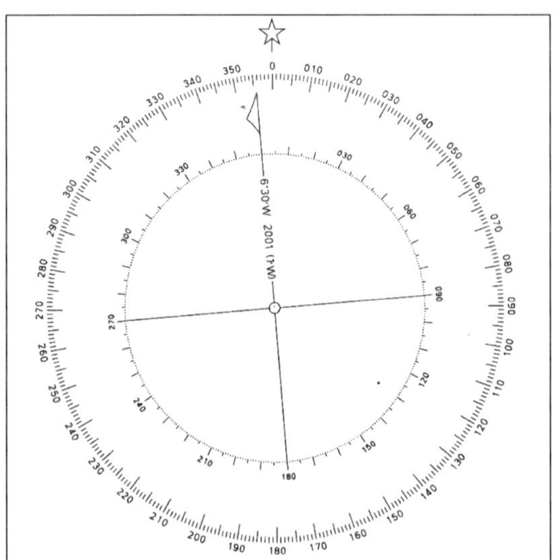

[그림 2-5] Compass Rose

편차는 자침이 가리키는 북(자북)이 진자오선(진북)의 오른쪽에 있을 때를 편동편차(Easterly variation), 왼쪽에 있을 때를 편서편차(Westerly variation)로 구별하며, 각각 E 또는 W를 붙여 표시한다. 편차를 구하려면 해도에 그려져 있는 나침도에서 그 중앙에 기재된 편차에 연차를 가감하거나 자침 편차도(variation chart)를 이용한다.

2. 자 차

마그네틱 컴퍼스는 원래 자기 자오선의 방향을 가리켜야 되지만, 선체, 선내 철기류 등의 영향을 받아 지자극의 방향을 가리키지 않고 자기 자오선과 약간의 교각을 이루는 방향을 가리키게 된다. 이 교각을 자차(自差, Deviation, Dev.)라 한다.

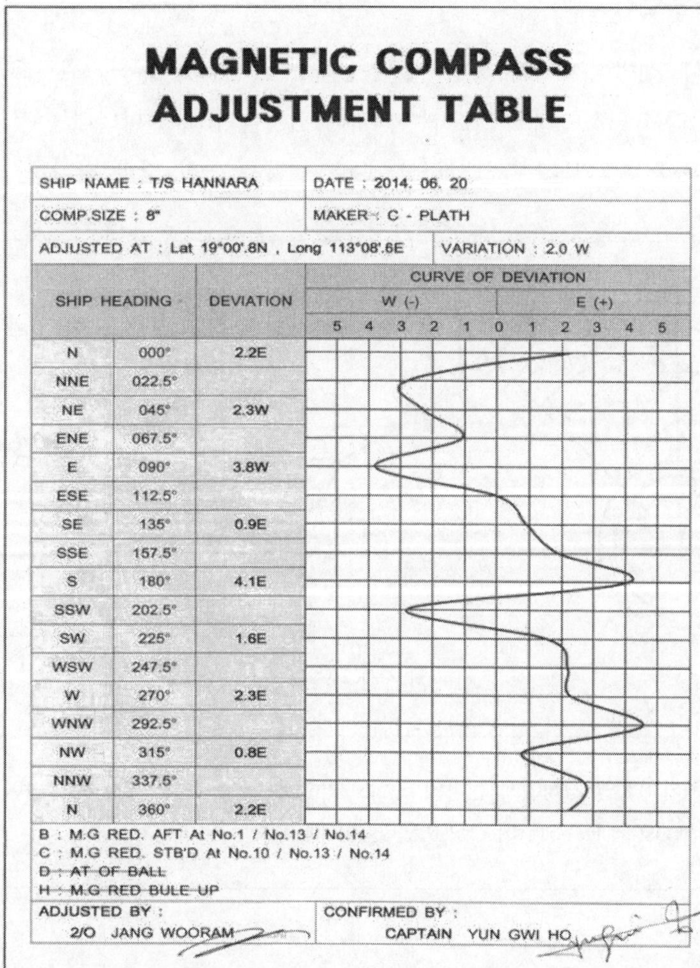

MAGNETIC COMPASS ADJUSTMENT TABLE

SHIP NAME : T/S HANNARA	DATE : 2014. 06. 20
COMP.SIZE : 8"	MAKER : C - PLATH
ADJUSTED AT : Lat 19°00'.8N , Long 113°08'.6E	VARIATION : 2.0 W

SHIP HEADING		DEVIATION
N	000°	2.2E
NNE	022.5°	
NE	045°	2.3W
ENE	067.5°	
E	090°	3.8W
ESE	112.5°	
SE	135°	0.9E
SSE	157.5°	
S	180°	4.1E
SSW	202.5°	
SW	225°	1.6E
WSW	247.5°	
W	270°	2.3E
WNW	292.5°	
NW	315°	0.8E
NNW	337.5°	
N	360°	2.2E

B : M.G RED. AFT At No.1 / No.13 / No.14
C : M.G RED. STB'D At No.10 / No.13 / No.14
D : AT OF BALL
H : M.G RED BULE UP

ADJUSTED BY : 2/O JANG WOORAM	CONFIRMED BY : CAPTAIN YUN GWI HO

[그림 2-6] Deviation Curve

즉, 자차는 선내 컴퍼스가 선체나 선내 철기류의 영향을 받아서 생기는 오차이다. 자차도 편차의 경우와 같이 컴퍼스의 북(나북)이 지자기의 북(자북)의 오른쪽에 있으면 편동자차(Easterly deviation), 나북이 자북의 왼쪽에 있으면 편서자차(Westerly deviation)로 구별하며, 각각 E 또는 W를 붙여 표시한다.

항해 중에 자이로 컴퍼스가 고장이 나면 마그네틱 컴퍼스로 항해를 하게 되는데, 마그네틱 컴퍼스는 자차와 편차로 인하여 컴퍼스 오차를 가지고 있다. 따라서 침로를 유지하여 목적지에 도달하기 위해서는 자차와 편차를 가감하여 진방위로 변환하는 과정이 필요하다.

3. 컴퍼스 오차

진자오선과 선박의 마그네틱 컴퍼스의 남북선이 이루는 교각을 컴퍼스 오차(Compass Error, C.E.)라고 말하는데, 이 오차는 자차와 편차를 합쳐서 나타내는 것으로 그 크기는 자차와 편차의 부호가 같으면 합하고, 다르면 차를 구한 것과 같다.

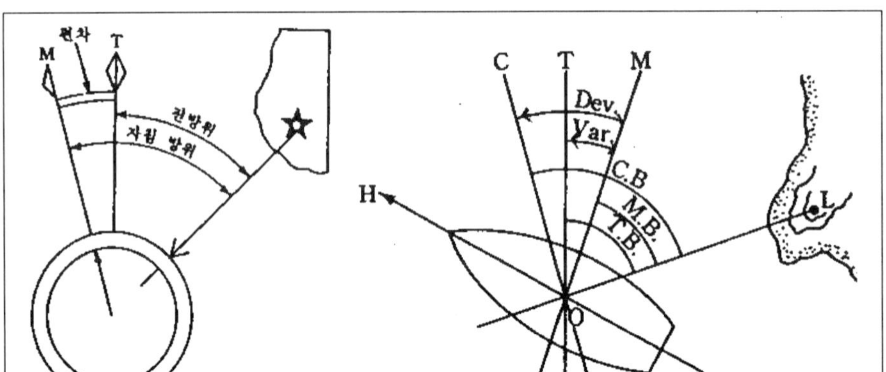

[그림 2-7] 컴퍼스 오차

[예제] 편차가 8°E인 지점에서 선박의 Gyro Compass(Error : 없음)의 Heading이 060°, Magnetic Compass의 Heading이 062°이었다. 이 때 자차는 얼마인가?

---

4. 자이로 컴퍼스의 에러

여러 가지 이유로 인하여 자이로 컴퍼스는 항상 정확히 진북을 가리키는 것은 아니다. 따라서 정확하게 목적지에 도달하기 위해서는 항해 전에 반드시 자이로 에러를 확인하고 있어야 한다. 자이로 에러의 정확도는 적도 부근에서는 최대 0.5°정도, 60°부근에서는 최대 1.0°정도로 인정하며 사용되고 있다.

[예제] 만약 해도에 위치가 정확히 표시된 두 물표를 항해 중에 Gyro Compass로 중시된 시점에 확인한 결과 077°이었고, 반면에 해도에서 확인한 진방위(True Bearing)가 075°이었다면 Gyro Error는 얼마인가?

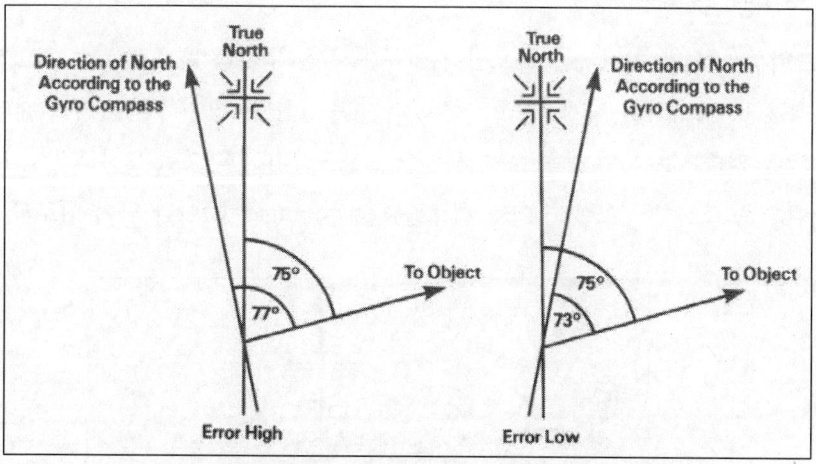

[그림 2-8] 자이로 컴퍼스의 에러(Error of the Gyro Compass)

[예제] Gyro Compass의 Error를 확인하는 또 다른 방법에 대하여 알아보자.

---

5. 방위와 방위각

어느 기준선과 관측자 및 물표를 지나는 대권이 이루는 교각은 북을 000°로 하여 시계 방향으로 360° 까지 측정한 것을 방위(Bearing, Bn)라 하고, 북 또는 남을 0°로 하여 동쪽 또는 서쪽으로 180°이내의 각으로 표시한 것을 방위각(Bearing angle, B)이라 한다.

① 진방위(True Bearing, T · B) : 진자오선과 물표 및 관측자를 지나는 대권이 이루는 교각이다.

② 자침방위(Magnetic Bearing, M · B) : 자기자오선과 물표 및 관측자를 지나는 대권이 이루는 교각이다.

③ 나침방위(Compass Bearing, C · B) : 나침의 남북선과 물표 및 관측자를 지나는 대권이 이루는 교각이다.

④ 상대방위(Relative Bearing, R · B) : 북쪽과 관계없이 자선의 선수를 0°로 하여 시계방향으로 360°까지 재거나, 좌현 우현으로 180°씩 측정하는 방법이 있으나 통상 180°식을 많이 사용한다.

※ 상대방위는 견시보고나 닻줄의 방향을 보고할 때 편리하게 사용된다.

6. 침로와 침로각

선수미선과 선박을 지나는 자오선이 이루는 각을 침로(針路, Course, Co.)라고 하며, 보통 북을 000°로 하여 시계방향으로 360°까지 측정한다. 방위각을 나타낼 때와 같이 북 또는 남을 기준으로 하여 동쪽 또는 서쪽으로 180°까지의 각으로 표시하거나, 또는 90°까지의 각으로 표시한 경우에는 이를 침로각(Course angle, C)이라 하여 침로와 구별한다.

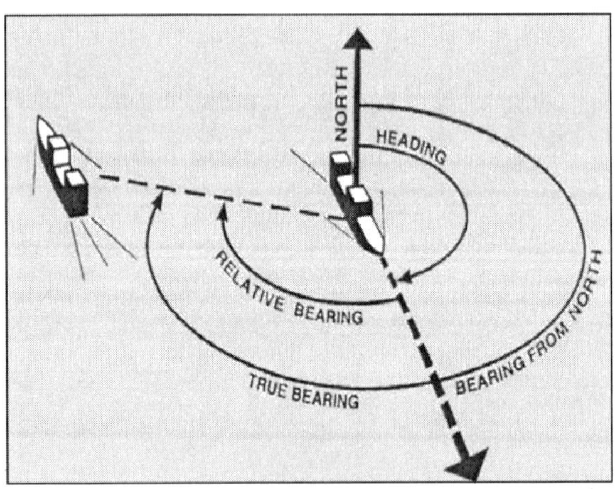

[그림 2-9] 침 로

1) 진침로

진자오선(진북)과 선수미선이 이루는 각을 진침로(True Course, T.Co.)라 하며, 풍압 · 유압차가 없을 때는 항적과 선수미선이 일치하므로 진자오선과 선수미선이 이루는 각이다.

2) 시침로

풍압차나 유압차가 있을 때의 진자오선과 선수미선이 이루는 각을 시침로(Apparent Course, App. Co.)라 하므로, 따라서 풍압·유압차가 없을 때는 진침로와 같으므로 시침로는 생기지 않는다.

3) 자침로

자기자오선(자북)과 선수미선이 이루는 각을 자침로(Magnetic Course, M.Co.)라 한다.

4) 나침로

나침의 남북선과 선수미선이 이루는 각을 나침로(Compass Course, C.Co.)라 하고, 자차와 편차, 즉 나침의 오차 때문에 진침로와 이루는 교각이다.

7. 풍압차와 유압차

일반적으로 선박에서는 풍압차와 유압차를 구별하지 않고 이들을 합쳐서 풍압차라고 하는 경우가 많다.

1) 풍압차(風壓差, Lee Way, LW)

선박이 항행할 때 어느 한쪽 현에서 바람을 받으면 풍하쪽으로 떠밀려 실제로 선박이 지나온 항적은 선수미선과 일치하지 않고 이루는 교각이다.

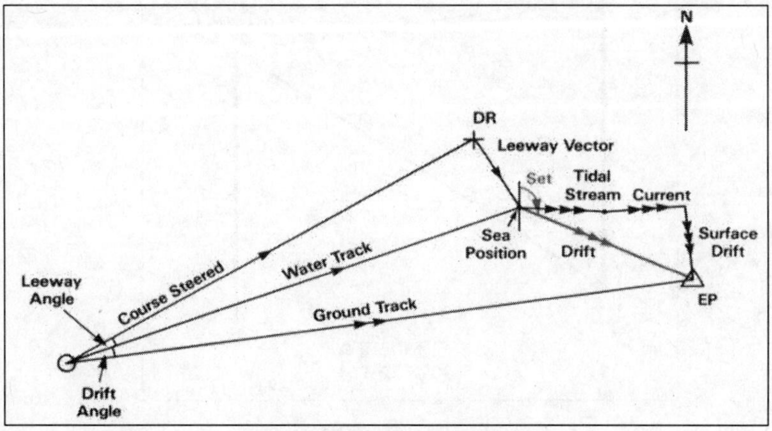

[그림 2-10] 풍압차와 유압차

2) **유압차(流壓差**, Tide way, Current way)

해류나 조류에 떠밀리는 경우에도 항적과 선수미선 사이에 생기는 교각이다.

8. 항 로

지구위의 어느 출발지점에서 도착지점까지 선박이 실제로 지나온 대지항적을 실항로라 하는데 실항로는 불규칙한 선을 그리게 된다. 또 출발지점에서 목적지까지 항해하려고 할 때 선박이 지나가리라고 예상되는 일련의 대지침로선들을 예정항로라고 한다.

2.1.5 진풍향 및 진풍속

선박의 항해 중 풍향풍속계(anemometer)를 통해 확인한 값은 상대풍향과 상대풍속이다. 상대풍향과 상대풍속은 실제로 불고 있는 바람과 선박의 진행으로 인해 선박 진행 반대방향으로 생기는 바람이 합성되어 만들어진다.

[그림 2-11]에서 000°로 15kts로 항행하는 선박에서 풍향풍속계로 확인한 결과 060° 방향에서 24kts의 바람이 불고 있다고 가정할 때, 이 상대풍향, 풍속은 선박의 진행 때문에 그 반대방향으로 생긴 바람(방향은 반대이나 속도는 동일, 180°, 15kts)과 실제로 불고 있는 바람(098°, 21kts)의 벡터 합성으로 생긴 것이므로 벡터 분해를 통하여 쉽게 구할 수 있다.

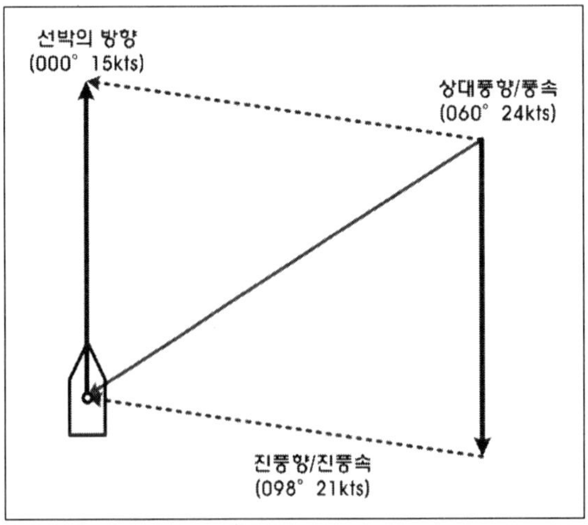

[그림 2-11] 진풍향과 진풍속의 계산

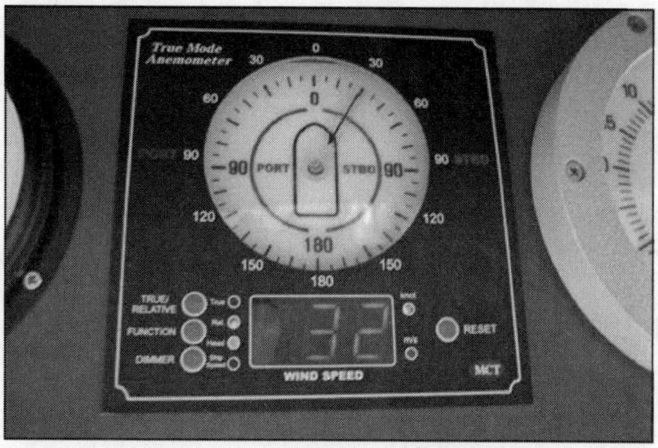

[그림 2-12] 풍향풍속계(상대풍향)

[예제] 선박이 Course 030°, Speed 16kts로 항해 중이며, 바람이 325도(절대방위), 20kts로 불 때, 플로팅 용지를 활용하여 진풍향, 진풍속을 정확히 구하시오.

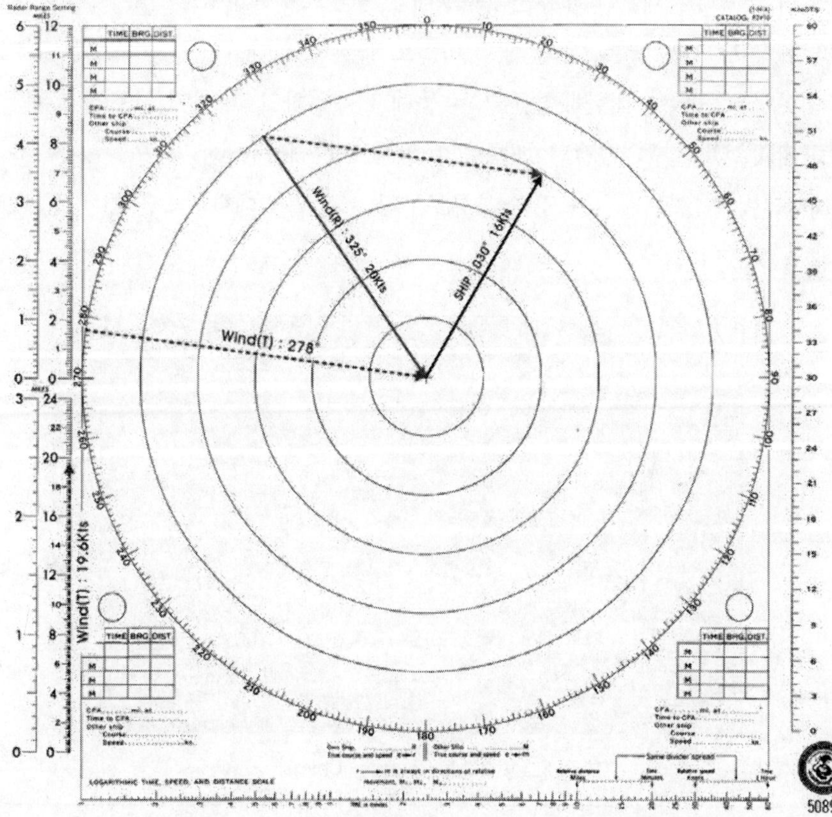

[그림 2-13] 플로팅 용지를 이용한 진풍향과 진풍속 계산

2.1.6 해 류

1. 세계의 해류

전 세계의 주요 해류를 보면 각 해양마다 아주 비슷한 형상을 나타내고 있다. 해류는 크게 저위도의 편동 무역풍에 의한 적도 해류와 중위도의 편서풍에 의한 서풍피류, 그리고 고위도의 아한대 편동풍에 의해 형성되는 해류의 흐름이 있다. 이는 지구의 풍계가 연중 거의 같은 방향으로 부는 바람으로 적도를 중심으로 남북대칭 분포를 가지기 때문에 해류 또한 남북대칭 분포를 이룬다.

2. 우리나라 주변의 해류

우리나라의 각 해역에 가장 큰 영향을 미치는 해류는 일본 큐슈(九州) 남쪽에서 북상하는 쿠로시오(黑潮)이다. 이 해류는 루존, 대만 동방으로부터 유구열도 서방의 동중국해의 외연을 유속 1.0~2.5노트, 유폭 수십 마일로 북상하여 아마미오시마의 북서방에서 분리하여 대마난류가 되어 대한해협을 거쳐 동해로 유입하고, 나머지 쿠로시오 주류는 큐슈, 시코쿠(四國) 및 혼슈(本州)의 대륙붕의 외연을 따라 북상한다.

우리나라에 영향을 미치는 것은 본류로부터 분리되어 대한해협을 거쳐 동해로 유입되는 대한난류이다.

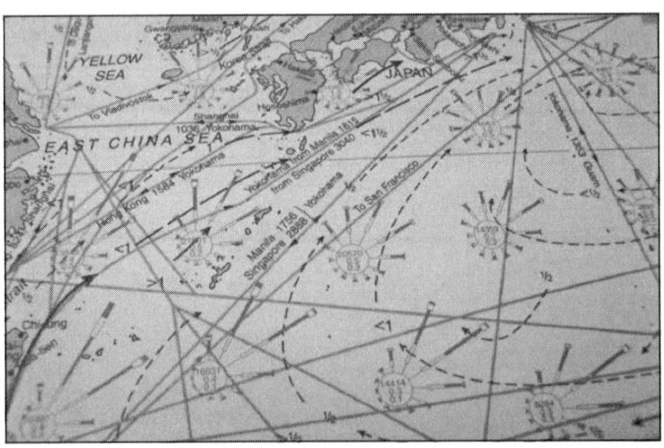

[그림 2-14] Routeing Chart

[예제] Routeing Chart를 해역별로 확인하여 그 내용을 살펴보자.

2.1.7 조석 및 조류

해수는 달과 태양의 인력작용에 의하여 상하운동과 수평운동을 한다. 지구, 달, 태양간의 인력인 기조력(tidal force)에 의한 해수면의 완만한 주기적 상하운동을 조석(tide)이라 하고, 조석에 따라 일어나는 주기적 수평운동을 조류(tidal current, tidal stream)라 한다. 만조부터 다음 만조까지, 또는 간조부터 다음 간조까지의 시간을 의미하는 조석 주기는 약 12시간 25분인데 만조와 간조는 각각 하루에 2회 정도씩 일어나면 매일 50분씩 늦어진다.

이는 지구가 하루 1회 자전하는 동안 달이 동쪽으로 약 13° 공전하기 때문이다. 달이 다음날 동일한 위치에 오기 위해서는 지구가 50분정도 더 자전을 해야 하고, 이에 따라 조석도 날마다 약 50분씩 늦어지게 된다. 즉, 2회의 만조가 일어나기 위해서는 24시간 50분이 걸리는 셈이므로 조석주기는 약 12시간 25분이 된다.

우리나라 서해안 특히, 인천의 경우는 조차가 10m에 가까워 부두 정박선들은 계류삭(mooring line)에 주의를 기울여야 한다. 또한 수심이 낮은 지역을 항해 시 고조를 이용하는 경우가 많다.

① 고조(高潮, High Water, H.W.) : 조석으로 인하여 해면이 가장 높아진 상태임
② 저조(低潮, Low Water, L.W.) : 해면이 가장 낮아진 상태임
③ 창조(flood tide) : 저조에서 고조로 되기까지 해면이 점차 높아지는 상태임
④ 낙조(ebb tide) : 고조에서 저조로 되기까지 해면이 점차 낮아지는 상태임
⑤ 정조(stand of tide) : 고조와 저조 시에 해면의 승강운동이 순간적으로 거의 정지한 것과 같이 보이는데 이러한 상태임
⑥ 월조간격(lunitidal interval) : 달이 어느 지점의 자오선을 통과한 때로부터 그 지점에 고조가 되기까지 걸리는 시간을 고조간격, 저조가 되기까지 걸리는 시간을 저조간격이라 하고, 이들을 통틀어 월조간격이라 함
⑦ 대조(大潮, spring tide) : 삭과 망이 지난 뒤 1~2일만에 생긴 조차가 극대인 조석으로 '사리'라고 함
⑧ 소조(小潮, neap tide) : 상현 (上弦) 및 하현(下弦)이 지난 뒤 1~2일만에 생긴 조차가 극소인 조석으로 '조금'이라고 함
⑨ 게류(憩流, slack water) : 조류는 창조류에서 낙조류, 또는 낙조류에서 창조류로 변할 때에는 흐름이 잠시 정지하는데 이것을 게류 또는 전류라고 함

[그림 2-15] 조석표

[예제] 부산항 및 싱가포르항의 오늘의 조석을 찾아보고, 또한 기타 항만의 조석을 표준항의 조시 및 조고를 이용하여 구해보자.

2.2 해도 및 수로서지

해도(海圖, chart)는 해상의 지형뿐만 아니라 바다 밑의 암초 등과 같은 장애물 등을 기호 및 약어로 표시한 바다의 안내도이며, 수로서지(navigation publication)는 해도에서 모두 나타낼 수 없는 내용들을 제동하는 안내서를 총칭하는 것이다. 이러한 해도 및 수로서지는 최신의 항행통보(notice to mariner)의 정보로 개정한 상태를 유지하여야 한다. 최근에는 전자해도 및 서지가 개발되어 있다.

2.2.1 해도의 구성

1. 해도도법에 의한 분류

해도의 사용목적에 따라 실용상 지장이 없을 정도로 오차를 작게 하여 둥근 지구표면을 평면에 나타내는 방법을 도법(map projection)이라 한다.

1) 평면도법

평면도(平面圖)는 지구표면의 좁은 한 구역을 평면으로 간주하고 그린 축척이 큰 해도이다. 주로 항박도(港泊圖, harbour plan)에 많이 이용되고 있다.

2) 점장도법

지구표면상의 항정선을 평면 위에 직선으로 나타내기 위해서 고안된 도법이 점장도법(漸長圖法, mercator projection)이며, 이 도법으로 그린 해도를 점장도라 한다. 항정선은 모든 자오선과 같은 각으로 만나므로, 항정선을 직선으로 나타내기 위해서는 자오선을 거등권에 직교하는 평행선으로 나타내어야 한다.

그런데, 자오선을 평행선으로 나타내면 자오선 사이의 거등권의 호의 길이는 실제보다 늘어나며, 그 비율은 위도가 높을수록 심하다. 점장도 위에서의 항정선은 모든 자오선과 같은 각도를 이루므로, 항해자가 어느 자오선을 이용해서도 손쉽게 침로와 방위를 측정할 수 있으므로 항해에 편리하게 이용할 수 있다.

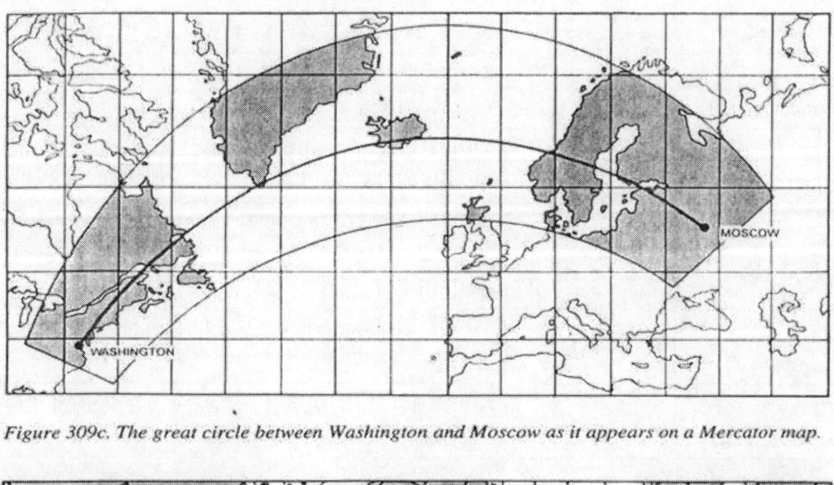

Figure 309c. The great circle between Washington and Moscow as it appears on a Mercator map.

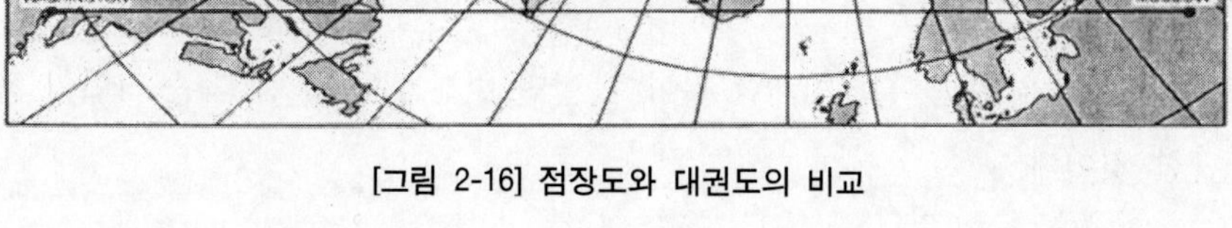

[그림 2-16] 점장도와 대권도의 비교

3) 대권도법

지구표면상의 한 점에 접하는 평면에 지구의 중심으로부터 지구표면의 점들을 투영한 해도를 대권도(大圈圖, great circle chart)라 하며 이는 심사도라고도 한다. 이 해도는 지구상의 모든 대권이 직선으로 표현되므로 대권 항법에 이용되며, 접점에서 멀리 떨어질수록 실제와 모양이 달라져서 일반적으로 방위, 거리 등을 직접 구할 수가 없다.

대권도법으로 작성된 해도 위에서는 대권이 직선으로 표현되므로 2점 사이의 최단 거리를 구하기가 편리하여 원양항해계획을 세울 때에 이용되며, 특히 거리가 긴 대양항해의 경우 대권 거리는 항정선 거리보다 훨씬 짧아지게 된다.

[예제] 대권도를 찾아보고 점장도와 비교하여 보자.

---

2. 사용목적에 의한 분류

사용목적에 따라 항해용 해도와 잡용 해도, 수로 특수도, 잠정판 해도 등으로 나눈다.

1) 항해용 해도

종 류	축 척	내 용
총도 (nautical chart)	1/400만 이하	세계 전도와 같이 넓은 구역을 나타낸 것으로 장거리 항해와 항해계획 수립에 이용됨
항양도 (sailing chart)	1/100만 이하	원거리 항해에 쓰이며, 먼바다의 수심, 주요 등대, 등부표, 먼 바다에서도 보이는 육상의 물표 등이 해도에 표시됨
항해도 (coastal chart)	1/30만 이하	육지를 바라보면서 항해할 때 사용하는 해도로서, 선위를 직접 해도상에서 구할 수 있도록 육상물표, 등대, 등표 등이 비교적 상세히 표시됨
해안도 (approach chart)	1/5만 이하	연안 항해에 사용하는 것이며, 연안의 상황이 상세하게 표시됨
항박도 (harbour plan)	1/5만 이상	항만, 정박지, 협수로 등 좁은 구역을 상세히 그린 평면도임

2) 잡용 해도

항해용 해도를 얇은 종이에 인쇄한 것이며, 항만 변경, 수산업 등 각종 작업의 참고용으로 출판한 것이다.

3) 수로 특수도

종 류	내 용
해저지형도 (bathymetric chart)	해안의 저조선을 포함한 해저면의 지형을 그린 해도로서 어장개발, 자원개발, 해저통신선 부설, 침몰선 인양, 해양오염 방제, 군사작전, 해양의 과학적 이용 및 환경보존 등에 중요한 자료를 제공함
어업용 해도 (fishery chart)	일반 항해용 해도에 각종 어업에 필요한 제반자료를 기재하여 제작한 해도로서 해도 번호앞에 'F'자가 표기됨
해류도 (ocean current chart)	일정한 방향과 유속을 가진 해수의 흐름을 나타낸 해도임
조류도 (tidal current chart)	조석현상에 따라 발생하는 해수의 수평적인 흐름인 조류의 상황을 그림으로 표시한 해도로서 창조류(밀물때의 조류), 낙조류(썰물때의 조류), 시간별 조류의 방향, 유속 등이 표시됨
해도도식(chart symbols and abbreviations)	해도상에 기재된 건물, 항만시설물, 등부표, 수중장애물, 조류, 해류, 해안선의 형태, 등고선, 연안지형 등의 기호 및 약어를 수록함
기타 특수도 (special charts)	위치기입도, 영해도, 세계 항로도 등임

4) 잠정판 해도

정규해도가 간행되기 이전에 임시조치로 간행된 해도를 말하며, 해도의 번호 앞에 P라는 기호를 붙여서 구별하고 있다.

3. 해도의 발행처에 의한 분류

세계의 여러 나라에서 발행되는 해도는 대부분 자국의 연안에 관한 사항만을 기록하고 있으므로, 여러 나라를 항해할 경우에는 각 국가가 발행한 해도뿐만 아니라 전세계를 모두 나타내는 미국 및 영국의 해도를 조합하여 적절히 이용할 필요가 있다. 우리가 자주 이용하고 있는 해도는 우리나라 해도 이외에 일본, 영국 및 미국 해도가 있다.

1) 한국해도(Korean charts)

한국해도는 국립해양조사원(National Oceanographic Research Institute, NORI)에서 발간하며, 간행구역은 한국 근해 및 일본 및 동남아시아 일부이다. 국립해양조사원에서 간행하는 모든 해도 및 수로서지의 종류와 내용은 서지 제810호 '수로도서지목록(catalogue of charts and publications)'을 참조하면 된다.

2) 일본해도(Japan charts)

일본해도는 해상보안청(Japan coast guard)에서 발간하며, 일본 근해를 중심으로 동남아시아는 물론 동쪽으로는 알래스카, 서쪽으로는 홍해, 남쪽으로는 호주, 북쪽으로는 베링해까지를 대상 해역으로 하고 있다. 따라서 우리가 특히 일본 근해를 항해할 경우에는 이 해도를 자주 이용하게 된다. 일본해도 및 수로서지의 종류와 내용은 해상보안청에서 서지 제901호로 발간되는 '수로도서지목록(Catalogue of Charts and Publications)'을 참조하면 된다.

3) 영국해도(Admiralty Charts)

영국해도는 해군수로국(Hydrographic Office)에서 간행하며 정식 명칭은 Admiralty Charts이나 보통은 B.A.(British Admiralty의 준말) Charts로 하고 있다. B.A. 해도 및 수로서지는 전세계 거의 모든 해역을 포함하고 있으며 그 종류와 내용은 'NP131'로 간행되는 '수로도서지목록'를 참조하면 된다.

4) 미국해도(DMA charts)

미국해도는 국방부(Department of Defence)의 National Imagery and Mapping Agency에서 발간하고 있으며, 일반적으로 DMA Charts라고 하고 있다. 전세계 거의 모든 해역의 해도 및 수로서지를 발간하고 있으며, 9개 지역(region)으로 나누어 구분하고 있다.

[그림 2-17] Catalogue of Charts and Publications

[예제] 부산항을 출발하여 홍콩항에 입항하는데 필요한 해도를 찾아보자.

4. 해도상의 정보

1) 해도의 축척

두 지점 사이의 실제 거리와 해도에서 이에 대응하는 두 지점 사이의 길이의 비를 축척(scale)이라 하는데, 예를 들어 1 : 25,000이라 함은 해도상의 길이 1단위는 지구표면상에서는 25,000단위에 해당한다.

항박도(대표적인 축척은 1 : 50,000)와 같이 작은 지역을 상세하게 표시한 해도를 대축척 해도라 하고 항양도(축척 100만분의 1 이하) 및 총도(축척이 400만분의 1 이하)와 같이 넓은 지역을 작게 나타낸 해도를 소축척 해도라 한다.

2) 해도의 번호 및 표제

모든 해도는 분류번호와 그 해도의 내용을 단적으로 표시한 표제가 기록되어 있으며 이는 임의로 변경할 수 없다. 해도의 번호 및 표제는 가급적 해면을 살려서 육지부분의 여백을 이용하여 눈에 잘 띄는 자리를 택하여 기재한다. 표제에는 지명, 도명, 축척, 수심과 높이의 단위, 필요한 경우에 도법명, 측량년도 및 자료의 출처, 조석에 관한 기사, 측지계, 기타 주의 기사가 기재되어 있다.

2.2.2 해도도식

해도상에 항로표지의 특질, 등심선, 수심, 해저 저질의 종류, 해류 및 조류의 크기와 방향, 해안선, 높은 산과 같이 잘 나타나는 육상물표의 등고선 등을 표시하기 위하여 사용되는 특수한 기호와 약어를 해도도식(海圖圖式, chart symbol)이라 한다.

[예제] 해도도식을 설명한 간행물(publication) 'CHART 5011'을 찾아보고, 한국해도 No. 204에서 해도도식에 해당되는 내용을 20개 이상 찾아보자.

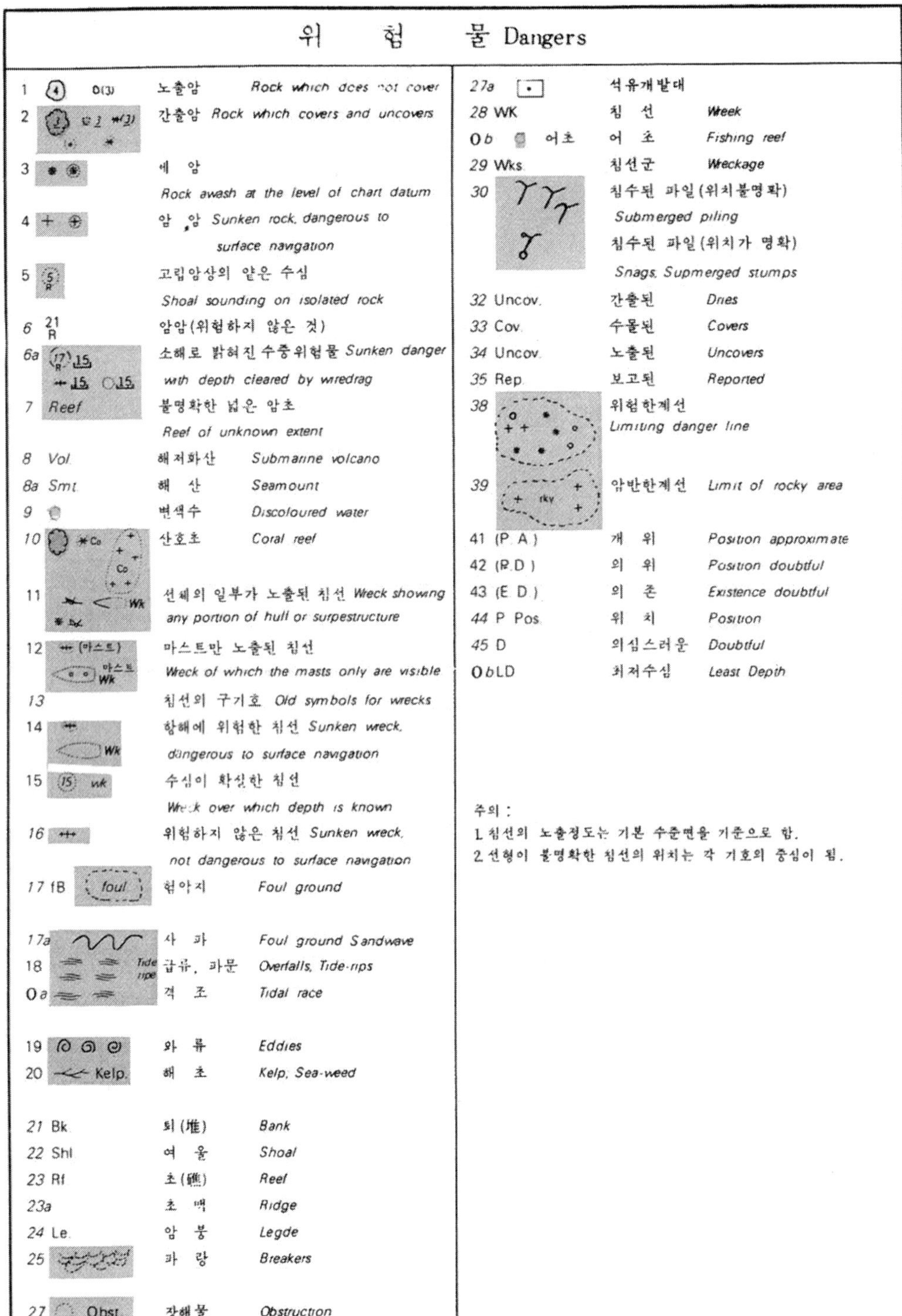

위 험 물 Dangers

No.	기호	명칭	Description
1	O(3)	노출암	Rock which does not cover
2		간출암	Rock which covers and uncovers
3		세 암	Rock awash at the level of chart datum
4		암, 암	Sunken rock, dangerous to surface navigation
5		고립암상의 얕은 수심	Shoal sounding on isolated rock
6	21 R	암암(위험하지 않은 것)	
6a		소해로 밝혀진 수중위험물	Sunken danger with depth cleared by wiredrag
7	Reef	불명확한 넓은 암초	Reef of unknown extent
8	Vol.	해저화산	Submarine volcano
8a	Smt.	해 산	Seamount
9		변색수	Discoloured water
10		산호초	Coral reef
11		선체의 일부가 노출된 침선	Wreck showing any portion of hull or surpestructure
12		마스트만 노출된 침선	Wreck of which the masts only are visible
13		침선의 구기호	Old symbols for wrecks
14		항해에 위험한 침선	Sunken wreck, dangerous to surface navigation
15		수심이 확실한 침선	Wreck over which depth is known
16		위험하지 않은 침선	Sunken wreck, not dangerous to surface navigation
17	fB	험악지	Foul ground
17a		사 파	Foul ground Sandwave
18		급류, 파문	Overfalls, Tide-rips
Oa		격 조	Tidal race
19		와 류	Eddies
20	Kelp.	해 초	Kelp, Sea-weed
21	Bk.	퇴(堆)	Bank
22	Shl	여 울	Shoal
23	Rf	초(礁)	Reef
23a		초 맥	Ridge
24	Le.	암 붕	Legde
25		파 랑	Breakers
27	Obst.	장해물	Obstruction

No.	기호	명칭	Description
27a		석유개발대	
28	WK	침 선	Week
Ob	어초	어 초	Fishing reef
29	Wks.	침선군	Wreckage
30		침수된 파일(위치불명확)	Submerged piling
		침수된 파일(위치가 명확)	Snags, Supmerged stumps
32	Uncov.	간출된	Dries
33	Cov.	수몰된	Covers
34	Uncov.	노출된	Uncovers
35	Rep.	보고된	Reported
38		위험한계선	Limiting danger line
39		암반한계선	Limit of rocky area
41	(P.A.)	개 위	Position approximate
42	(P.D.)	의 위	Position doubtful
43	(E.D.)	의 존	Existence doubtful
44	P. Pos.	위 치	Position
45	D	의심스러운	Doubtful
Ob	LD	최저수심	Least Depth

주의 :
1. 침선의 노출정도는 기본 수준면을 기준으로 함.
2. 선형이 불명확한 침선의 위치는 각 기호의 중심이 됨.

[그림 2-18] 위험물의 해도도식상의 표시

2.2.3 해도사용법

두 지점간의 거리를 구하는 방법은 두 지점에 디바이더의 발을 각각 정확히 맞추어 두 지점간의 간격을 재고, 이것을 그들 두 지점의 위도와 가장 가까운 위도의 눈금에 대어 거리를 구한다. 두 지점과 너무 떨어진 위도의 위도척에 의하여 거리를 구하면 정확하지 않게 된다.

1. 해도작업에 필요한 도구

해도작업에 필요한 기본적인 도구로는 삼각자, 디바이더, 컴퍼스, 지우개 및 연필 등이 있다.

삼각자는 해도상에서 방위를 재는 도구로 이용되는데 우리나라 선박에서는 대부분 이 노우에식 삼각자를 사용하고 있으나, 유럽의 경우에는 평행자(parallel ruler)를 많이 이용하고 있다. 일반적으로 해도용 삼각자는 2개를 한 세트로 사용하고 있다.

디바이더(divider)는 해도상에서 거리를 재는 도구인데, 일반적으로 많이 사용하는 직선형(straight type)과 활모양(bow type)이 있다. 디바이더를 크게 열어 사용하면 측정오차가 발생하기 쉬우므로 그러한 경우에는 몇 번에 나누어 측정하는 것이 좋다.

컴퍼스(compass)는 디바이더와 유사한 형태이나 한쪽에 연필심이 부착되어 레이더 등을 이용하여 물표까지의 거리를 파악하고 이를 해도위에 작도할 때 사용한다. 반원 등을 그리는 경우에는 편리하지만 디바이더로 사용하기에는 적당하지 않다. 만약 디바이더로 사용하려는 경우에는 항상 연필심을 가늘게 유지할 필요가 있다.

해도에 항해계획 등을 기입하는 경우 되도록 부드러운 연필로 기입을 하고 항해를 마치면 지우개로 지워 해도를 다시 사용할 수 있게 한다.

2. 해도사용시의 주의할 점

1) 해도의 선택

대축척의 해도는 지형, 지물, 수심 등이 상세하게 표시되어 있는 대신 좁은 구역만을 나타내므로 항해목적에 따른 적합한 축척의 해도를 선택하는 것이 좋다. 해도는 최근까지의 변경사항이 수정된 후 판매되므로 구입시 확인하고 그 이후에 변경된 사항은 항행통보를 이용하여 수정한 후 사용하여야 항해안전을 확보할 수 있다.

2) 측지계의 변환

위성항법장치를 이용하여 선위를 측정한 경우 항법장치와 사용해도의 측지계가 일치하는지 여부를 반드시 확인하고 일치하지 않을 경우는 측지계에 따른 거리오차를 수정하여 해도에 기입한다.

3) 해도의 관리

① 해도를 해도대의 서랍에 넣을 때에는 반드시 펴서 넣어야 하며, 부득이 접어야 할 때에는 구김살이 생기지 않도록 주의해야 한다.
② 해도는 발행기관별 번호순서로 정리하고 항해 중에는 사용할 것과 사용한 것을 분리하여 정리하면 편리하다.
③ 서랍의 앞면에는 그 속에 들어 있는 해도번호나 구역을 표시해 두어야 한다.
④ 해도를 사용할 때에는 삼각자, 끝이 뾰족한 디바이더, 누르개, 질이 좋은 지우개와 연필을 준비해야 한다. 연필은 2B나 4B를 사용하되 끝은 납작하게 깎아야 한다.
⑤ 해도에는 필요한 선만을 긋도록 하며 여백에 낙서하는 일이 있어서는 안된다.
⑥ 해도를 운반할 때는 반드시 말아서 가지고 다녀야 하며 비바람에 맞지 않도록 하여야 한다.

3. 해도에 선위를 기입할 때까지의 주의사항

① 선수가 지정된 침로에 정침되어 있는지 여부, 주변상황에 대한 정밀한 파악 및 경계를 실시한다. 해도작업(chart work) 중에는 경계가 중단되므로 선위를 기입할 때는 항로부근에 타선 및 기타의 장애물이 없는가에 특히 주의한다.
② 야간에 해도실(chart room)의 조명은 작업에 적당하도록 조절하고 경계에 방해되는 빛이 외부에 새어나가지 않도록 주의한다.
③ 선위부근의 해도면을 청결히 하여 새롭게 기입되는 위치선이 확실히 판명되도록 한다.
④ 비나 습기에 젖은 손, 안경, 모자 등으로 인해 해도가 젖지 않도록 주의한다.

4. 해도에 선위를 기입할 때의 주의사항

① 물표를 정확히 식별한 후 해도 위에 신속하게 작도한다. 본선의 예측위치 부근의 좁은 범위에만 위치선을 작도한다.

② 위치선을 작도할 때 연필의 각도에 주의한다. 물표 혹은 경도·위도선의 위치선을 같은 연필각도로써 그으면 보다 정확성을 기할 수 있다.

5. 해도에 선위를 기입한 후의 주의사항

① 선위를 표시하고 선위를 측정한 시각을 표시한다. 표시된 위치를 '전자해도 표시정보시스템(ECDIS)'의 본선 위치와 비교한다.

② 이전의 선위와 비교해서 본선의 이동거리 및 속력을 확인하여 예정대로 항행하고 있는지 확인한다.

③ 천후, 풍랑, 해조류, 기관회전 등의 영향을 고려해서 선위에 이상이 있다고 판단되면 선장에게 즉시 보고한다.

④ 해도 위는 항상 깨끗하게 정돈해 둔다.

6. 삼각자를 이용한 침로나 방위의 기입법

① 일반적으로 사용하는 이노우에(井上) 삼각자는 각도 45°식(A자, 일명 받침자), 30°와 60°식(B자, 일명 작업자)으로 구성되어 있다.

② 그림과 같이 B자(운용자)의 각도 눈금을 경도선에 맞춘 후 A자와 B자를 평행이동시켜 방위선 및 침로를 해도에 작도한다. 이는 이동범위를 줄이면서 편리하게 침로나 방위선을 기입할 수 있다는 장점이 있다. 작도시에는 눈금이 기재되어 있는 면을 해도측에 밀착시킨다. 두 삼각자를 항상 붙여서 사용해야 한다.

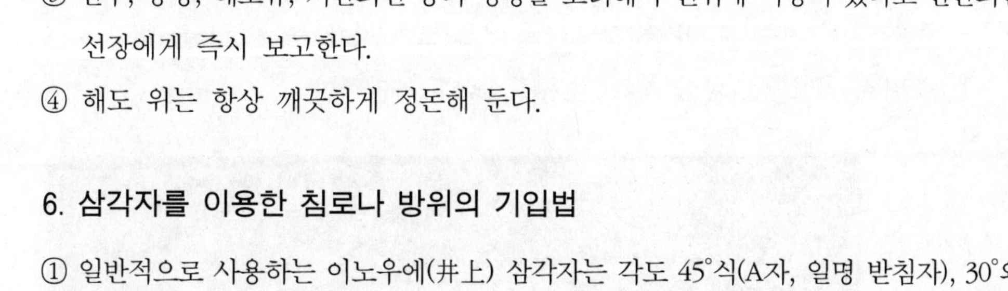

[그림 2-19] 이노우에 삼각자

7. 기 타

① 해도의 선위 기입은 가능한 한 빨리 하고 즉시 경계에 임한다. 당직사관이 Chart Room에 머무르는 기간에 충돌사고 등의 해양사고가 많이 발생하기 때문이다.

② 해도가 오손되거나 구겨지지 않도록 주의한다.

③ 디바이더를 직각 또는 직각에 가까운 각도로 이용하는 것은 해도에 구멍을 내므로 피하도록 한다.

④ 연필은 항상 사용이 가능하도록 여러 개를 준비해 둔다.

⑤ 해도 위에서 지우개를 사용한 후에는 지우개 털이를 이용하여 쓸어 모아 휴지통에 버린다.

⑥ 선체의 동요로 차트 테이블(chart table)에서 떨어지기 쉬운 연필, 지우개 등은 디바이더(divider) 또는 삼각자를 이용하여 고정시킨다. 해도는 문진(weight)을 이용한다.

⑦ 삼각자는 수시로 비눗물로 닦아주어 해도에 때가 묻지 않도록 한다.

⑧ 차트 테이블은 항상 정리 정돈된 상태를 유지한다.

[그림 2-20] 해도대의 정리상태

2.2.4 해도의 기준면

현재 우리나라에서 사용하고 있는 해도에서 높이와 수심을 나타내는 단위는 미터이며, 그 기준면은 다음과 같다.

2014년21호307항
NM(307)2014
Chart 001(H20)

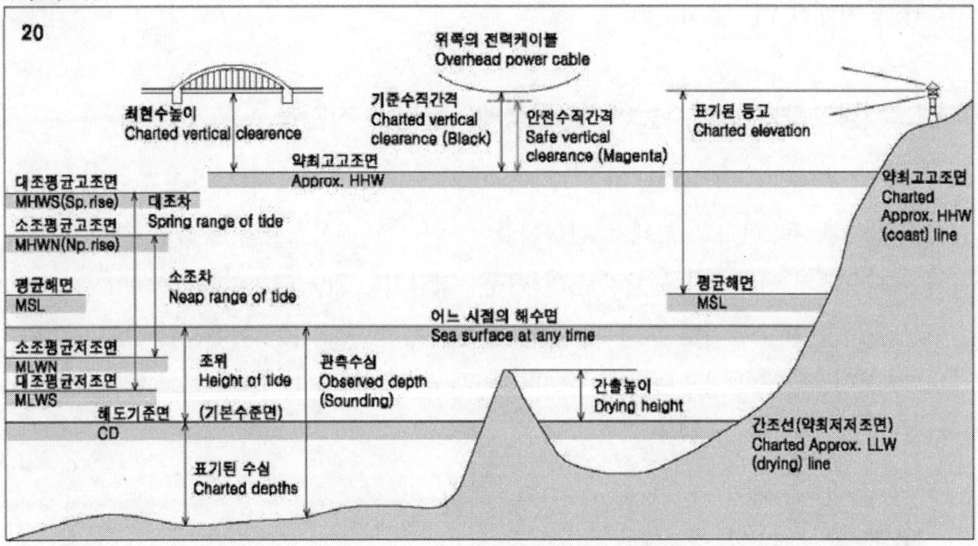

[그림 2-21] 수심 및 높이의 기준면

1. 수 심

연중 해면이 그 이상으로 낮아지는 일이 거의 없다고 생각되는 수면을 해도기준면(Chart Datum, CD) 또는 약최저저조면(Approximate Lowest Low Water, Approx.LLW)이라 하며, 우리나라 해도의 수심(水深)은 이 수면을 기준으로 하여 나타낸다.

2. 물표의 높이

등대 등과 같은 고정물표의 높이는 장기간 관측한 해면의 평균 높이에 있는 수면을 평균해면(平均海面, Mean Sea Level, MSL)이라 한다.

3. 조고와 간출암

조석에 의하여 변동하는 수면의 높이를 조고라 하고, 조석의 간만에 따라 수면 위에 나타났다 수중에 감추어졌다 하는 바위를 간출암(干出岩)이라 한다. 그런데, 이들의 높이는 해도기준면을 기준으로 하여 측정된다.

4. 안 선

일반적으로 해안선(海岸線)을 줄여서 안선이라고도 하며, 고조면(high water)에서의 수륙의 경계선으로 표시한다.

5. 해상교량의 최현수높이 및 전력케이블의 수직간격

최근 우리나라 부산, 인천, 여수, 울산, 마산 등 주요 항만에는 항만을 횡단하는 해상교량이 사장교 및 현수교 형태로 건설되었다. 이러한 해상교량의 형하고 결정에 필요한 최현수거리 및 전력케이블(해월선)의 기준수직간격(charted vertical clearance) 등은 더 이상 높아질 일이 없다고 판단되는 약최고고조면(Approx. HHW)을 기준으로 하여 나타낸다. 전력케이블의 경우 케이블 처짐이 발생하므로 여유간격(clearance)을 고려한 안전수직간격도 별도로 표시한다.

2.2.5 해도의 측지계

GPS에서 기본적으로 사용하는 좌표계는 지구의 중심을 원점으로 하는 3차원 직교좌표계(세계측지계라고 하며, 약어로는 WGS-84)로서, 2001년 초까지 우리나라에서 해도 등에 사용하였던 Bessel 타원체를 이용한 도쿄측지계(Tokyo Datum)와 약 300~400미터의 차이가 있었으며, 이러한 측지계에 대한 인식부족은 해도상에 잘못된 선위를 기입하도록 하여 선박의 안전운항을 위협하였다.

따라서 국제수로기구(IHO), 국제해사기구(IMO)에서는 WGS-84 측지계로 작성된 해도를 사용할 것을 권고하였다. 이에 따라 우리나라는 기존의 도쿄측지계와 세계측지계의 좌표계를 혼용하여 사용하였던 해도를 2004년까지 모두 세계측지계 해도로 교체하여 사용하고 있다.

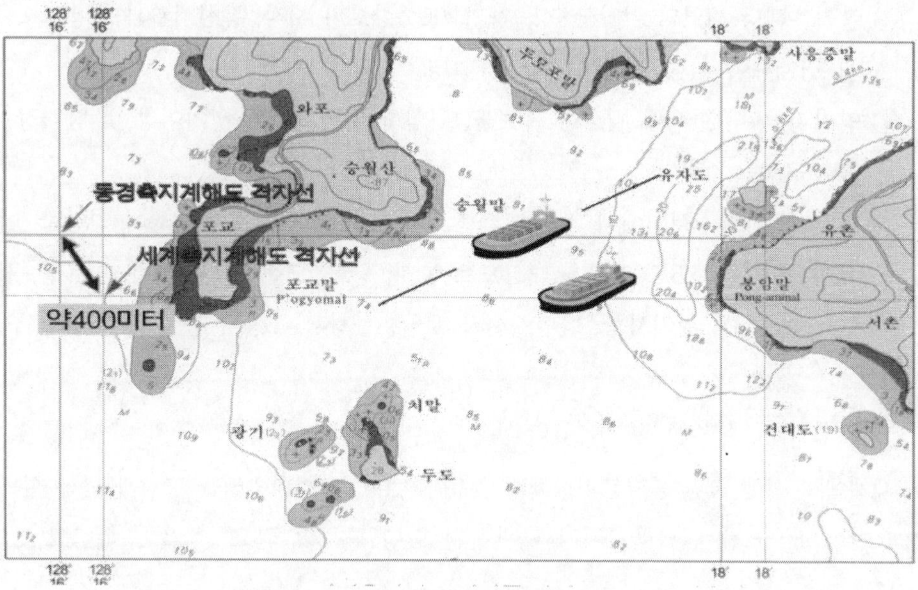

[그림 2-22] 도쿄측지계와 세계측지계의 차이

2.2.6 수로서지

해도 이외에 항해에 도움을 주는 모든 간행물을 통틀어 수로서지(nautical publications)라 하며, 일반적으로 이러한 수로서지는 항해시 또는 항만 입출항시에 안내기능을 담당하는 항로지와 기타 참고자료를 제공하는 특수서지(등대표, 조석표, 천측력, 국제신호서 등)의 2가지로 분류된다.

1. 수로서지의 사용

주로 사용하는 수로서지(publication)는 한국, 일본, 미국(Defence Mapping Agency, DMA), 영국(British Admiralty, BA)판이다.

2. 수로서지의 종류

1) 항로지(Coast Pilot ; Sailing Direction)

해당 해역의 기상, 해상, 항로와 항만의 사정 등 일반사항(총기), 목표물 등(연안기) 및 주요항의 항계등(항만기)을 구분하여 현장을 파악하고 조사한 자료이다.

① 총기 : 해당 해역의 기상, 해상, 자기, 표준항로와 시각, 통신기관과 이용방법, 특별히 경계해야 할 사항과 관련 법규설명
② 연안기 : 연안항해에 필요한 목표물, 위험한 지역, 닻밭, 정치어장 및 양식장, 침선 등에 관하여 설명
③ 항만기 : 개항질서법에서 정하는 개항, 항만법에서 정하는 무역항과 연안항, 어항법에서 정하는 1·3종 어항을 대상으로 항계, 항로, 도선구간과 구역, 침로법, 정박구역, 적용법규, 검역사항, 항만시설과 보급, 관광과 교통편 등에 관한 내용을 설명

2) 항로지정(Ships' Routeing)

세계에서 시행 중인 통항분리방식과 깊은 수심항로, 피항수역을 그림과 함께 설명하고 있으며, IMO가 채택한 항로지정과 관련된 항행규칙 및 권고사항을 수록하고 있다.

3) 국제신호서(International Code of Signal)

선박의 의사 및 상태를 전달하기 위해, 특히 의사소통의 문제가 발생한 경우 항해안전 및 인명안전과 관련된 의사소통의 소통을 제공하기 위해 국제적으로 약속된 부호와 그 부호의 의미를 해설한 것으로 일반부문(single letter & two letter signals)과 의료부문(medical section, three letter signals)으로 나누어 있으며 수신호 또는 기류신호로서 표시한다.

4) 해상거리표(Distance Tables)

주요항까지의 항로상 거리를 해리로 표시하였으며 선박의 크기, 계절 등의 요인에 따라 여러 항로를 선택할 수 있을 경우 그 경유지도 함께 표기하였다.

5) 천측력(Nautical Almanac)

선박의 위치 및 컴퍼스 오차를 측정하기 위해 필요한 천체 관련 자료를 수록하였다.
① 1년간 매일, 매시각의 수성, 금성, 지구, 달, 목성, 토성의 본초시각 및 적위
② 1년간 매일의 일출몰 시각, 월출몰 시각 및 박명시각
③ 57개 항성의 항성시각 및 적위

6) 천측계산표(Sight Reduction Tables for Marine Navigation)

측정한 천체의 고도와 방위각을 이용하여 선위 및 컴퍼스의 오차는 물론 대권항해에 필요한 침로, 거리 등을 얻는데 이용되는 것으로 지방시각, 위도, 적위에 대한 계산고도 및 방위각을 수록하였다.

7) 색성판(Star Finder)

계산을 하지 않고 천측력에 수록된 57개 항성의 춘분점까지의 지방시각을 가지고 개략적인 항성의 고도와 방위를 얻고자 할 때 이용된다.

8) 태양방위각표(Azimuth of the Sun)

지방시각을 시시로 환산한 값, 적위, 위도를 통해 태양의 고도 및 방위각을 제공하여 선위 및 컴퍼스 오차 산출에 이용될 뿐 아니라 일출몰 시각을 제공한다.

9) 등대표(List of Light)

항로표지에 일련번호를 부여하여 수록한 서지로 항로표지의 명칭과 위치, 등질, 등고, 광달거리, 색상과 구조에 관한 자료와 함께 항로표지에 대한 해설, 항로표지에 관한 법규 등을 게재하고 있다. 항행통보를 이용하여 자료를 최신화하여야 한다.

10) 조류도(Tidal Current Charts)

주요 수역의 조류현황을 설명과 함께 그림으로 나타낸 서지로서 경제항로 선정 및 입출항 시각결정에 유용하게 활용할 수 있다.

11) 조석표(Tide Tables)

주요 항구의 고조 및 저조시각 및 높이가 기록되고 협수로 등의 전류시, 최강류시 및 유속 등을 날짜별로 수록하고 있다.

12) Guide to Port Entry

전세계 모든 개항에 대하여 항만사정, 입출항 서류, 하역설비, Pilotage, Tug, 보급,

의료지원 등 항해사에게 유용한 각종 자료를 제공하고 있다.

13) 해양환경도(Marine Environmental Atlas of Korea Waters)

해저지형, 수온 및 염분분포, 용존산소량 등의 해양환경에 관한 여러 사항을 그림으로 표시한 서지이다.

3. 수로도서지목록(Catalogue of Charts and Publication)

해도와 서지를 색인도와 함께 번호별로 나타낸 목록이다.

1) Pilot Chart

항해사에게 유용한 아래의 각종 자료를 해역별, 월별로 도표의 형태로 제공하고 있다.

① 등편차 곡선(Magnetic variation)
② 해류(Currents)
③ 각 지점의 탁월풍(Prevailing winds and calms)
④ 폭풍의 빈도(Percentage of gale)
⑤ 폭풍의 진로 및 지역별 발생횟수(Tracks of tropical and extra tropical cyclones)
⑥ 안개발생분포(Percentage of fog)
⑦ 표면대기압(Surface barometric pressure)
⑧ 유빙의 한계(Ice and iceburg limit)
⑨ Ocean weather station ship의 위치
⑩ 추천항로 및 주요 지점간의 항해거리(Recommend routes)
⑪ 등온선(Surface air and water temperatures)
⑫ 해역의 파고(Wave height)
⑬ 기상·천후의 일반해설 등

2) Radio Signal

항만의 입출항에 필요한 통신내용 및 보고내용 등을 국가별, 항구별로 제공하고 있다.

〈표 2-3〉 수로서지의 종류

형태	서지번호	서 지 명	내 용
항로지	110	동해안 항로지	두만강 하구에서 승두말
	120	남해안 항로지	승두말에서 해남각
	130	서해안 항로지	해남각에서 압록강 하구
	140	중국연안 항로지	중국연안 전체
	150	말라카 해협	수마트라섬의 북서단에서 싱가포르 해협
	210	근해 항로지	한국근해, 일본, 동남아의 항로선정 자료제공
	220	대양 항로지	대양에 있어서의 항로선정을 위한 자료제공
	230	항로지정	연안, 해협, 진입로 등의 통항분리 방식과 주의사항
표형태	240	국제신호서	시각 및 무선전신으로 육 · 해 · 공간의 통신방법 수록
	250	해상거리표	한국 및 세계 각국의 주요 항간의 거리를 수록
	310	천측력	주요 행성의 적위, 항성의 항성시각, 해와 달의 출몰시간이 기록된 것을 천문항법용으로 사용
	320	태양방위각표	위도, 적위 및 시시에 의하여 태양의 진방위를 알아내는 것으로서 컴퍼스 오차 확인 등에 사용
	331~336	천측계산표	천문항해에서 천측위치를 계산하는데 필요한 제원을 수록한 책
	410	등대표	한국연안의 항로표지를 수록
	510	조석표(한국연안)	한국연안의 주요 항만과 표준항에 대한 조석 및 조류의 자료를 수록
	520	조석표(태평양)	태평양의 주요 항만과 표준항에 대한 조석 및 조류의 자료를 수록
그림형태	620~634	조류도	우리나라 주요 19개 지역의 조류 현황 제공
	710	한국해양환경도	한국 근해의 해양환경 자료를 수록

[예제] 수로서지의 종류에는 어떤 것들이 있으며, 그 중 실습 중인 선박에 비치된 수로서지는 어떤 것들이 있는지 찾아보고 그 내용을 정리해 보자.

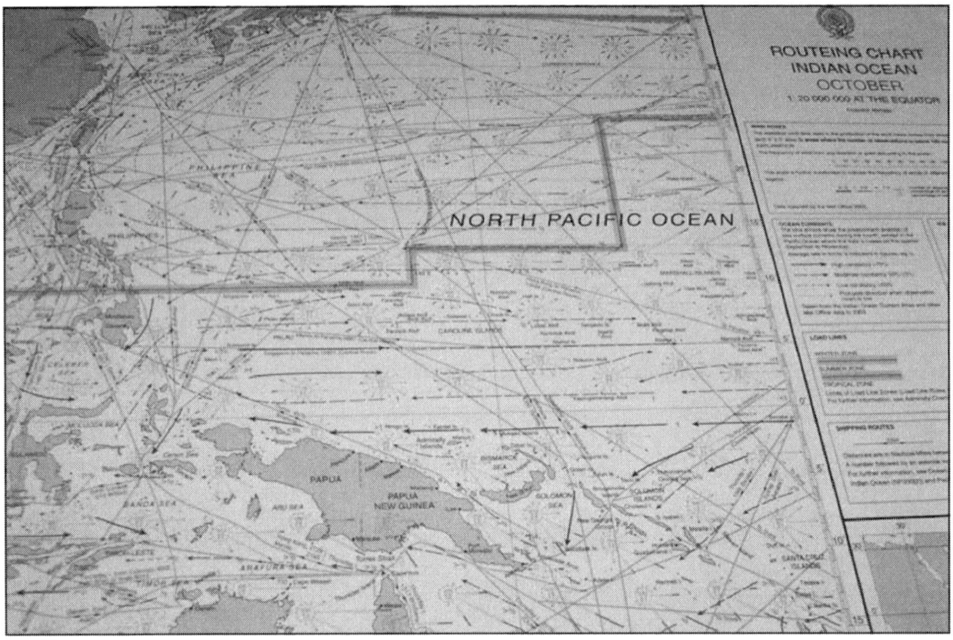

[그림 2-23] Routeing Chart(Pilot Chart)

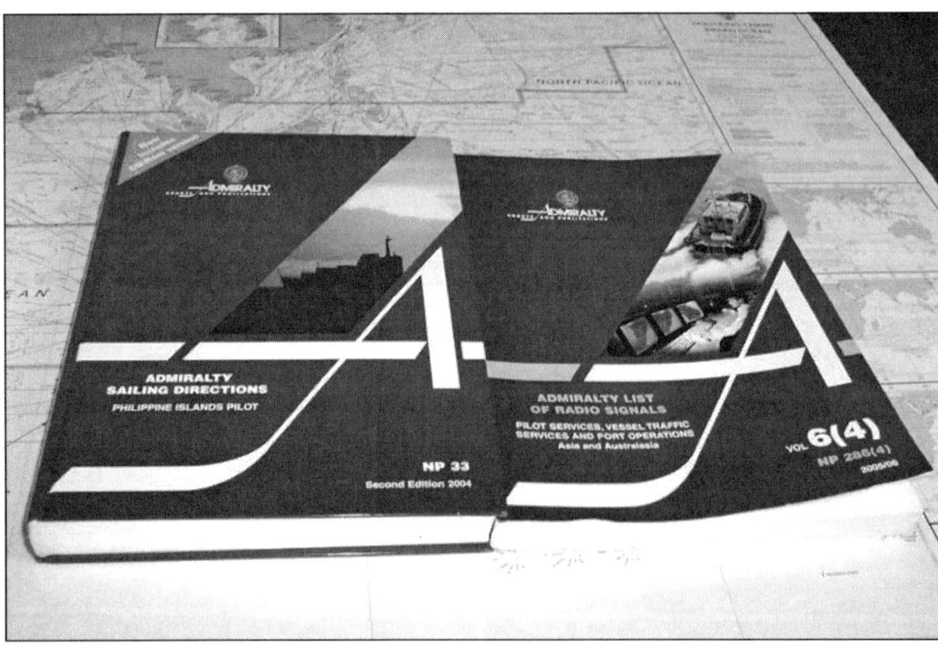

[그림 2-24] Sailing Directions & Radio Signals

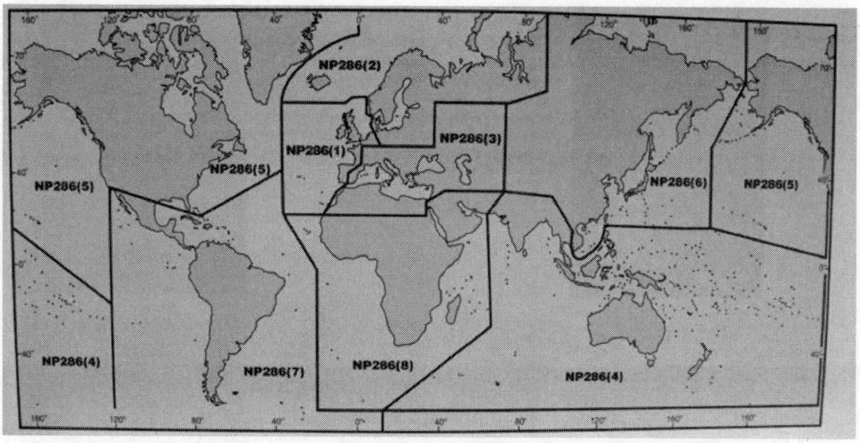

[그림 2-25] Limits of Admiralty List of Radio Signals NP286

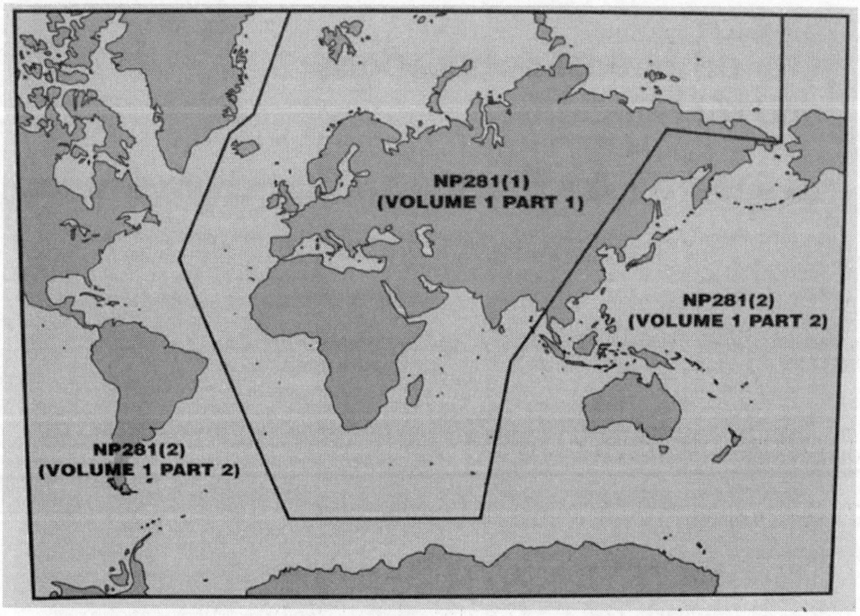

[그림 2-26] Limits of Admiralty List of Radio Signals NP281

2.2.7 수로도서지의 개정

수로도지 및 서지는 항상 현재 상태와 일치되어 있어야 하므로, 그 내용에 변경이 있거나, 간행 이후에 새로운 자료를 얻었을 때에는 항행통보에 의하여 개정해야 한다.

1. 항행통보

선박의 안전항행에 필요한 각종 정보를 수록한 해도 및 수로도서의 내용을 최신자료로 유지하기 위해 항로표지의 신설 및 변경, 암초, 침선, 수중장애물, 항만공사, 해상사격훈련, 항로변경, 해저케이블, 표류위험물, 해적행위, 수로도서지의 수정 및 보완사항 등에 관한 사항을 정기적으로 통보한다.

2. 항행통보의 구성

① 지역별 항수색인

② 해도 · 서지별 항수색인

③ 제 I 부 : 간행예고, 일반, 해도 · 서지 수정사항, 일시사항 등

④ 제II부 : 해상사격훈련 등 항해안전을 위한 정보 및 주의사항

⑤ 별지 : 수로도 · 서지 수정사항 및 항해안전을 위한 참고사항 등

⑥ 보정도

3. 항행통보 제공방법(한국)

① 홈페이지 게재(www.nori.go.kr) : 매주 수요일

② 인쇄물에 의한 항행통보 제공 : 매주 금요일

③ E-mail 정보배달 서비스 : 신청자에 한해 전자메일로 발송

4. 항행통보 제공방법(UKHO)

인쇄물에 의한 항행통보도 이루어지고 있으나 모든 항행통보를 전자파일 형태로 제공받는 추세이다.

6|0|6 - 8|0|6 부산광역시 영도구 해양로 351
전화 051-400-4331~4
팩스 051-400-4194
인터넷 http://www.khoa.go.kr
전자우편(E-Mail) : ntmkhoa@korea.kr

2015년 4월 3일 발행
(매주 금요일 발행)

제 Ⅱ 부 162 ~ 174항
제 Ⅲ 부 (58) ~ (66)항

항 행 통 보

제 14 호

Ⅰ. 목 차
Ⅱ. 항행통보
Ⅲ. 항행정보
Ⅳ. 항해서지 개정

▶ 일 러 두 기 ◀

항해자는 항행통보를 받으면 즉시 관련 해도와 항해서지를 개정하여야 하며,
안전항해에 영향을 주는 위험물 또는 항로표지의 이상을 발견하거나
해도와 항해서지의 표기에 의문이 있을 때는 국립해양조사원으로 알려주시기 바랍니다.

---

1. 항행통보 항 번호 앞에 ★표시는 통보사항의 출처가 대한민국인 것을 의미한다.
2. 항행통보 항 번호 뒤에 표시한 (T)는 일시정보, (P)는 예고정보를 뜻하며, 항 번호 뒤에 ※가 있는 것은 이전에 일시정보 또는 예고정보로 통보된 적이 있는 것을 의미한다.
3. 경위도 위치는 세계측지계(WGS-84) 좌표를 표기한다.
4. 침로와 방위는 진방위로, 시계방향 000°~359°로 표시한다. 명호나 지도선의 방위는 해상에서 목표물을 바라본 방위로 표시한다.
5. 시각은 4자리 숫자(00:00~24:00)로 표시하며, 별도 표시가 없는 것은 한국 표준시(KST)이다.
6. 서지번호 앞에 †가 있는 것은 개정사항이 별지에 있는 것을 의미한다.
7. 해도번호 뒤에 붙은 대괄호는 이전에 마지막으로 통보한 항행통보의 연도와 항 번호를 표시한 것이다.

국 립 해 양 조 사 원

[그림 2-27] 항행통보(국립해양조사원)

Notices
2074-2215/15
T & P Notices in Force

United Kingdom
Hydrographic Office

ADMIRALTY NOTICES TO MARINERS

Weekly Edition 18

30 April 2015
(Published on the UKHO Website 20 April 2015)

CONTENTS

For information on how to update your ADMIRALTY products using ADMIRALTY Notices to Mariners, please refer to NP294 How to Keep Your ADMIRALTY Products Up-to-Date.

Mariners are requested to inform the UKHO immediately of the discovery of new or suspected dangers to navigation, observed changes to navigational aids and of shortcomings in both paper and digital ADMIRALTY Charts or Publications.
The Hydrographic Note Form (H102) should be used to forward this information and to report any ENC display issues.
H102A should be used for reporting changes to Port Information.
H102B should be used for reporting GPS/Chart Datum observations.
Copies of these forms can be found at the back of this bulletin and on the UKHO website.

The following communication facilities are available:

Notices to Mariners Website:	Web:	www.ukho.gov.uk/msi
Searchable Notices to Mariners:	Web:	www.ukho.gov.uk/nmwebsearch
Urgent navigational information:	e-mail: Phone: Fax:	navwarnings@btconnect.com +44(0)1823 353448 +44(0)1823 322352
H102 forms (see back pages of this Weekly Edition)	e-mail: Post:	sdr@ukho.gov.uk UKHO, Admiralty Way, Taunton, Somerset, TA1 2DN, UK
All other enquiries/information	e-mail: Phone:	customerservices@ukho.gov.uk +44(0)1823 337900 (24/7)

Printed in the United Kingdom for the UKHO

For UKHO use only 215418

[그림 2-28] 항행통보(UKHO)

5. 해도의 개정 및 소개정

해도를 최신 상태로 유지하기 위한 방법으로는 해도 발행 당국에서 직접 개판(new edition)이나 재판(reprint)을 통해 개정하는 방법과 항해자 자신이 항행통보에 의하여 정정하는 소개정(small correction) 방법이 있다.

1) 개 판

새로운 자료에 의해 해도의 내용을 전반적으로 개정하거나 또는 해도의 포함구역이나 크기 등을 변경하기 위해 해도 원판을 새로 만드는 것으로 이전의 해도는 폐기 처분한다.

2) 재 판

현재 사용 중인 해도의 원판을 약간 수정하여 다시 발행하는 것이다.

3) 소개정

항행통보에 의해 직접 해도상에 수정 보완하거나 또는 보정도로서 개보하는 것으로 소개정 후에도 좌측 하단에 있는 《소개정》란에 통보 연도수와 통보항수를 기입하여 어느 시점까지 소개정이 되었는지를 누가 보아도 알 수 있도록 한다. 소개정의 방법은 다음과 같다.

① 수기로 하는 개보 : 통보내용을 그리거나 써서 개정하는 것으로 불필요한 부분은 두 줄을 그어 지운다.

㉠ 개보할 때에는 붉은색 잉크를 사용해야 한다.

㉡ 기사는 해도의 여백(보통 좌측 하단)에 간결하고 알기 쉽게 가로로 써야 한다.

㉢ 해도 도식에 기호가 정해지지 않는 지물의 위치는 ⊙ 또는 ○ 로 표시하고, 그 옆에 명칭을 기입해야 한다.

㉣ 수심은 수심을 나타내는 숫자의 정수 부분의 중앙이 되도록 기재해야 한다.

㉤ 침선, 암초 등의 바로 위에 표지로써 설치된 부표를 기입할 때에는 침선, 암초 등을 삭제하지 말고 거기에서 가장 가까운 항로쪽이나 외해쪽에 기입한다.

② 보정도에 의한 개보 : 지형, 해안선 또는 광범위하게 수심이 변화된 경우 또는 개보사항이 좁은 구역에 밀집된 경우 등은 손으로 직접 적어 개보하기 곤란하므로 변경사항이 그림으로 제공되는 보정도를 항행통보에서 오려서 해도의 개정위치에 붙인다.

보정도를 사용할 때 주의할 사항은 다음과 같다.

㉠ 보정도를 붙이기 전에 해도와 겹쳐서 대조하고 개보내용을 확인한 다음 정확하게 붙인다.

㉡ 개보할 부분만 정확하게 자르고, 불필요한 부분은 절단하여도 좋다.

㉢ 신축으로 인하여 차질이 없도록 확실하게 붙인다.

6. 수로서지의 개정

수로서지도 그 내용에 변경이 있을 때에는 해도와 마찬가지로 항행통보에 의하여 개정한다. 항행통보에는 해도와 수로서지별 개정내용이 따로 기록되어 있으므로 해당 수로서지를 찾아 항행통보의 개보내용대로 정정하거나 오려 붙여 개정한다. 수로서지에 개정을 하였을 경우에는 수로서지의 앞부분에 있는 '개정이력'란에 통보연도수와 통보항수를 기입하여 어느 시점까지 개정이 되었는지를 누가 보아도 알 수 있도록 한다.

2.3 항로표지

선박이 안전항해를 하기 위해서는 항상 자선의 위치를 정확히 확인할 수 있어야 한다. 날씨가 좋은 낮 시간에 연안항해 혹은 항만을 출입항하는 경우에는 섬이나 육지의 산봉우리 등을 이용하여 자선의 위치확인이 가능하겠으나, 밤이 되거나 안개가 끼거나 주변의 지형이 비슷비슷해 눈으로 쉽게 식별할 수 있는 목표물들을 발견할 수 없을 경우는 위치를 확인하는 것이 쉽지 않을 것이다.

따라서 선박통항량이 많은 항로, 항만, 협수로, 연안해역 등에서는 주·야간에 상관없이 짧은 시간 안에 식별이 가능하여 자선의 위치를 정확하게 알 수 있도록 하는 인위적 시설물을 설치할 필요가 있다. 이러한 시설물을 항로표지(aids to navigation)라 하며 등광, 형상, 색채, 음향 또는 전파를 이용하여 식별한다.

2.3.1 주간표지

주간표지(주표)는 점등장치가 없는 표지로 그 모양과 색깔로써 식별하므로 형상표지라고도 한다. 주간에 선위를 결정할 때 이용되며, 암초, 침선 등을 표시하여 항로를 유도하는 역할을 한다.

1. 주간표지의 종류

① 입표(beacon) : 암초, 사주(모래톱) 등의 위치를 표시하기 위하여 그 위에 세워진 경계표이다. 여기에 등광을 함께 설치하면 등표가 된다.

② 부표(buoy) : 비교적 항행이 곤란한 장소나 항만의 유도표지로서 항로를 따라 설치하며 변침점에 설치한다. 여기에 등광을 함께 설치하면 등부표가 되는데, 특별한 경우가 아니면 주로 등부표를 사용한다.

③ 육표(land mark) : 육상에 설치된 간단한 기둥표로서 여기에 등광을 함께 설치한 것은 등주라 한다.

④ 도표(leading mark) : 좁은 수로의 항로를 표시하기 위하여 항로의 연장선 위에 앞뒤로 2개 이상의 육표로 된 것과 방향표로 되어 선박을 인도하는 것을 말하며, 여기에 등광이 함께 설치된 것을 도등이라 한다.

2. 국제해상부표 시스템

국제항로표지협회(International Association of Marine Aids to Navigation and Lighthouse Authorities, IALA)에서는 각국의 부표식의 형식과 적용 방법을 통일하여 적용하도록 하였으며, 우리나라도 이에 따라서 설치하여 운용하고 있다.

IALA 해상부표식은 세계를 A와 B지역으로 구분하여 측방표지의 도장 및 등화의 색상이 서로 반대로 되어 있는 것으로, 캐나다와 미국을 비롯한 북미, 중미 및 남미 지역과 카리브해 지역, 일본, 한국 및 필리핀 등의 지역은 B방식을, 나머지 지역은 A방식을 따르고 있다.

우리나라에서 적용하고 있는 B방식의 경우를 보면, 항구의 입항 · 출항 항로에서는 입항자를 기준으로 우현쪽의 측방표지인 등부표는 짝수번호에 적색(赤, red)으로 하고, 좌현의 측방표지인 등부표는 홀수번호에 녹색(綠, green)으로 배치하도록 하여 항만에 입항하는 선박을 기준으로는 현등색깔과 반대색깔의 배치를 하고 있다.

국제항해에 종사하는 선박은 위와 같이 상이한 부표방식을 채택하고 있는 국가간을 입·출항할 수 있으므로, 항해사는 해당 국가가 채택한 부표방식을 확인하는 것뿐만 아니라 항내나 항구의 진입수로 및 협수로를 항행할 때 매우 중요하게 이용되는 항로표지가 의미하는 바를 정확하게 이해하고 있어야 된다. IALA 해상부표식은 측방표지, 방위표지, 고립장해표지, 안전수역표지 및 특수표지의 다섯 가지로 이루어진다.

1) 측방표지(Lateral Marks)

측방표지는 우리가 항행하는 수로의 좌·우측 한계를 표시하기 위해 설치된 표지로서 좌우를 정하는 방법은 다음의 두 가지 방법 중 한 가지에 의해 결정된다.

① 항해자가 외해에서 항구, 강, 하구, 수로에 접근할 경우 취하는 일반적인 방향을 기준으로 한다.

② 육지를 중심으로 하여 시계방향으로 정하는 것을 원칙으로 하되 필요하면 인접국가와 협의하여 결정한다.

그러나 대부분의 해도에는 이러한 측방표지의 방향(conventional direction of buoyage)을 보라색의 화살표로 표시해주고 있으므로 이것을 참조하면 된다. 측방표지의 표시방법이 A와 B지역에서 서로 다른데, A지역에서는 좌현부표의 색깔과 등화의 색상이 적색이고, 우현부표의 색깔과 등화의 색상이 녹색으로 되어 있어 선박의 좌우를 표시하는 현등의 색상과 일치한다. 그러나 B지역의 경우에는 이와 반대로 되어 있다.

좌현부표는 이 부표의 위치가 항로의 왼쪽 한계에 있음을 의미하며 따라서 이 부표의 오른쪽에 가항수역이 있고, 왼쪽에는 암초나 침선 또는 수심이 얕아 항행에 위험이 있음을 나타낸다. 반대로 우현부표는 이 부표의 위치가 항로의 오른쪽 가장자리에 있음을 말하며, 따라서 이 부표의 왼쪽을 항행하면 안전하고, 오른쪽에는 항행에 위험이 있음을 나타낸다.

하나의 목적지에 이르는 항로가 2개로 나누어졌을 때 좌측항로가 일반적인 항로임을 나타낼 때는 좌항로 우선표지를, 우측항로가 일반적인 항로일 때는 우항로 우선표지를 설치한다.

2) 방위표지(Cardinal Marks)

방위표지는 장해물을 중심으로 하여 주위를 4개 상한으로 나누고 그들 상한에 각각 북, 동, 남, 서라는 이름을 붙이고, 이 각각의 상한에 설치된 표지를 북방위(north mark), 동방위(east mark), 남방위(south mark) 및 서방위(west mark) 표지라 한다. 이러한 방위 표지가 있을 때 항해자는 이 표지를 기준으로 하여 북방위 표지의 북쪽, 동방위 표지의 동쪽, 남방위 표지의 남쪽, 서방위 표지의 서쪽으로 항행하면 안전하다.

방위표지에 부착하는 두표(top mark)는 반드시 2개의 원추형을 사용하는데, 각 방위에 따라 서로 연관이 있는 모양으로 부착한다. 방위표지의 색상은 흑색과 황색 밴드를 사용하고, 이 표지에 부설되는 등화의 색상은 백색이며, 그 등질 역시 각 방위에 따라 서로 다르게 하고 있다.

3) 고립장해표지(Isolated Danger Marks)

고립장해표지는 그 전 주위가 가항수역인 암초나 침선 등 고립된 장해물의 위에 설치 또는 계류하는 표지를 말한다. 이 표지의 두표는 2개의 흑구를 수직으로 부착하며, 표지의 색상은 흑색 바탕에 한 개 또는 그 이상의 적색 띠를 둘러 표시한다. 등화는 흰색을 사용하며 2회의 섬광등{Fl(2)}을 사용한다.

4) 안전수역표지(Safe Water Marks)

안전수역표지는 이 표지의 모든 주위가 가항수역임을 알려주는 표지로서, 중앙선이나 수로의 중앙을 나타내 준다. 이 표지의 두표는 1개의 적색구를 부착하며, 표지의 색상은 적색과 백색의 세로방향 줄무늬로 되어 있다. 등화는 흰색을 사용한다.

5) 특수표지(Special Marks)

특수표지는 항행원조를 주요 목적으로 하는 것이 아니고 해당되는 수로도지에 기재되어 있는 공사구역, 토사채취장 등 특별한 구역 또는 특별한 시설이 있음을 표시하는 표지이다. 이 표지의 두표는 노랑색으로 된 1개의 X자 모양의 형상물을 부착하며, 표지의 색상 및 등화도 노랑색을 사용한다.

[그림 2-29] 방위표지(남방위)

[그림 2-30] 고립장해표지(인천항)

[그림 2-31] 안전수역표지(부산항)

[그림 2-32] 측방표지(좌) 및 특수표지(우)

2.3.2 야간표지

1. 항로표지의 구조적 분류

① 등대(light house) : 해양으로 돌출한 곶(갑), 섬 등 선박의 물표가 되기에 알맞은 장소에 설치된 탑과 같이 생긴 구조물이다. 다른 표지와 쉽게 구별할 수 있도록 등질을 서로 다르게 하고 있다.

② 등부표(light buoy) : 등대와 함께 가장 널리 쓰이고 있는 야간표지인 등부표는 해저의 일정한 지점에 체인으로 연결되어 해면에 떠있는 구조물이다. 이와 같이 해면에 떠있기 때문에 일정한 선회 반지름을 가지고 이동하게 되며, 강한 파랑이나 조류에 의해 위치가 이동되거나 유실되는 경우도 있으므로 주의를 요한다.

③ 등주(staff light) : 쇠나 나무 또는 콘크리트와 같이 기둥 모양의 꼭대기에 등을 달아 놓은 것으로 광달거리가 별로 크지 않아도 되는 항구, 항내 등에 설치한다.

④ 등표(light beacon) : 항로, 항행에 위험한 암초, 항행 금지구역 등을 표시하는 지점에 설치되는 표지이다.

⑤ 등선(light ship) : 육지에서 멀리 떨어진 해양, 항로의 중요한 위치에 있는 모래 기둥 등을 알리기 위하여 일정한 지점에 정박하고 있는 특수 구조의 선박을 말하며, 이들 등선을 대신하여 설치되고 있는 대형 등부표를 LANBY(Large Automatic Navigational Buoy)라 한다.

[그림 2-33] 등부표

[그림 2-34] 등표(싱가포르항)

[그림 2-35] 등선(미얀마 양곤항)

[그림 2-36] LANBY(부산 신항)

2. 용도에 의한 분류

① 도등(leading light) : 통항이 곤란한 좁은 수로, 항만 입구 등에서 안전항로의 연장선 위에 높고 낮은 2~3개의 등화를 앞뒤로 설치하여 그들의 중시선에 의해서 선박을 인도하도록 되어 있다.

[그림 2-37] 도등

② 조사등(projector) : 풍랑이나 조류로 등부표를 설치하거나 관리하기가 곤란한 지점에서 근처의 등대에 강력한 투광기를 설치하여 그 구역을 비추어 위험을 표시하는 등화를 말한다.

③ 지향등(sector light) : 좁은 수로, 항구, 만 입구 등에서 선박에게 안전한 항로를 알려주기 위하여 항로 연장선상의 육지에 설치한 분호등을 말한다.

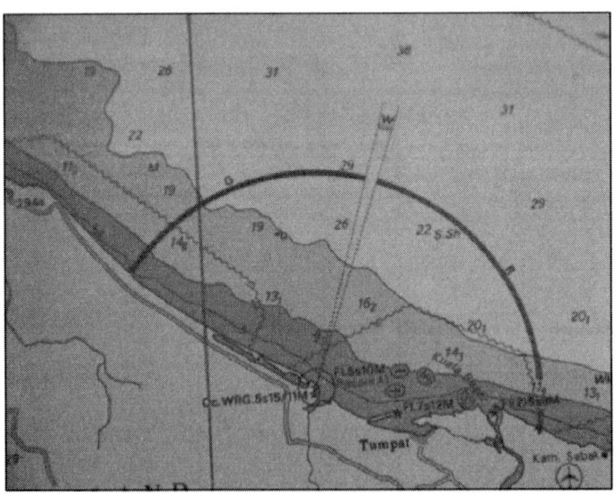

[그림 2-38] 지향등

3. 등 질

야간표지는 일반 등화와 혼동되지 않고 부근에 있는 다른 야간표지와도 구별될 수 있도록 등광의 발사상태를 달리하고 있는데 이것을 등질(character of light)이라 하며 다음과 같은 종류가 있다.

① 부동광(Fixed light, F.) : 등색이나 광력이 바뀌지 않고 일정하게 계속 빛을 내는 등이다.

② 섬광(Flashing light, Fl.) : 빛을 비추는 시간이 꺼져 있는 시간보다 짧으며 일정한 간격으로 섬광을 내는 등이다.

③ 명암광(Occulting light, Oc) : 한 주기 동안에 빛을 비추는 시간이 꺼져 있는 시간보다 길거나 같은 등이다.

④ 등명암광(Isopharse light, Iso) : 한 주기 동안에 빛을 비추는 시간과 꺼져 있는 시간이 같은 등이다.

⑤ 호광(Alternating light, Al.) : 색깔이 다른 종류의 빛을 교대로 내며 그 사이에 등광은 꺼지는 일이 없이 계속 빛을 낸다.

⑥ 분호등(sectored lights) : 한 가지 이상의 색깔을 비추는 등화는 위에서 설명한 호광등과 분호등의 두 가지이다. 그러나 분호등은 호광등처럼 등광의 색상이 바뀌는 것은 아니고 서로 다른 지역을 다른 색상으로 비추는 등화를 말한다.

⑦ 모르스 부호등(Morse code light, Mo.) : 모르스 부호를 빛으로 발하는 것으로 어떤 부호를 발하느냐에 따라 그 등질이 달라진다.

4. 주기, 등색, 등고 및 점등시간

① 주기(period) : 정해진 등질이 반복되는 시간을 말하며 초(sec)단위로 표시한다. 즉, 섬광의 경우 1 섬광이 최초에 시작되는 시각으로부터 그 다음 섬광이 시작될 때까지의 시간을 말한다.

② 등색(colour of light) : 등화에 이용되는 색은 백색(White, W), 적색(Red, R), 녹색(Green, G), 황색(Yellow, Y) 등이 있다.

③ 등고(height of light) : 해도 및 등대표에 표기된 등고(charted elevation)는 평균해면(MSL)에서 등화의 중심까지를 등대의 높이(미터)로 표시하고 있다.

④ 점등시간 : 특정한 것을 제외하고는 일몰시부터 일출시까지이다. 그러나 무인등대

의 경우에는 항시 점등되어 있는 것도 많다.

⑤ 광달거리 : 등광을 알아볼 수 있는 최대 거리를 광달거리(visibility of light)라 하며 해도상에는 해리(M)로 표시된다. 광달거리에는 지리학적 광달거리(등광은 충분히 밝으나 그 빛이 지표면에 의해 가려져 항해사의 눈에 들어가지 않기 때문에 제한)와 광학적 광달거리(빛이 발산이나 대기 중에서의 흡수, 산란에 의해 감쇠되어 항해사의 눈에 감지되지 않기 때문에 제한)가 있는데 고시되는 광달거리는 이 둘 중 작은 값을 기재한다.

5. 항로표지 도식의 이해

해도상에 표시되어 있는 각각의 항로표지의 도식은 다음과 같은 의미를 포함하고 있다.

① 유인등대 : 오륙도 항로표지관리소 - Fl. 10s 52m 22M HORN(1) 40s

등질은 10초의 주기내에 섬광등(flashing light)을 1차례 점등하며, 표기된 등고(평균해면(MSL)에서 광원까지의 높이)는 52m이고, 광달거리는 22마일이고, 무중신호는 Horn을 설치하고 있는데 40초 주기내에 1번 울린다.

② 무인등대 : 목도등대 - Fl(2) 6s 68m 11M Racon(B)

등질은 6초의 주기내에 섬광등을 2차례 점등하며, 등고 68m, 광달거리는 11마일이고, 레이콘(B, — ···) 이 설치되어 있다.

③ 등표 : 생도등표 - Fl(2) 5s 56m 11M

고립장애표지로서 등질은 5초의 주기내에 섬광등을 2차례 점등하며, 등고는 56m이고, 광달거리는 11마일이다.

④ 등부표 : 부산신항 가덕수도 제1호 등부표 - Fl(4) G. 8S No.1

부산신항 가덕수도 출항 · 입항 항로에 있어서 좌현쪽으로 설치되어 있는 등부표로 8초 주기내의 녹색 섬광등이 4차례 점등되는 1호 등부표이다.

⑤ 등부표 : 부산항 제6호등부표 - Iso. R 4s 10M

부산항 출항 · 입항 항로에 있어서 우현쪽으로 설치되어 있는 등부표로서 등질은 등명암광(燈明暗光, Isophase)의 의미로 일정한 광도의 빛이 일정한 간격으로 발하며, 불이 켜져 있는 시간과 꺼져 있는 시간이 동일하다는 의미이다. 그러므로 이 등부표의 경우에는 홍색등이 2초 동안 켜져 있고, 2초 동안 꺼져 있는 4초의 주기로 작동한다는 것을 나타내며, 광달거리는 10마일이다.

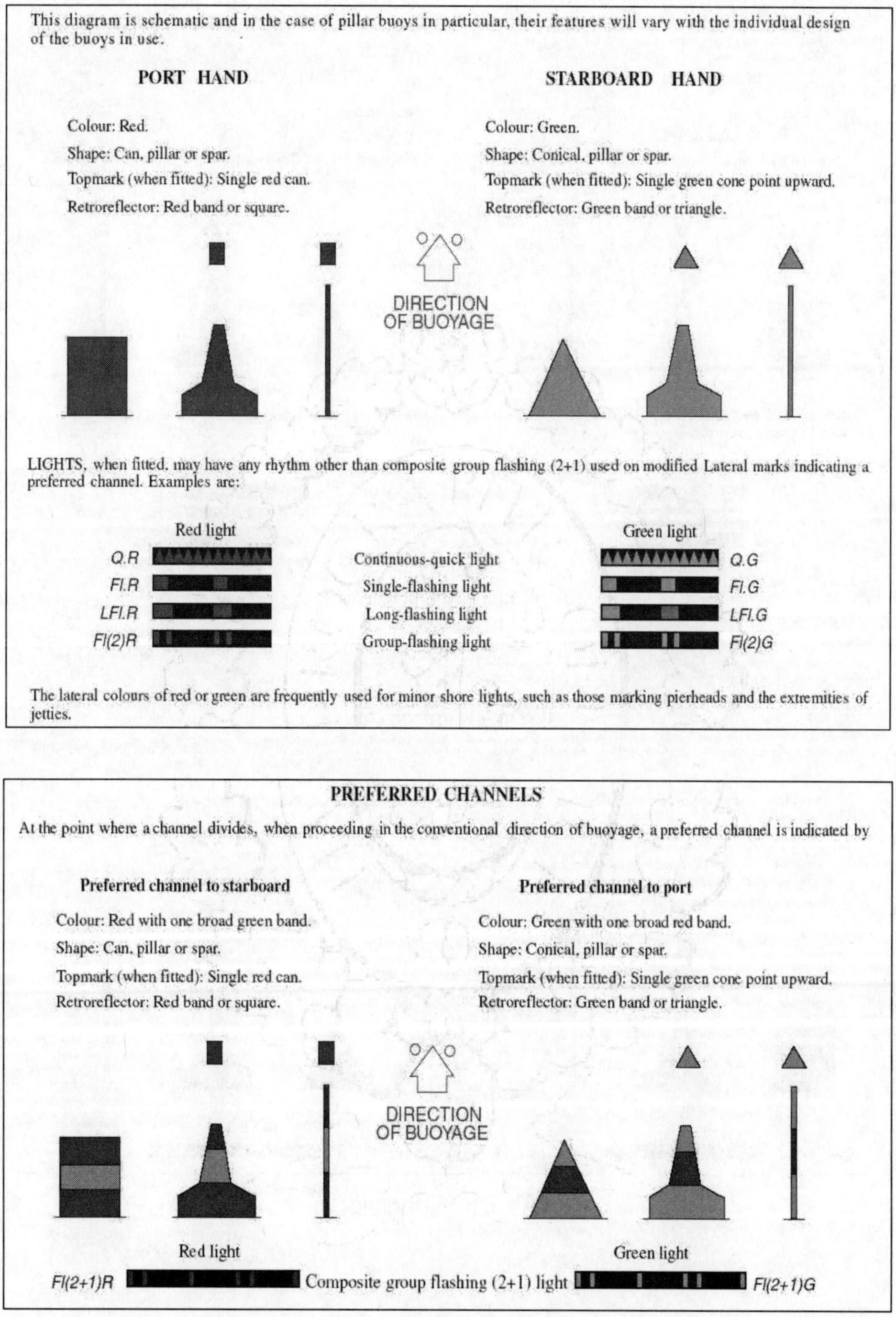

[그림 2-39] 측방표지(A방식)

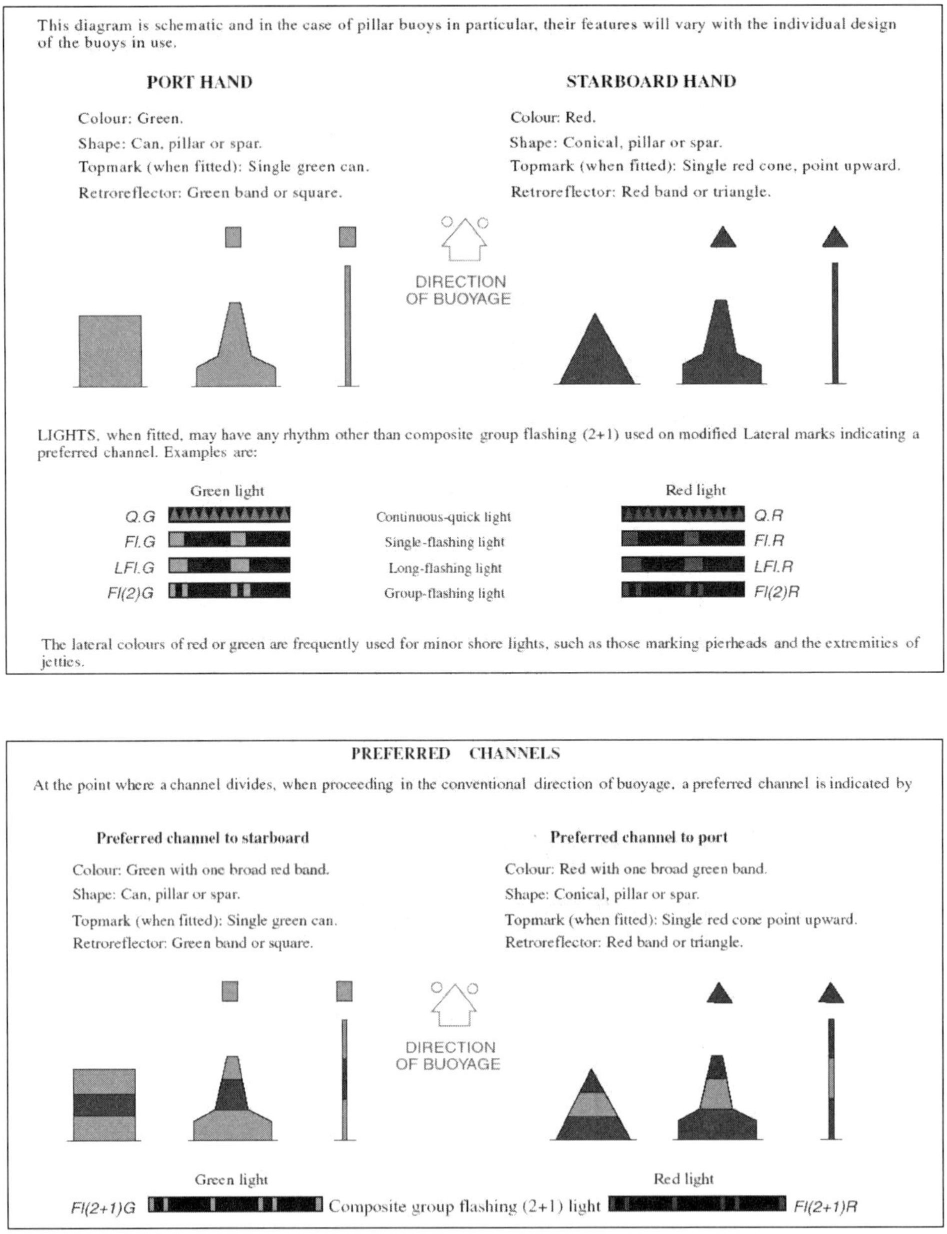

[그림 2-40] 측방표지(B방식)

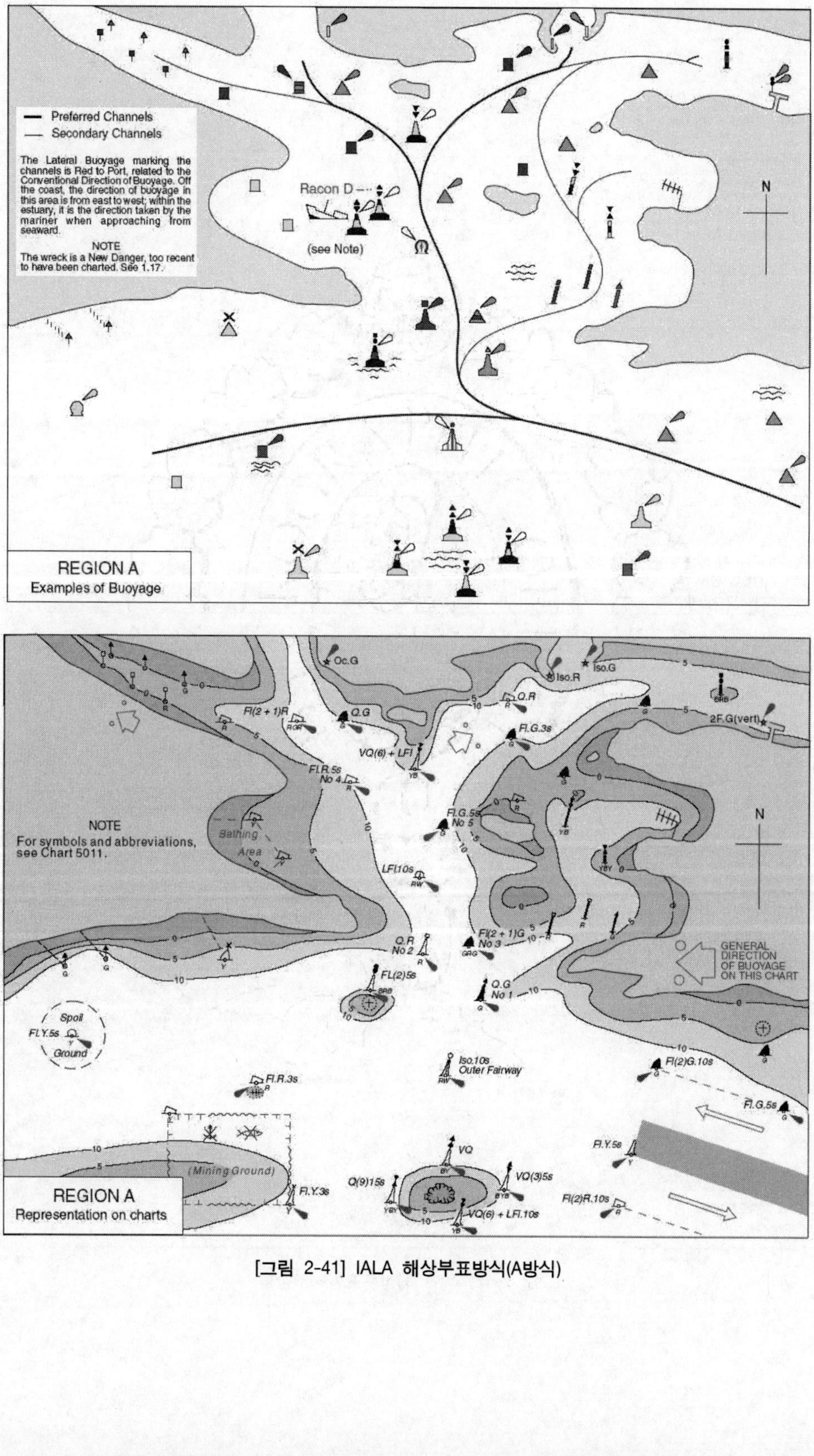

[그림 2-41] IALA 해상부표방식(A방식)

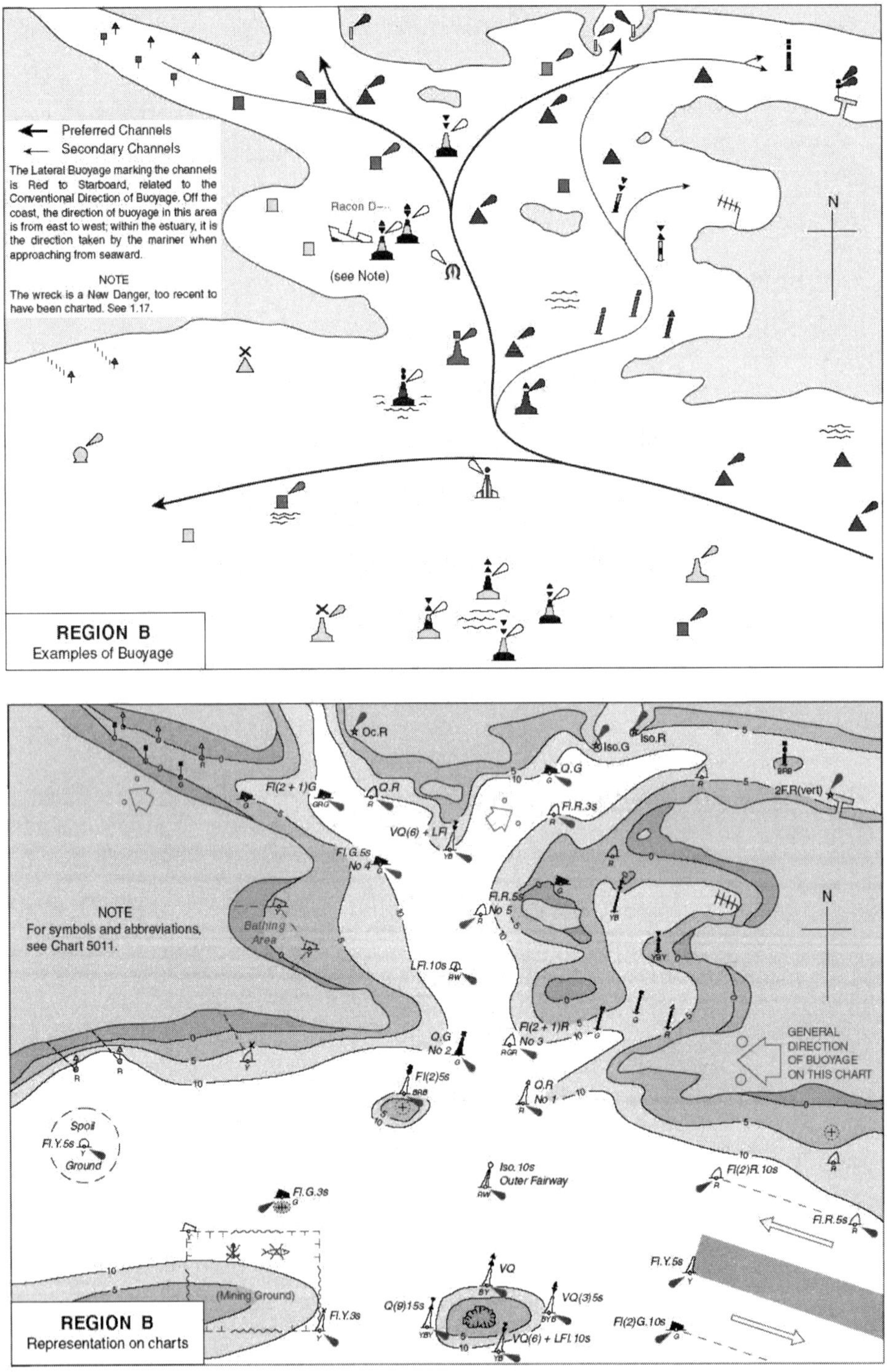

[그림 2-42] IALA 해상부표방식(B방식)

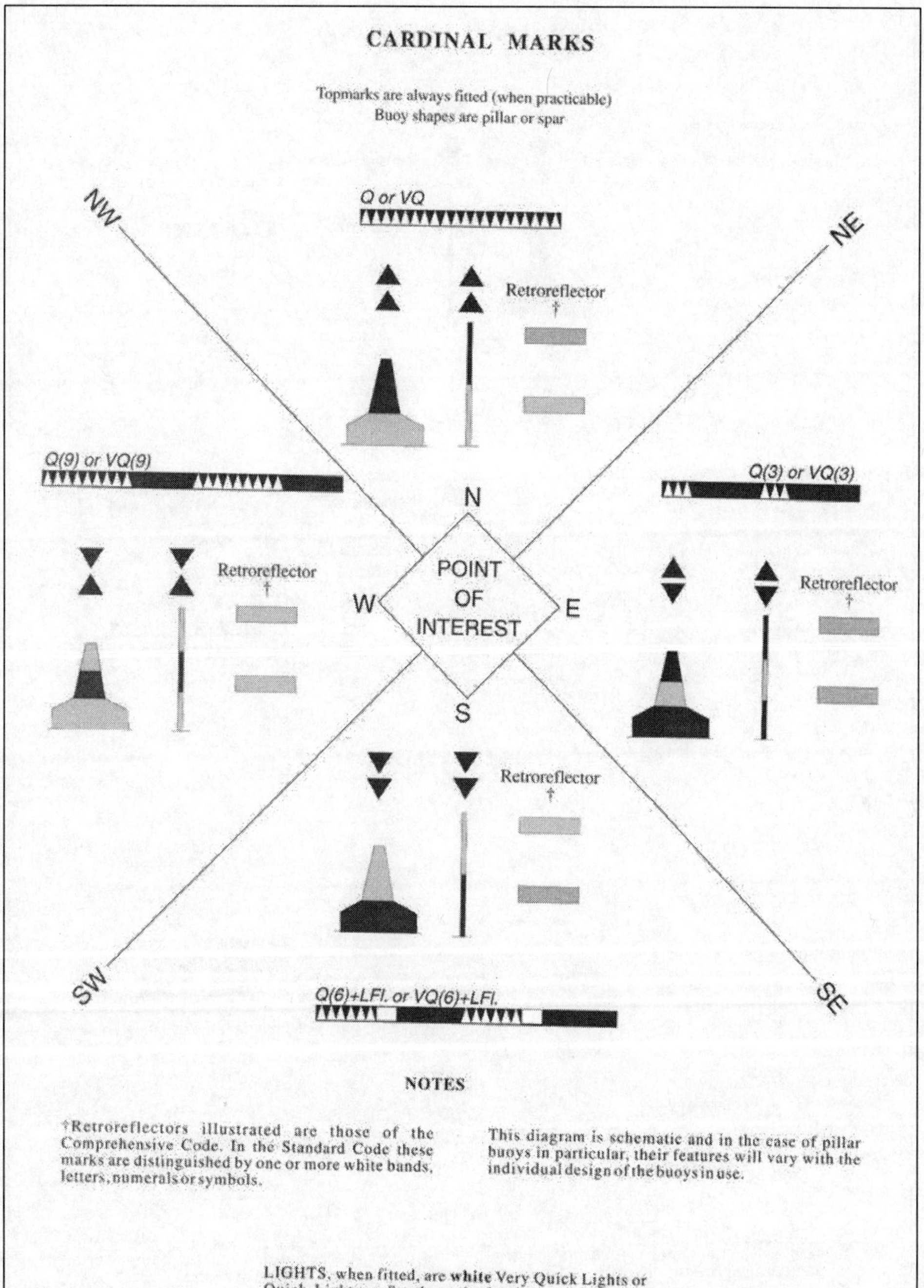
CARDINAL MARKS
Topmarks are always fitted (when practicable)
Buoy shapes are pillar or spar
NW
NE
SW
SE
Q or VQ
Retroreflector
†
Q(9) or VQ(9)
Q(3) or VQ(3)
N
W
E
S
POINT
OF
INTEREST
Retroreflector
†
Retroreflector
†
Retroreflector
†
Q(6)+LFl. or VQ(6)+LFl.
NOTES
†Retroreflectors illustrated are those of the Comprehensive Code. In the Standard Code these marks are distinguished by one or more white bands, letters, numerals or symbols.
This diagram is schematic and in the case of pillar buoys in particular, their features will vary with the individual design of the buoys in use.
LIGHTS, when fitted, are white Very Quick Lights or Quick Lights; a South mark also has a Long Flash immediately following the quick flashes.

[그림 2-43] 방위표지

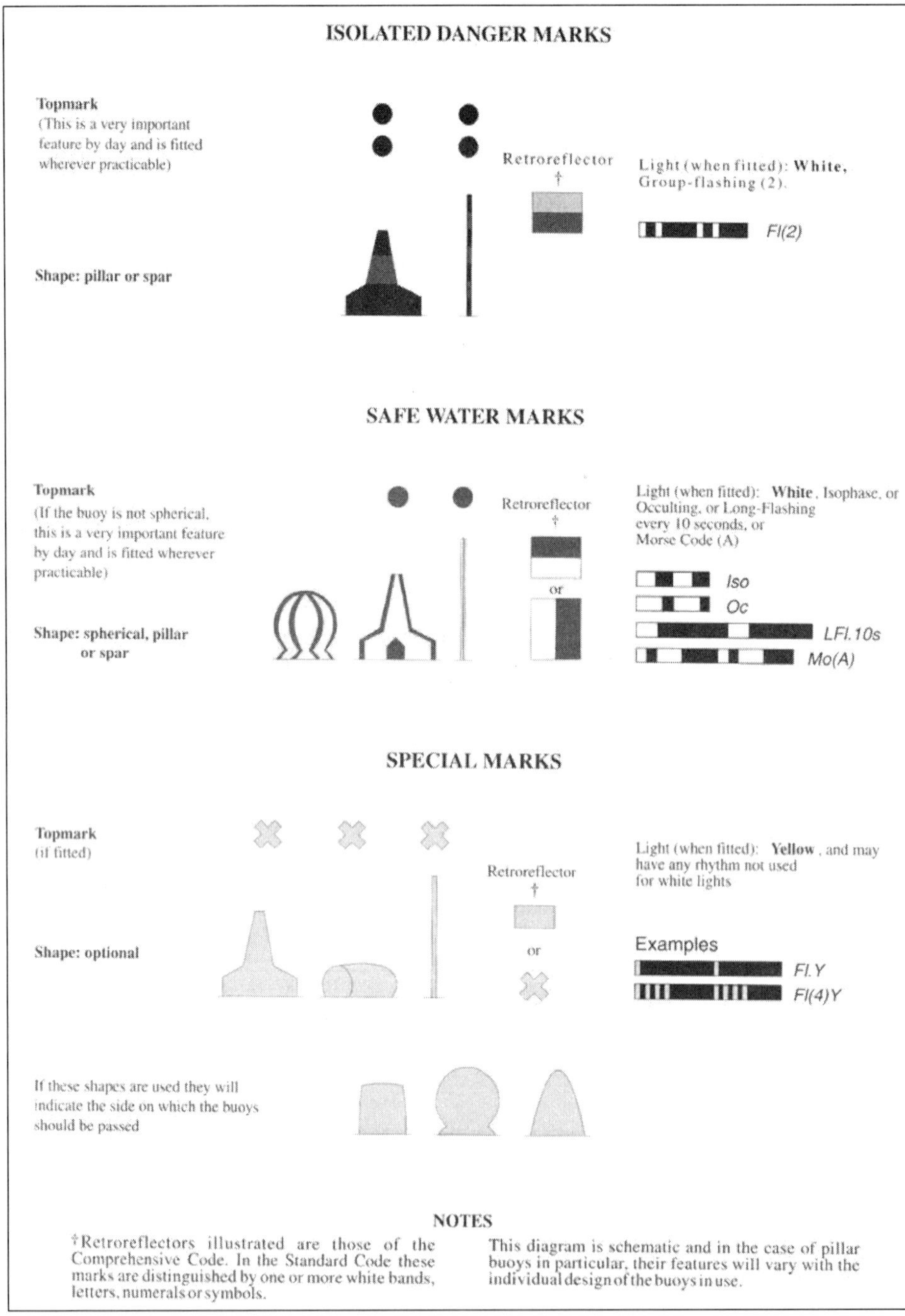

[그림 2-44] 고립장해표지, 안전수역표지, 특수표지

2.3.3 음향표지

안개가 끼거나 눈 또는 비 등으로 시계가 나빠서 육지나 등화를 발견하기 어려울 때에 부근을 항해하는 선박에게 항로표지의 위치를 알리거나 경고할 목적으로 설치된 표지이며, 이 음향신호를 무신호(fog signal)라고도 한다. 대개는 등대나 다른 항로표지에 부설되어 있으며, 소리가 전달되는 경로에 따라 공중음 신호와 수중음 신호로 구별하는데, 현재는 거의 대부분 공중음 신호만 이용되고 있다.

1. 음향표지의 종류

① 에어사이렌(air siren) : 압축공기에 의하여 사이렌을 울리는 장치이다.

② 모터 사이렌(motor siren) : 전동기에 의하여 사이렌을 울리는 장치이다.

③ 다이어폰(diaphone) : 압축공기에 의해서 발음체인 피스톤을 왕복시켜서 소리를 내는 장치이다.

④ 다이아프램 혼(diaphragm horne) : 전자력에 의해서 발음판을 진동시켜 소리를 울리는 장치이다.

⑤ 무종(fog bell) : 가스의 압력 또는 기계장치로서 종을 쳐서 소리를 내는 장치를 말한다. 또한 부표의 꼭대기에 종을 달아 파랑에 의한 흔들림을 이용하여 종이 울리게 된 것을 타종 부표라 한다.

2. 음파표지 이용에 있어서의 주의사항

① 무신호는 안개 · 눈 · 기타에 의해서 시계가 나빠 항행에 지장을 초래할 우려가 있을 경우에 한하여 행한다.

② 그러나 해상에서는 이미 안개가 발생하였더라도 무신호소에서 이러한 상황을 알지 못하여 신호를 행하지 않는 경우가 있으며, 알았더라도 신호를 시작하기까지에는 시간이 걸린다.

③ 무신호의 음향 전달거리는 대기의 상태나 지형에 따라 변하는 수가 있으므로, 신호음의 방향 및 강약만으로 신호소의 방위나 거리를 판단해서는 안된다. 또, 소리의 특성 때문에 신호소 가까이에서도 소리가 들리지 않는 경우가 있으며, 소리를 듣는 사람의 위치에 따라서도 다르다. 더구나 갑판상에서는 들리지 않는 소리가 마스트

꼭대기에서는 들리는 경우도 있다.

④ 이와 같이 무신호는 불규칙하기 때문에, 무중 항해시 선내를 정숙하게 하고 경계원을 여러 곳에 나누어 배치하는 등 특별한 주의가 필요하다. 따라서, 무신호에만 의지할 것이 아니라 측심의나 레이더의 활용에 노력해야 한다.

2.3.4 전파표지

전파의 특성인 직진성, 등속성, 반사성 등을 이용하여 선박의 위치를 측정할 수 있도록 개발된 항로표지를 무선표지 또는 전파표지라고 한다. 전파표지는 전파를 이용하므로 천후에 관계 없이 항상 이용이 가능하고, 넓은 지역에 걸쳐서 이용할 수 있는 이점이 있다.

육상이나 등선과 같이 일정한 장소에 설치된 고정국에서 표지 전파를 발사하면 관측자가 무선방위측정기(Radio Direction Finder, RDF)로 그 방위를 측정하는 무선표지국과 선박에서 발사한 전파를 육상에 장치한 무선방향탐지기로 방위를 측정하여 이를 선박에 통보하는 무선방향탐지국이 있다.

이 중 무선방향탐지국은 육상의 고정된 위치에서 선박의 송신전파를 측정하므로 정확도가 무선표지국에 비해 좋다는 이점이 있으나, 위치를 낼 때마다 탐지국에 측정을 의뢰하여야 하는 불편함 때문에 현재는 거의 사라지고 무선표지국으로 대체되었다. 무선표지국은 전파의 발사방식에 의하여 중파 표지와 마이크로파 표지로 분류된다.

1. 중파 표지

1) 무지향식 무선표지(Circular Radiobeacon, RC)

선박이 방향탐지기에 의해 송신국의 방위를 측정할 수 있도록 무지향식 전파(중파)를 발사하는 시설을 말한다. 항상 전파를 발사하는 표지가 많으므로 선박은 필요할 때마다 전파를 수신하여 선박위치를 구할 수 있다. 유효거리 2도 이내의 오차로 방위를 측정할 수 있는 거리는 주간 150M, 야간 50M 정도이다.

2) 지향성 회전식 무선표지(Rotating loop radiobeacon, R.W. or Directional Radiobeacon, R.D.)

선박이 중파 수신기에 의해 송신국에서 방위를 측정할 수 있도록 정시에 방위수를 나타내 단점신호의 지향성 전파(중파)를 발사하는 시설을 말한다. 유효거리는 주간 100M, 야간 50M 정도이다.

3) 지향성, 회전식 · 무지향식 무선표지

지향성 회전식 표지전파를 발사하는 사이에 몇 분간 무지향성식 전파로, 표지부호 및 장음을 수회 발사하는 표지이다. 따라서 이 사이에 국명을 확인하고 전파를 청취하면서 감도나 음량을 조정할 수가 있어 편리하다.

2. 마이크로파 표지

파장이 수 cm, 수 mm와 같이 아주 짧은 전파를 발사하는 표지로써 오차가 적고, 정확한 방위를 구할 수 있으며, 수신기의 취급이 간편한 반면 중파 표지에 비해 이용범위가 좁은 단점이 있다. 마이크로파 표지에는 다음과 같은 종류가 있다.

1) 레이더 트랜스폰더(Radar Transponder)

레이더 반사파를 강하게 하고 방위와 거리 정보를 제공하는 면에서 레이콘과 유사하나 정확한 질문을 받거나 송신이 국부명령으로 이루어질 때 다른 관련 자료를 자동적으로 송신할 수 있다는 점에서 레이콘과 구별된다. 송신내용에는 부호화된 식별신호 및 데이터가 들어 있으며 이것이 레이더 화면상에 나타난다.

2) 레이더 반사기(Radar Reflector)

부표와 작은 배 특히 목선 등은 레이더에 잘 나타나지 않으므로, 3개의 금속면을 상호 수직으로 접합하여 어느 방향에서 전파가 오더라도 강하게 반사시킬 수 있는 코너 리플렉터(corner reflector)를 부착한다. 국제해사기구는 총톤수 100톤 이하의 선박이 적절한 최소 성능을 가진 레이더 반사기를 장치하도록 권고하고 있다.

3) 레이 마크(Radar Marker Beacon, Raymark)

선박의 레이더 영상에 송신국의 방향이 휘선(밝은선)으로 나타나도록 전파(마이크로파)를 발사하는 것으로 표지국의 방향과 거리를 쉽게 알 수 있어 편리하다. 유효거리는 주간 · 야간 20마일 정도이다.

4) 레이콘(Radar Transponder Beacon, Racon)

선박 레이더에서 발사된 전파를 받은 때에만 응답하여 레이더 화면상에 일정 형태의 신호가 나타날 수 있도록 전파를 발사한다. 레이콘의 신호로는 표준신호와 모르스 부호가 이용된다. 유효거리는 주간 · 야간 10마일 정도이다.

2.3.5 특수신호표지

대표적인 특수신호표지가 조류신호소(tidal current signal station)인데 인천, 진도와 같이 조류가 강한 항만이나 수로에 설치하여 조류의 방향과 속력을 실시간으로 측정하여 항해사에게 제공하는 신호표지이다.

[그림 2-45] 인천항 갑문입구의 조류신호소

2.4 연안 항법

연안 항법(Coastal Navigation 또는 Coasting)이란 육지나 섬 또는 항로표지(Aids to Navigation, 航路標識) 등과 같이 육상 또는 해상 물표의 방위와 거리에 의한 위치선을 이용하여 선박의 위치를 결정하면서 안전하고 경제적인 항해를 계속하는 항법이다. 이 항법의 항로에는 선박 항해에 위험을 주는 암초, 부유물, 양식장 등의 많은 장애물이 존재하고 있으므로 항해자의 신속하고 정확한 판단과 많은 경험을 필요로 한다.

2.4.1 위치선

1. 위치와 위치선

위치선(Line Of Position, LOP)이란 어떤 물표를 관측하여 얻은 방위, 거리, 협각, 고도 등을 만족시키는 점의 자취로서 관측을 실시한 시점에 선박이 그 자취 위에 있다고 생각되는 특정한 선을 말한다. 따라서 하나의 위치선은 그 위치선 상의 어딘가에 선박이 존재한다는 것만을 말해줄 뿐이므로, 선박의 실제 위치는 이러한 위치선들의 2개 또는 그 이상의 교점을 구하여 나타낼 수 있다. 이러한 위치선들을 구할 수 있는 방법으로는 다음과 같은 것들이 있다.

2. 위치선을 구하는 방법

1) 방위에 의한 위치선

선박에 장치된 Gyro Compass 등의 방위측정장치를 이용하여 물표의 방위를 측정하고, 그 측정방위에 컴퍼스의 오차를 개정하면 진방위를 구할 수 있다. 또한 레이더의 EBL(Electronic Bearing Line)을 이용하여 방위를 측정할 수 있다.

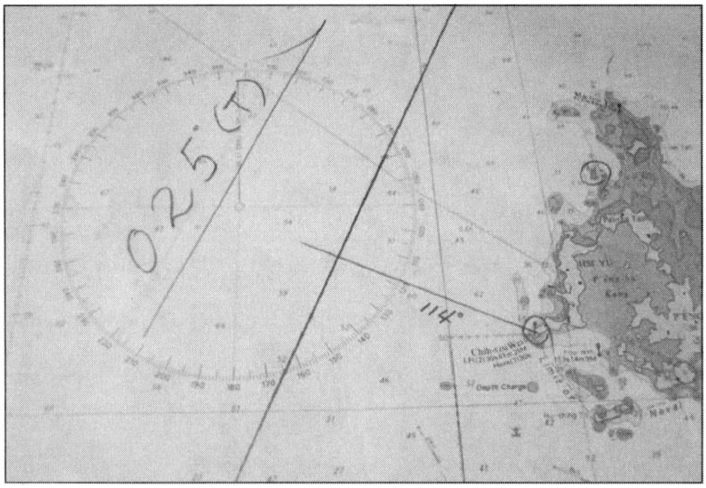

[그림 2-46] 방위에 의한 위치선

2) 중시선에 의한 위치선

두 물표가 일직선상에 겹쳐 보일 때 그들 물표를 연결한 직선을 중시선(Range Line 또는 Transit Line)이라고 하며, 이와 같은 중시선은 위치선이 된다. 특히, 관측자와 가까운 물표 사이의 거리가 두 물표 사이의 거리의 3배 이내이면 매우 정확한 위치선이 된다. 중시선은 선위를 측정하는 이외에도 좁은 수로를 통과할 때의 피험선, 컴퍼스 오차의 측정 등에 이용된다.

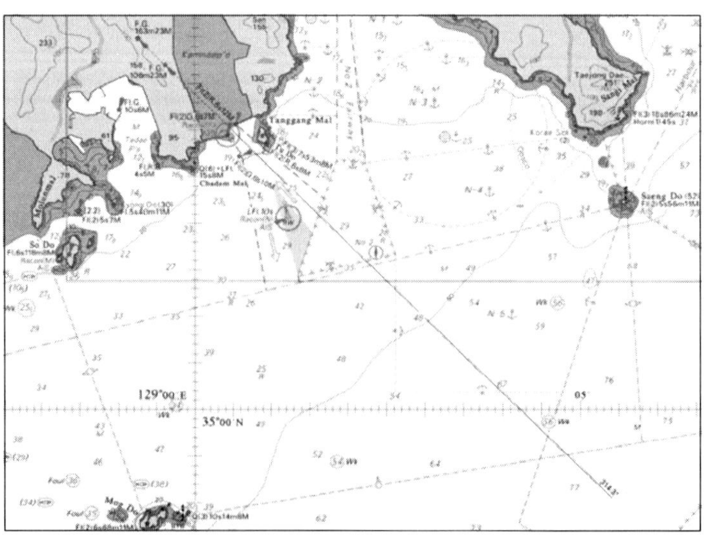

[그림 2-47] 중시선 예시(부산 감천항)

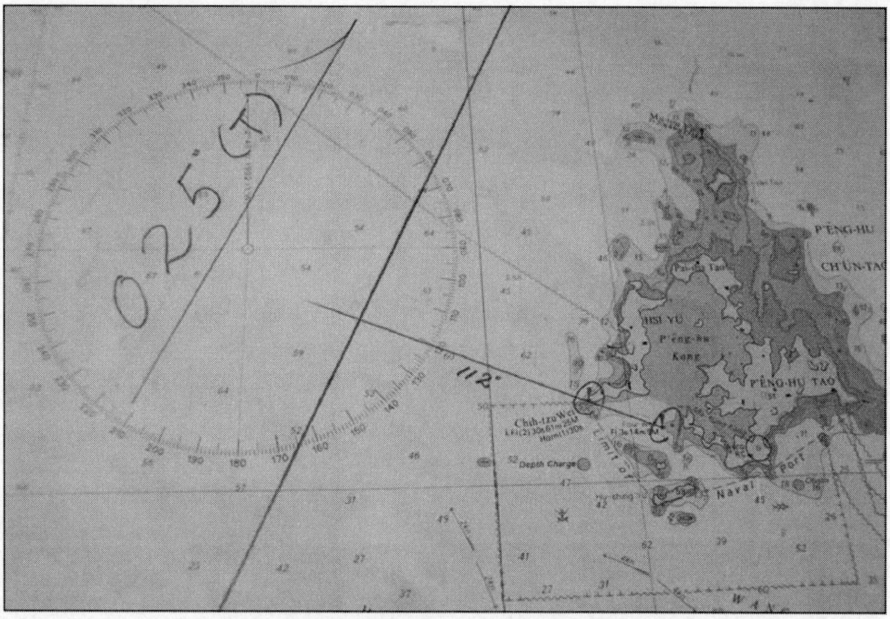

[그림 2-48] 중시선에 의한 위치선

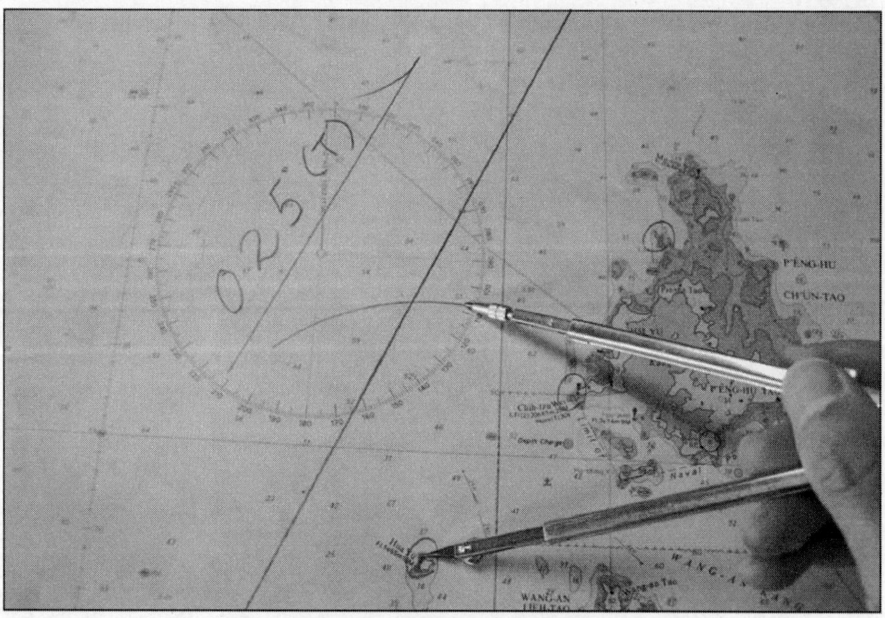

[그림 2-49] 수평거리에 의한 위치선

3) 수평거리에 의한 위치선

어떤 선박으로부터 물표까지의 거리를 레이더와 같은 항행장치로 측정하여 그 거리를 반지름으로 하고, 물표를 중심으로 하는 원을 그리면 선박은 그 원주 위에 있게 되므로 이 원은 위치선이 된다.

4) 수평협각에 의한 위치선

두 물표 사이의 수평 협각을 육분의(Sextant)로 측정한 다음, 해도 위에서 두 물표를 지나고 측정한 각을 품는 원을 작도하면 선박은 이 원주 위에 있게 되므로 이 원은 위치선이 된다.

5) 수심에 의한 위치선

수심의 변화가 규칙적이고 측량이 잘 된 해도를 사용할 때에는 직접 측정하여 얻은 수심과 같은 수심인 점들을 연결하여 등심선을 그려 넣을 수 있다. 이 등심선도 위치선으로 이용할 수 있으나 다른 방법에 비하여 부정확하다.

6) 천체의 고도 측정에 의한 위치선

주간에 태양 및 달의 고도나 박명시에 행성이나 항성의 고도를 육분의로 측정하면 위치선을 얻을 수 있다.

2.4.2 추측위치와 추정위치

1. 추측위치

1) 추측위치의 정의

가장 최근에 얻은 실측위치를 기준으로 그 후에 조타한 진침로와 속력계(Speed Log) 또는 주기관의 회전수로 구한 항정에 의하여 선위를 결정하는 것을 선위의 추측(Dead Reckoning)이라 하고, 이와 같이 하여 결정된 선위를 추측위치라고 한다. 해상 실무에서는 일반적으로 추측위치를 DR 위치(Dead Reckoning Position, DRP)라고 하며, 선박의 침로와 속력의 두 가지 요소를 이용하여 구하게 된다.

2) 추측위치의 활용

추측위치는 침로와 속력(또는 거리) 이외의 조류, 해류 및 바람 등의 자연환경의 외력 요소는 전혀 고려하지 않은 것이지만, 때로는 놀랄 정도로 정확한 위치가 얻어지기도 한다. 그러므로 실측위치를 얻을 수 없을 때에는 이러한 추측위치를 매 시간마다 기점하는 것이 좋고, 또한 전 항해구간을 통해서 계속 기점할 필요가 있다. 이렇게 함으로써 본선의 위치가 어디쯤에 있을 것인가를 언제든지 알 수 있게 된다.

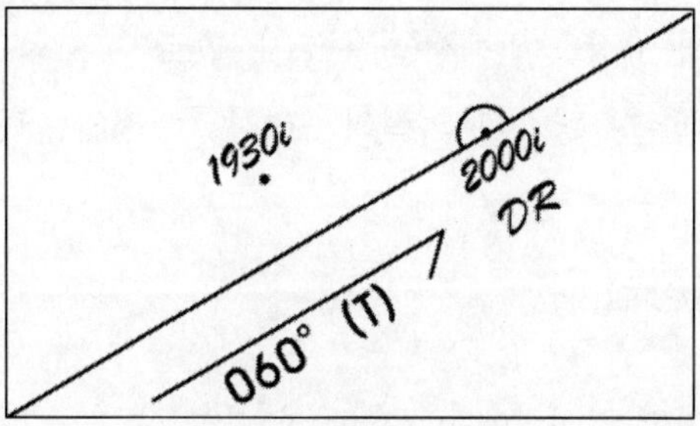

[그림 2-50] 추측위치의 표시

2. 추정위치

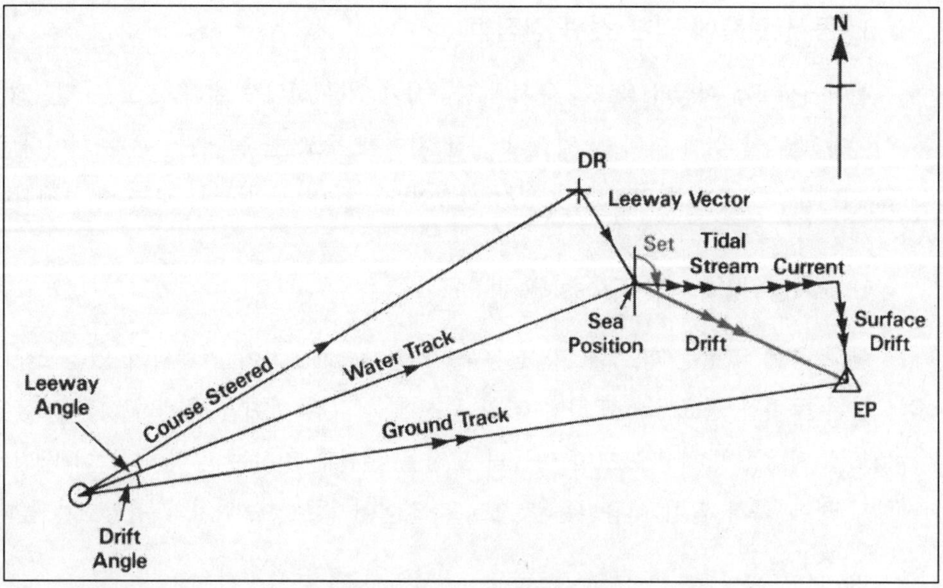

[그림 2-51] 추정위치를 구하는 방법

추정위치(Estimated Position, EP)는 추측위치에 대하여 조류(Tidal current 또는 Tidal stream), 해류(Current) 및 풍압차(Leeway) 등의 외력의 영향을 고려하여 구한 위치를 말한다.

만일 해류 · 조류의 방향(Set)과 속력(Rate, Drift)을 알고 있다면 먼저 추측위치를 기점하고, 여기에 해류 · 조류의 방향과 속력의 영향을 수정해 주면 추정위치가 얻어진다. 물론 여기에 풍압차를 고려해 주면 더욱 신뢰성이 높은 위치가 얻어질 것이다.

외력에 의한 선위의 변화는 선박의 크기, 흘수, 트림, 침로, 해역, 계절 등에 따라 다르기 때문에 정확한 선위 추정을 위해서는 수로지, 항로지, 조석표, 해류도, 조류도, 파일럿 차트 등을 참고하는 한편, 평소에 외력의 크기를 추정하는 능력을 길러 두어야 한다.

2.4.3 선위 결정법

물표 관측에 의한 선위 결정법에는 2개 이상의 물표를 동시에 관측하여 선위를 구하는 동시 관측법과 시간차를 두고 동일 물표나 다른 물표를 2회 이상 관측하여 위치선과 전위선을 이용하여 선위를 구하는 격시 관측법이 있다.

1. 동시 관측에 의한 선위 결정법

선박에서 2개 이상의 물표를 동시에 관측하여 얻은 선위는 실제적으로는 정확히 동일한 시각에 관측한 것이 아니고, 약간의 시간 차이를 두고 관측하여 얻은 위치이다. 그러나 그 차이는 극히 미소하기 때문에 동시 관측이라고 하여도 실무적으로는 별 문제점이 없다.

1) 교차방위법(Fix by Cross Bearing)

항해 중에 고정된 항로 표지, 산봉우리, 섬 등 해도상에 기재되어 있는 2개 이상의 고정된 뚜렷한 물표를 선정하여 거의 동시에 각각의 방위를 측정하고, 해도상에서 각 물표를 지나는 방위선을 그어 이들의 교점을 선위로 정하는 방법이다. 이 방법은 연안 항해 중에 가장 많이 이용되는 것으로 측정법이 쉽고 또 위치의 정밀도가 높다.

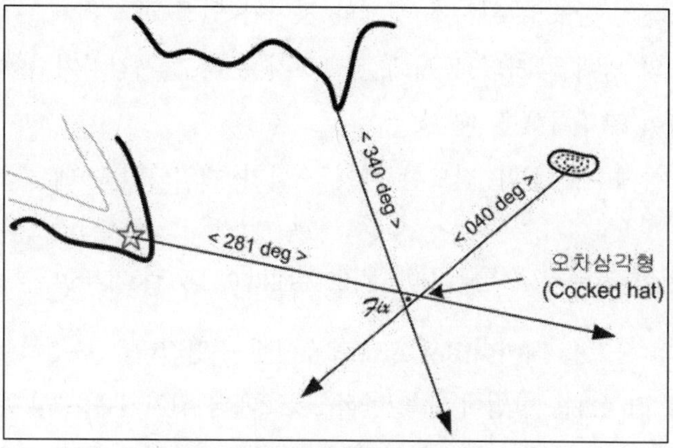

[그림 2-52] 교차방위법을 이용한 선위 결정법

① 물표의 선정에 있어서의 주의사항

㉠ 해도상의 위치가 명확하고 뚜렷한 목표를 선정한다.

㉡ 먼 물표보다는 적당히 가까운 물표를 선택한다.

㉢ 물표 상호간의 각도는 가능한 한 30°~150°인 것을 선정해야 하며, 두 물표일 때에는 90°, 세 물표일 때에는 60°정도가 가장 좋다.

㉣ 물표가 많을 때에는 2개보다 3개 이상을 선정하는 것이 정확도가 높다.

② 방위측정시의 주의사항

㉠ 방위변화가 빠른 물표는 나중에 측정해야 한다. 즉, 선수미 방향이나 먼 물표를 먼저 재고, 정횡 방향이나 가까운 물표는 나중에 측정한다.

㉡ 물표가 선수미선의 어느 한쪽에만 있을 경우 앞에서부터 뒤로 또는 뒤에서부터 앞으로 차례로 측정하는 경우가 많은데, 이 때에는 선위가 예정 침로의 오른쪽 또는 왼쪽으로 편위될 수 있으므로 주의하여야 한다.

㉢ 방위측정은 빠르고 정확하게 해야 하며, 또 해도상에 방위선을 작도할 때에도 신속히 해야 한다.

㉣ 위치선을 기입할 때에는 전위할 때를 고려하여 관측시각과 방위를 기입해 두도록 하며, 선위에도 그 관측시각을 항상 기입하여야 한다. 관측한 3개의 방위선이 1점에서 만나지 않고, 작은 삼각형을 이룰 때에는 그 삼각형의 중심을 선위로 하며 이 삼각형을 오차 삼각형(Cocked hat)이라 한다.

③ 오차 삼각형이 생기는 이유

㉠ 자차나 편차에 오차가 있을 때

㉡ 해도상의 물표의 위치가 실제와 차이가 있을 때
㉢ 물표의 방위를 거의 동시에 관측하지 못하고 시간차가 많이 생겼을 때
㉣ 관측이 부정확했을 때
㉤ 해도상에 위치선을 작도할 때에 오차가 개입되었을 때

2) 두 개 이상 물표의 수평거리에 의한 방법(Fix by Ranges)

두 개 이상의 물표를 레이더(RADAR)로 동시에 수평거리를 측정하여 각각의 위치권의 교점을 선위로 결정하는 방법이다. 일반적으로 2개의 위치권은 항상 두 점에서 만나지만 선박과 물표의 위치 관계에서 그 2개의 교점 가운데 어느 것을 선위로 정할 것인가는 쉽게 판단할 수 있다. 그러나 제3물표의 위치권을 이용하면 보다 확실한 선위를 얻을 수 있으며, 물표가 가깝고 위치권의 교각이 90°에 가까울수록 선위의 정밀도가 높다.

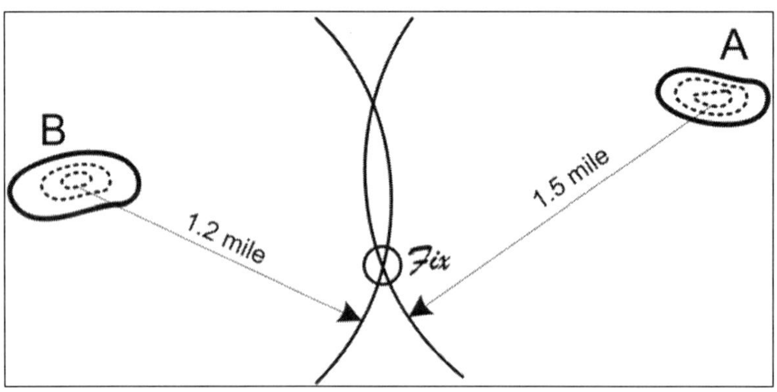

[그림 2-53] 레이더를 이용한 수평거리에 의한 방법

3) 물표의 방위와 거리에 의한 방법(Fix by bearing & Range)

물표의 방위와 거리를 동시에 측정하여 그 방위에 의한 위치선과 수평거리에 의한 위치권과의 교점을 선위로 정하는 방법이다. 특히, 이 방법은 물표가 하나밖에 없을 때에 유용하게 사용할 수 있는 방법으로 이 방법을 사용하기 위해서는 거리를 정확히 측정하여야 한다.

거리측정 방법에는 레이더에 의한 방법, 시달거리에 의한 방법, 앙각에 의한 방법 및 소리의 전파속도에 의한 방법 등이 있다. 그러나 선박에서 실무적으로는 레이더의 작동에 의하여 정확한 거리를 측정하는 방법을 가장 많이 이용하고 있다.

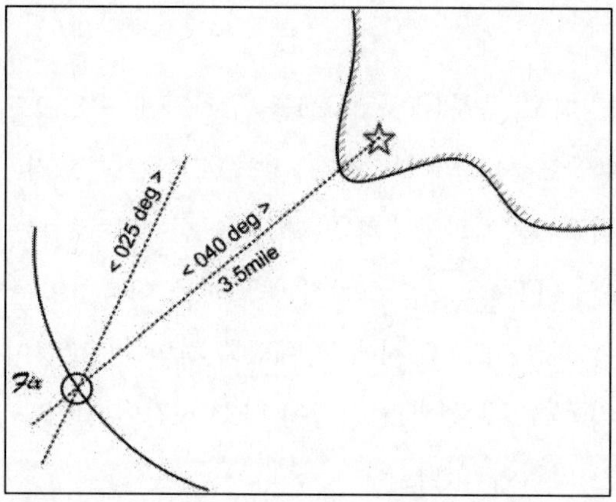

[그림 2-54] 물표의 방위와 거리에 의한 방법

4) 두 중시선에 의한 방법

두 개의 물표 A, B와 C, D가 서로 각각 겹쳐서 보일 때의 시각을 관측하면 그 선박은 그 때 두 중시선의 교점에 위치하게 된다. 이 방법은 아주 정확한 위치를 구할 수 있는 관측법으로, 협수로 통과시의 변침점을 구하거나, 투묘위치를 선정하는 경우와 같이 목표물은 많으나 선위를 측정할 시간적인 여유가 없는 경우에 유용하게 이용할 수 있다. 이 방법을 이용하기 위해서는 미리 계획을 세워서 실행해야 하며, 자선의 자차 측정에도 아주 좋은 방법으로 활용할 수 있다.

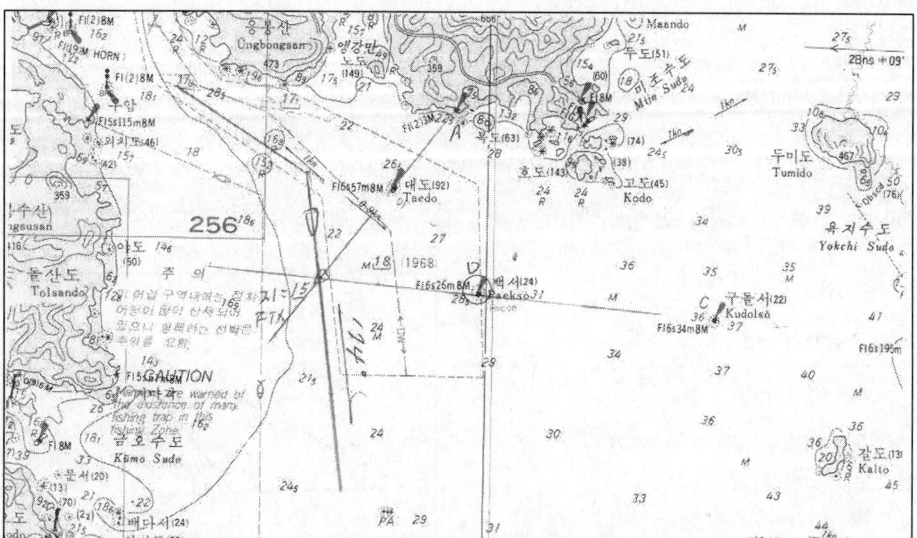

[그림 2-55] 두 중시선에 의한 방법

5) 수평협각법

3개의 물표를 선정하고 육분의(Sextant)를 정확하게 수평으로 쥔 다음, 중앙에 있는 물표와 좌우 각각의 물표 사이의 협각을 측정하고, 삼간 분도기(Three arm protractor)를 이용하여 이들 두 각을 품는 원둘레의 만난 점을 선위로 정하는 방법이다. 그러나 삼간 분도기가 없으면 투사지(Tracing paper)와 분도기를 이용하여 관측한 사이각을 갖는 세 직선을 그어, 관측한 세 목표를 동시에 지나도록 하면 그 교점이 관측시의 선위가 된다.

그러나 수평협각의 측정 및 선위의 결정에 다소 시간이 걸리는 점과 반드시 3개 이상의 물표가 있어야 하는 단점이 있다.

2. 격시 관측에 의한 선위결정

선박이 야간에 연안을 항행할 때 관측이 가능한 등광이 1개뿐인 경우와 같이 물표가 1개뿐이거나 방위와 거리 중 어느 한 가지 밖에 구할 수 없을 경우에 개략적인 선위를 구하기 위하여 사용하는 방법이다. 이 방법은 시간차를 두고 두 번 이상 같은 물표나 또는 다른 물표를 관측하여 그들의 전위선과 위치선을 이용하여 구한 선위를 격시 관측 위치(Running fix)라고 한다.

특별한 경우에는 항해표 제7표를 사용하지 않고도 격시 관측 삼각형을 풀어서 쉽게 제2 관측시에 있어서의 물표까지의 거리와 예상 정횡거리를 구할 수 있으며, 대표적인 경우는 다음과 같은 것들이 있다.

1) 선수 배각법

LA를 상대 선수각이 α인 제1 위치선이라 하고, LB를 상대 선수각이 2α 되었을 때의 방위선이라 하면, 관측점간의 항주거리는 제2 관측시의 관측자에서 물표까지의 거리와 같다.

△ABL에서

$\beta = 180^\circ - 2\alpha$

$\alpha + \beta = + \ \delta = 180^\circ$

$\alpha + (180^\circ - 2\alpha) + \delta = 180^\circ$

$\therefore \alpha = \delta$

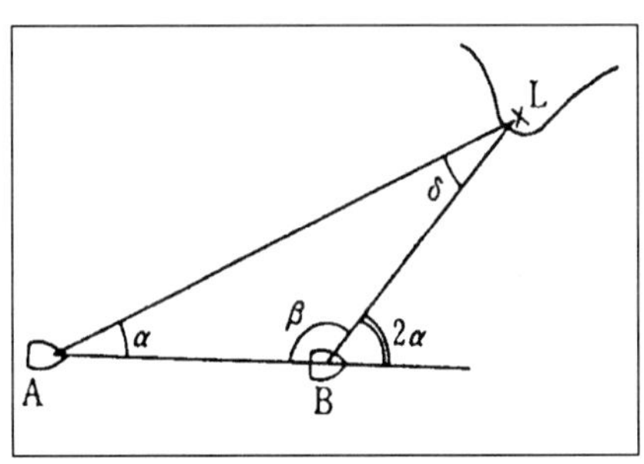

[그림 2-56] 선수 배각법

따라서, △ABL은 이등변삼각형이므로, AB = BL이다.

그러므로, 제2 위치선 상에서 항주한 거리와 같은 점을 물표로부터 취하면 선위가 된다. 이 방법은 본선의 침로와 속력이 일정한 경우에 정확하며, 해류 · 조류, 그 밖의 외력의 영향이 있을 때에는 결정된 선위가 부정확하게 된다.

2) 4점 방위법

물표 L의 상대 선수각이 45°로 될 때의 방위선을 측정하여 이것을 AL이라 하고, 이 침로를 계속 유지해 항행한 후 그 물표를 정횡에서 바라본 점을 B라고 하면, AB=BL이므로 B점은 제2관측시의 선위가 된다. 이 방법은 물표까지의 정횡거리를 구하는데 편리하므로 연안항해에서 많이 이용된다.

2.4.4 항해계획

여기에서는 항해계획을 수립할 때 고려하여야 될 사항, 항해당직 중 유념해야 될 점 및 연안해역의 항행안전을 위한 교통관리 시스템에 대한 내용을 익힌다. 항해자는 안전하고 성공적인 항해를 성취할 수 있도록 출항 전에 충분한 항해계획을 세워야 한다. 또한 당직항해자는 해상에서의 인명과 재산의 안전 및 해양환경의 오염을 방지하기 위하여 자신의 당직임무를 책임감있게 수행해야 한다.

1. 항해계획의 수립

항로의 상황은 해역 및 계절에 따라 다르기 때문에 항해자는 선박의 출항 전에 충분한 시간적 여유를 갖고 앞으로 항해하게 될 수역의 상황을 조사하여 면밀한 항해계획을 세워야 한다. 그 원칙은 안전한 항해를 첫째 목표로 하고, 다음으로 항해일수의 단축과 경제성을 고려하여야 한다.

특히, 바다에서는 기상과 해상상태에 의한 외력을 효과적으로 이용하거나 또는 경우에 따라서는 이를 피하는 것이 항행의 안전과 경제적인 면에서 유리할 때가 있다. 이러한 관점에서 항해계획을 수립할 때에는 [그림 2-57]과 같은 순서로 한다.

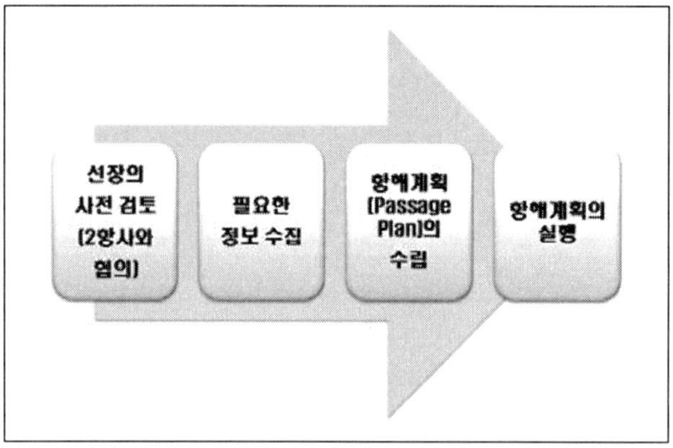

[그림 2-57] 항해계획의 수립절차

① 각종 수로도지에 의한 항행해역의 조사 및 연구와 자신의 경험을 바탕으로 가장 적합한 항로를 선정한다.
② 소축적 해도상에 선정한 항로를 기입하고 일단 대략적인 항정을 구한다.
③ 사용속력을 결정하고 실속력을 추정한다.
④ 대략의 항정과 추정한 실속력으로 항행할 시간을 구하여 출항 · 입항시각 및 항로상의 중요한 지점을 통과하는 시각 등을 추정한다.
⑤ 위에서 수립한 계획이 적절한가를 검토한다.
⑥ 항해에 사용하는 대축척 해도에 출항 · 입항 항로, 연안 항로를 그리고, 다시 정확한 항정을 구하여 예정 항해계획표를 작성한다.
⑦ 상세한 항행일정을 구하여 출항 · 입항시각을 결정한다.

2. 필요한 정보수집(정보의 출처)

① 해도목록(Chart Catalogue)
② 항해용 해도(Navigational Charts)
③ 세계 대양항로지(Ocean Passage for the World)
④ 항로지정 해도(Routeing Charts ; Pilot Charts)
⑤ 수로지 및 항해안내서(Sailing Directions ; Pilot Book)
⑥ 등대표(Light Lists)
⑦ 조석표(Tide Table)

⑧ 조류도(Tidal Stream Atlases)

⑨ 항행통보(Notice to Mariners)

⑩ 무선국 정보(Radio Signals)

⑪ 만재흘수선 해도(Load Line Charts)

⑫ 입항안내서(Guides to Port Entry)

3. 항해계획 작성시 고려할 요소

① 충분한 UKC

② 위험물로부터의 안전거리(Safe Distances off Dangers)

③ 확실한 변침목표

④ 해도의 변경(Management of Chart Changes)

⑤ 통항분리제도(Traffic Separation Schemes)

⑥ Tidal Information

⑦ Speed Plan

⑧ Reporting Point

⑨ Abort Position & Contingency Plan

⑩ Charts & Publications up-date

4. 연안 항로의 선정

선박이 연안을 항행할 때는 다음과 같은 항로를 선정하는 것이 안전하다.

1) 해안선과 평행한 항로

연안 항로에 있어서 도중에 뚜렷한 육상 물표가 없는 해안을 따라서 항행해야 될 경우에는 일반적으로 해안선과 평행한 항로를 선정하는 것이 좋다. 그러나 야간 항해의 경우나 육지로 향하는 해류 · 조류나 바람이 예상될 때에는 해안선과 평행한 항로에서 약간 바다쪽으로 벗어난 항로를 선정하는 것이 안전확보에 도움이 된다.

2) 우회 항로

위험물이 많은 연안을 항해하거나 또는 조종성능에 제한을 받는 상태에서 항해할 때에는 해안선에 근접한 항로를 선정하거나 장애물이 많은 지름길을 선정하지 말고 다소 우회하더라도 안전한 항로를 선정해야 한다.

3) 추천 항로

생소한 해역을 처음 항행할 때에는 수로지, 항로지, 해도 등에 추천 항로(Recommended Route or Track)가 설정되어 있으면 특별한 이유가 없는 한 그 항로를 따르도록 한다.

5. 이안거리 및 경계선

1) 이안거리

이안거리는 해안선으로부터 떨어진 거리를 말하며, 항정을 단축하고 항로표지나 자연물표를 충분히 이용할 수 있도록 육지에 접근한 항로를 선정하는 것이 원칙이지만, 다음과 같은 요소들을 충분히 고려하여 결정하되 지나치게 육지에 접근하지 않도록 하여야 한다.

① 선박의 크기 및 제반상태
② 항로의 교통량 및 항로길이
③ 선위측정 방법 및 정확성
④ 수심을 포함한 해도상에 표시된 각종 자료의 정확성
⑤ 해상, 기상 및 시정의 영향조건 및 본선의 통과시기(주간, 야간)
⑥ 당직자의 자질 및 위기대처 능력

안전한 이안거리는 숫자로 정해진 바가 없고, 위의 조건에 따라 서로 다르나, 보통 내해항로이면 1마일, 외양 항로이면 3~5마일, 야간 항로표지가 없는 외양 항로이면 10마일 이상 떨어져 항해하는 것이 좋다.

2) 경계선

경계선이란 어느 기준 수심보다 더 얕은 위험구역을 표시하는 등심선을 말한다. 보통 흘수가 얕은 선박은 10m 등심선을, 흘수가 깊은 선박이나 해저의 기복이 심하고 암초가 많은 해역에서는 20m 등심선을 경계선으로 선정하는 것이 좋다. 그러나, 상용 항로나 수심 측정이 잘 되어 있어서 신뢰할 수 있고, 저질도 양호하며, 수심의 변화가 단조로운 곳에서는 더 얕은 곳을 항행할 수도 있다. 경계선을 결정함에 있어서 무엇보다 중요한 요소는 본선의 흘수일 것이다.

6. 변침 물표의 선정

연안 항해 중에는 자주 변침하게 되며, 변침을 실시한 후에 선박이 예정된 항로상을 항행하도록 하기 위해서는 변침지점을 미리 선정해 둘 필요가 있다. 이러한 목적으로 선정한 물표를 변침 물표라 하는데, 이것을 선정할 때는 다음과 같은 점에 주의하여야 한다.

① 물표가 변침 후의 침로방향에 있고, 그 침로와 평행이거나 또는 거의 평행인 방향에 있으면서 거리가 가까운 것을 선정한다.

② 위와 같은 물표가 없으면 전타할 현쪽의 정횡 부근에 있는 뚜렷한 물표, 또는 중시 물표와 같이 정밀도가 높은 것을 선정한다.

③ 변침 물표로는 등대, 입표, 섬, 산봉우리 등과 같이 뚜렷하고 방위를 측정하기 좋은 것을 선정하고, 곶, 등부표 등은 불가피한 경우가 아니면 이용하지 않는 것이 좋다. 또, 산봉우리는 관측자의 위치에 따라 모양이 다르게 보이는 수가 있으므로 주의해야 한다.

④ 중요한 변침 지점이거나 뚜렷한 물표가 없는 곳이면 반드시 예비 물표를 선정해 두어야 한다.

7. 피험선

협수로를 통과할 때나 항만을 출항 · 입항할 때에는 자주 변침하여 마주치는 선박을 적절히 피해야 하므로, 선위측정을 자주 하거나 예정 침로를 계속 유지하기가 어려운 때가 많다. 이런 경우에는 미리 해도를 보고 조사해 둔 뚜렷한 물표의 방위, 거리, 수평협각 등에 의한 위치선을 이용하여 위험을 피하는 것이 좋다. 이와 같은 목적으로 준비한 위험예방선을 피험선이라 한다. 피험선을 선정하는 방법으로는 다음과 같은 것들이 있다.

1) 두 물표의 중시선에 의한 것

침로 양쪽에 점선으로 둘러싸여 있는 수역은 수심이 낮은 곳이다. 이 두 곳 사이를 안전하게 피해갈 수 있는 침로는 물표 L과 B, 또는 L과 P가 일직선으로 보이는 중시선 위를 따라 항행하면 침로 양쪽에 있는 위험지역을 안전하게 통과해 갈 수 있다. 그러나 만일 이들 두 물표가 일직선으로 보이지 않으면 위험지역을 항해하고 있다고 판단하면 된다.

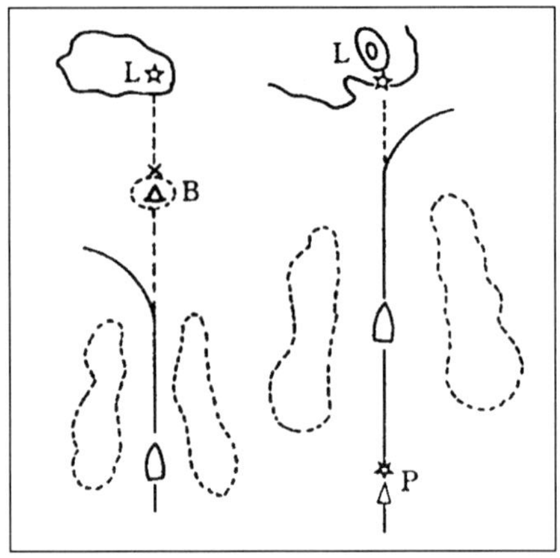

[그림 2-58] 중시선을 이용한 피험선

2) 선수 방향에 있는 물표의 방위선에 의한 것

C점과 D점 부근 수역은 항행에 위험이 존재하는 곳으로, 이곳을 안전하게 피해가기 위하여 선수 방향의 L등대를 이용하는 방법을 알아본다.

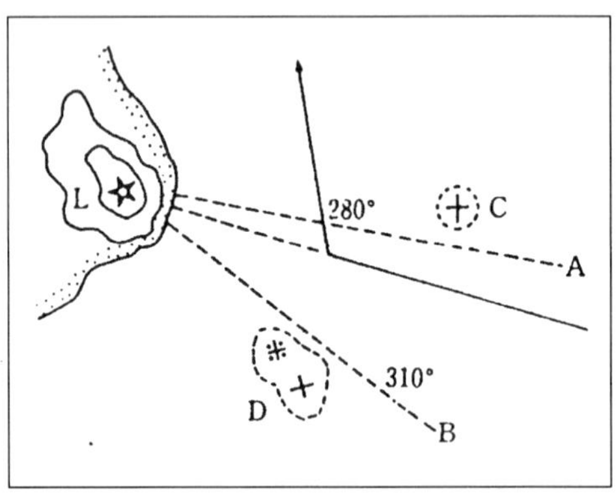

[그림 2-59] 선수 방위 물표의 방위선에 의한 것

물표 L의 방위가 280°보다 크게 보이는 곳을 항행하면 위험물 C를 피하게 되고, 310°보다 작게 보이는 곳을 항행하면 위험물 D수역에 대하여 안전하다. 따라서 등대 L이 280°보다 작은 방위각으로 측정되거나, 또는 310°보다 큰 방위각으로 측정되면 위험한 상황에 있게 된다.

3) 침로의 전방에 있는 한 물표의 방위선에 의한 것

선박이 000°의 침로로 항행하고 있을 때 침로 부근에 존재하는 위험지역 D를 피할 수 있는 방위선 LC(020°)를 그어 둔다. 따라서 선박에서 물표 L의 방위가 020°보다 크게 관측되면 안전구역에 있고, 020°보다 작게 관측되면 위험구역에 있음을 알 수 있다.

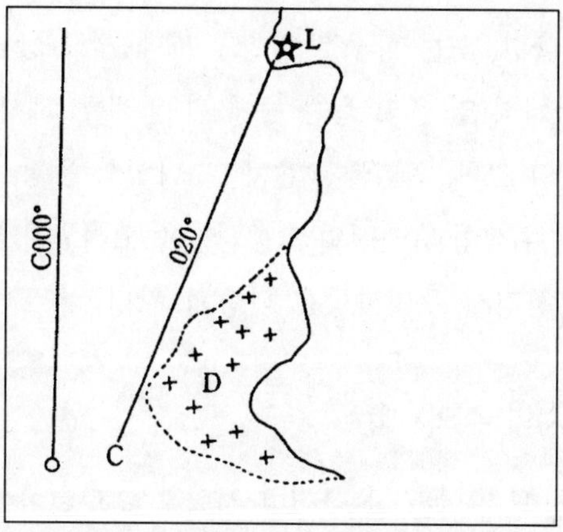

[그림 2-60] 침로의 전방에 있는 한 물표의 방위선에 의한 것

4) 측면에 있는 물표로부터의 거리에 의한 것

선박은 북쪽으로부터 남쪽을 향하여 항행하고 있고, 선박의 좌현쪽에 있는 섬과 우현쪽의 "O" 수역에 있는 몇 군데의 장해물 사이로 빠져나가고자 한다. 이 경우 선박이 원주 위를 따라 항행하면 위험지역을 안전하게 통과할 수 있음을 알 수 있다.

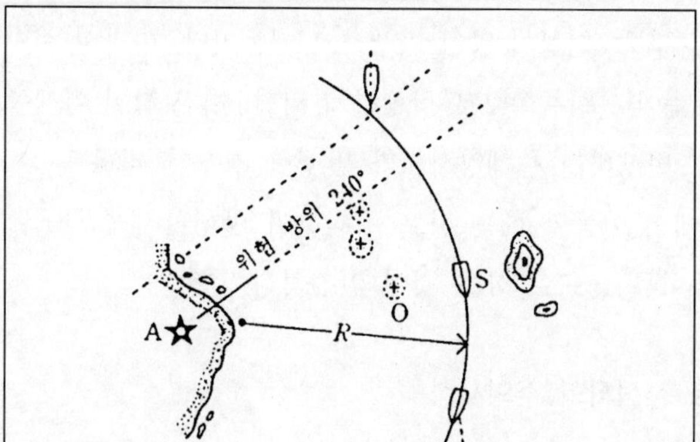

[그림 2-61] 측면에 있는 물표로부터의 거리에 의한 것

따라서 레이더를 이용하여 물표 A가 있는 곳으로부터 일정한 거리(R)를 유지하며 항행하도록 한다. 주로 레이더를 이용하면서 협수로 항해를 실시할 때 많이 선택하는 가장 쉬우면서도 안전한 방법이다.

8. 출항·입항 항로

선박이 항구로부터 출항, 또는 항구로 입항할 때에는 그 항만에 적용되는 각종 항행법규(우리나라의 경우 선박의 입항 및 출항 등에 관한 해류, 해사안전법 등)에 따라 항행하여야 한다. 또한, 항만의 크기, 위험물의 존재, 정박선의 동정, 다른 선박의 내왕 및 바람과 조류의 영향 등의 주위상황과 자선의 조종성능을 고려하여 가장 안전한 항로를 선정해야 하며, 이러한 항로선정시 주의사항은 다음과 같다.

1) 출항 항로 선정시의 주의사항

① 출항할 때에는 선박 주위를 잘 살피며 선박을 조종하여야 하기 때문에 선위측정에 전념할 여유가 없다. 그러므로 선수나 선미에 물표를 정하여 선박이 항로의 좌우로 벗어나는지를 쉽게 파악할 수 있도록 하여야 한다. 그리고, 예정 침로 주위에 위험물이 있으면 미리 상세하게 피험선을 선정해 놓도록 한다.

② 정박선 또는 장애물로부터는 될 수 있는 대로 멀리 떨어져 통항하도록 한다. 바람이나 조류가 있을 때에는 풍하쪽이나 흐름의 하류쪽을 지나도록 하고, 부득이 풍상쪽이나 상류쪽을 통과해야 할 때에는 충분한 거리와 적당한 속력을 유지해야 한다.
그 이유는 선박이 항내를 항해할 때는 저속으로 움직이고 있기 때문에 바람이나 조류에 압류되는 현상이 많이 발생하기 때문이다. 따라서 선박이 정박선이나 위험물의 풍상측이나 상류쪽으로 항행하게 되면, 이와 같이 아래쪽으로 압류되어 위험한 상황이 발생하거나 해양사고가 발생한 경우가 많다.

③ 정박지점에서 이동할 때는 바로 정침하여 항행할 수 있도록 계획하고, 야간에는 확실한 등화를 물표로 삼는 것이 안전하다.

2) 입항 항로 선정시의 주의사항

① 사전에 항만의 상황, 정박지의 수심 및 저질, 기상, 해상의 상태 등을 조사해 두어야 한다.

② 지정된 항로, 추천 항로 또는 상용의 항로가 있으면 이에 따른다.

③ 정박지로 들어갈 때에는 될 수 있는 대로 선수 물표를 정하고, 투묘시기를 알 수 있도록 정횡 방향 또는 항로 양쪽에 있는 물표의 방위를 미리 구해 둔다.

④ 해도의 정밀성을 고려하여 일반적으로 수심이 얕거나 고르지 못한 지역, 고립된 암초나 침선 등은 가급적 멀리 피하도록 한다.

⑤ 항로에 설치되어 있는 각종 항로표지의 특질 및 위치를 확인하고, 자선의 안전항행에 영향을 미치는 것에 대해서는 통과시에 항상 유의하도록 한다.

⑥ 특별히 항계내로 접근하는 항로에서는 일반 잡종선들의 출입이 잦은 연안 항로대 등은 멀리 피하여 접근하도록 한다.

2.5 대양 항법

출발지 및 도착지의 경도 · 위도를 알고서 두 지점 사이의 침로 및 항정을 구하거나, 출발지의 경도 · 위도와 침로 및 항정을 알고서 도착지의 경도 · 위도를 포함한 지구상의 각종 요소를 구하는 계산법을 항법(Sailing)이라 한다. 이러한 항법은 항정선 항법(Rhumb Line Sailing)과 대권 항법(Great Circle Sailing)으로 크게 나눌 수 있다.

항정선 항법은 침로를 일정하게 유지하면서 항해하는 것으로, 현재 이용되고 있는 것으로는 중분 위도 항법(Middle Latitude Sailing)과 점장 위도 항법(Mercator Sailing)이 있다.

대권 항법은 실제로는 대권 위에 변침점을 정해 놓고, 각 변침점 사이를 항정선 항법에 의해 항해하는 항법으로 결국 항정선 항법의 집합이라 볼 수 있다.

2.5.1 점장 위도 항법

점장 위도 항법(Mercator Sailing)은 점장도의 구성 이론에 근거를 두고 고안된 항법으로, 중분 위도 항법의 이용이 부적당하거나 정확한 계산이 필요할 때에 이용한다.

먼저 항법 계산에 앞서 점장 위도에 대하여 알아보면 모든 자오선은 극에 모이므로 자오선 사이의 간격은 극에 접근할수록 좁아진다. 점장도에서는 이와 같은 자오선이 평행선으로 나타나기 때문에, 해도에 있어서 거등권의 길이는 지구상의 실제 길이에 비하여 적도에서는 같으나, 위도가 높아짐에 따라 늘어나는 비율도 커진다.

따라서, 지구표면과 닮게 나타내기 위해서는 거등권이 늘어난 것과 똑같은 비율로 자오선의 길이, 즉 거등권 간의 간격도 늘려주어야 한다. 점장 해도에 있어서 위도가 L인 어느 지점까지의 자오선의 길이를 이와 같은 비율로 늘인 길이를 그 지점의 점장 위도(Meridional Parts, M)라 한다.

〈점장 위도 항법의 계산〉

점장 위도 항법의 공식이 중분 위도 항법과 다른 점은 변경을 계산하는 데 있어서 동서거를 이용하지 않고, 점장 변위(Meridional Difference of Latitude, m)를 이용한다는 점이다. 그리고 침로, 항정 및 변위의 계산은 평면 항법의 공식을 이용한다. 점장 변위는 두 지점의 점장 위도의 차이이며, 임의의 위도 L에 대한 점장 위도를 구하는 공식을 근사식으로 다음과 같이 나타낼 수 있다.

$$M \fallingdotseq 7{,}915.7054\, Log \tan\left(45^{\circ} + \frac{L}{2}\right) - 23.2689 \sin L$$

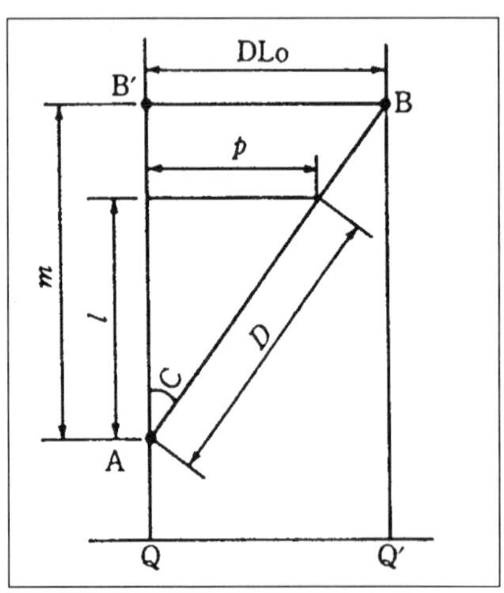

[그림 2-62] 점정도에서의 침로, 점장변위 및 변경과의 관계

그림에서 AB를 점장도에 그린 항정선, QQ′를 적도라 하면, QA는 A 지점의 점장 위도, Q′B는 B 지점의 점장 위도이다. AB′ 는 A, B 두 지점의 점장 변위, BB′는 변경이므로, 점장 위도, 변경, 침로 사이의 관계는 평면 직각삼각형을 푸는 것과 같이 다음의 식으로 나타낼 수 있다.

$$DL_0 = m \cdot \tan C_0 \ \rightarrow \ \tan C_0 = DL_0 \,/\, m$$

$$l = D \cdot \cos C_0 \ \rightarrow \ D = l / \cos C_0$$

2.5.2 대권 항법

지구표면에 있는 두 지점 사이의 최단 거리는 두 지점을 지나는 대권이다. 따라서, 대양을 횡단하는 등의 먼 거리를 항해할 때에는 항해거리를 단축하고 연료를 절약할 목적으로 두 지점의 대권을 따라 항해한다. 이러한 항법을 대권 항법(Great Circle Sailing)이라 하며, 항정선 위를 항해하는 항정선 항법과 구별하고 있다.

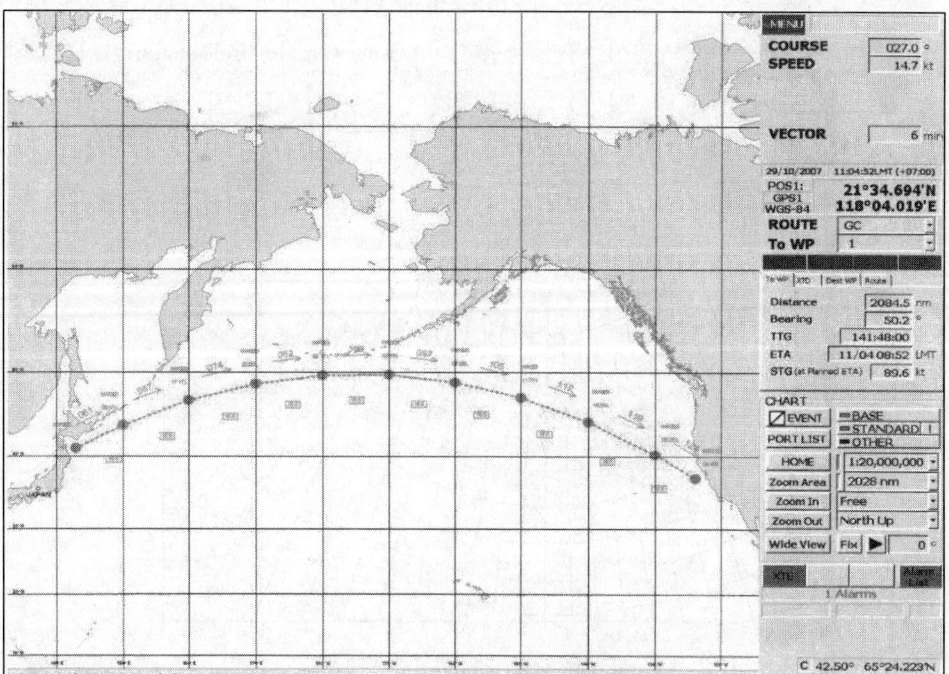

[그림 2-63] 점장도에 표시된 대권

대권 항법에서 필요한 요소, 즉 대권 거리, 침로, 변침점(way-point) 등은 계산에 의하여 구할 수 있으나, 대권도를 이용하면 간단히 구할 수 있다. 대권도에서는 모든 대권이 직선으로 나타나기 때문에, 출발지와 도착지를 직선으로 이으면 구하려는 대권 항로가 된다. 대권 위를 항해하려면 시시각각으로 침로를 바꾸지 않으면 안된다. 그러나, 실제로는 그렇게 할 수 없으므로 항로상의 경도를 5° 또는 10° 정도의 간격으로 나누고, 각 구간 내에서는 항정선 항법으로 항해하여 근사적으로 대권을 따라 항해하는 것이다.

2.5.3 집성 대권 항법

두 지점을 지나는 대권은 극쪽으로 볼록한 곡선이 되므로, 대권을 따라 항해하면 자연히 위도가 높은 지역을 항해하게 되어 바람, 파도, 유빙 등의 해상상태가 오히려 불리한 경우가 있다. 또 섬이나 육지 등에 차단되어 항해를 못할 수도 있다. 이런 경우에는 항해에 지장을 주지 않는 범위 내에서 최고 위도를 설정하고, 대권 항법과 거등권 항법을 함께 이용하여 목적지로 항해하는 것이다. 이 때 미리 선정한 최고 위도를 제한 위도라 하고, 이와 같은 항법을 집성 대권 항법(Composite Great Circle Sailing)이라 한다.

[그림 2-64]에서 제한 위도를 50° N이라 정하고, 출발지 A와 도착지 B에서 제한 위도의 거등권에 접하는 직선 AV_1 및 BV_2를 긋는다. 이 때 AV_1은 대권 항법으로, V_1V_2는 거등권 항법으로, V_2B는 다시 대권 항법으로 항해하면 된다.

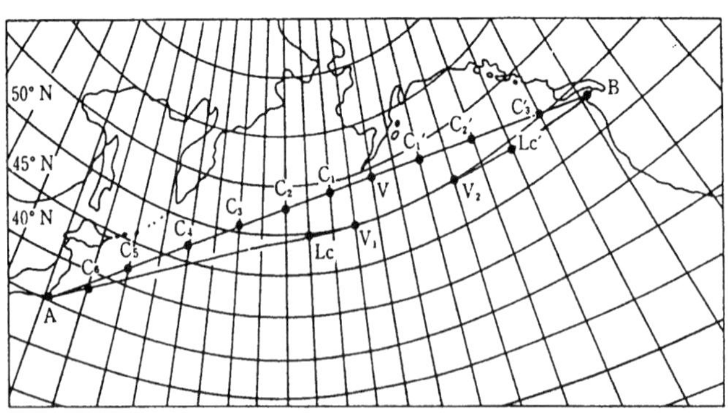

[그림 2-64] 대권도의 사용법

2.6 천체측정의 원리

2.6.1 천문항해학의 목적

① 천체의 고도 관측을 통한 천측 위치선의 도출한다.

② 천체의 천문 방위를 계산하고 Gyro compass 방위와 비교하여 Gyro compass Error를 측정한다.

③ 시의 개념을 이해하여 각종 시각(일출몰시, 박명시 등)을 계산한다.

2.6.2 천체측정의 원리

천체를 이용하여 어떻게 천측위치선을 도출하고 Gyro compass Error를 측정할 것인가를 이해해야 할 필요성이 있다.

1. 등거리권에 의한 위치결정

레이더의 가변거리환(Variable Range Marker)을 이용하면 A의 물표에서 3.5마일 떨어진 등거리권 1개를 얻을 수 있다. 이때 자선은 A의 물표에서 3.5마일 떨어진 등거리권 상에 위치할 것이다.

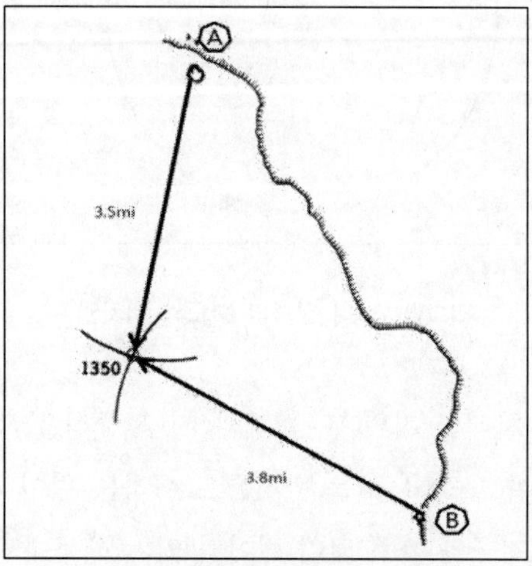

[그림 2-65] 수평거리의 방법

그러나 정확히 등거리권 위의 어느 점에 위치하는지는 알 수 없으므로 B물표까지의 거리(3.8마일)을 측정하여 등거리권 하나를 추가로 작도하게 되고 한 점에서 만나게 된다. 이때 자선의 위치는 A물표에서 3.5마일, B물표에서 3.8마일 떨어진 점에 있게 되며 이것은 [그림 2-65]과 같이 2개의 등거리권의 교차점이 된다.

2. 등고도권 및 천측위치선 작도의 원리

앞서 설명한 등거리권은 보통 레이더의 가변거리환을 이용하여 자선의 위치를 내고자 할 때 사용한다. 그러나 천체를 이용하여 위치를 내고자 할 때는 등고도권이 이용된다. 등고도권이란 천체를 일정한 고도로 관측할 수 있는 지구 표면상의 점들을 연결해서 생긴 하나의 원을 말한다.

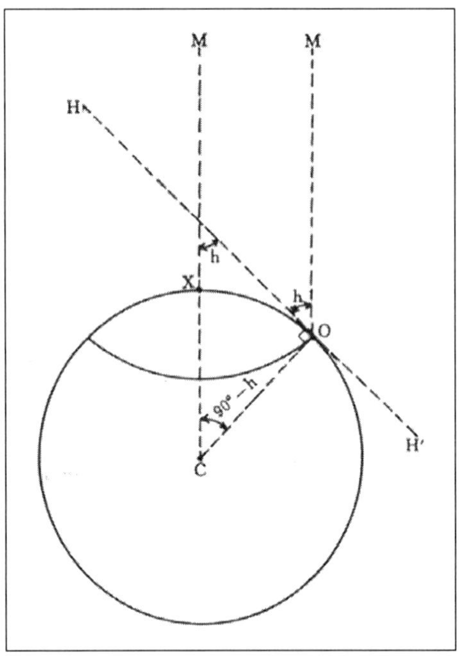

[그림 2-66] 천체의 고도와 위치권

이러한 등고도권은 천체의 지위(M, 천체와 지구 중심을 잇는 선이 지표면에 만나는 점)를 중심으로 하고 여고도(90°-h)를 반지름으로 하는 원이 된다. [그림 2-66]는 그것을 증명한다. 즉, 천체는 지구와 아주 멀리 떨어져 있기 때문에 천체의 지위(X)와 관측자의 위치(O)에서 천체 M까지 연결한 선이 평행선을 이룬다고 생각해도 무방하다. 수평선(HH')은 관측자의 위치에서 지표면과 직각을 이루고 삼각형 △CZO의 내각의 합은 180°이므로 지구중심(C)에서의 내각은 90°-h가 된다.

즉, [그림 2-66]를 통해 관측자(O)가 천체(M)을 일정한 고도(h)로 관측했다면 관측자는 지위(X)에서 90°-h 만큼 떨어진 등고도권의 어느 한 지점에 있게 되는 것이다. 따라서 천체의 고도를 관측하고 그 천체의 지위를 정확하게 안다면 하나의 위치권(등고도권)을 그릴 수 있게 되고, 만일 여러 천체를 관측하면 위치권들이 만나는 지점인 정확한 위치를 파악할 수 있을 것이다.

그러나 지위를 어떻게 알아낼 것이며, 몇 천마일(만일 천체의 고도가 60°라면 위치권의 반지름은 1800해리(30°×60해리))이나 되는 위치권을 어떻게 해도에 작도할 것인가 하는 것이 문제이다.

3. 항해삼각형

큰 반지름을 가진 위치권을 해도에 표시하기 위해 고안된 것이 항해삼각형이다. 즉, 천체(X), 관측자(Z), 천의 극(P)은 [그림 2-67]에서와 같이 삼각형을 이루고 있다고 가정한다. 이 삼각형의 세 변은 여고도(h′), 여적위(d′), 여위도(L′)로 이루어져 있으며, 세 각은 자오선각(ZPX), 방위각(XZP), 위치각으로 구성되어 있다.

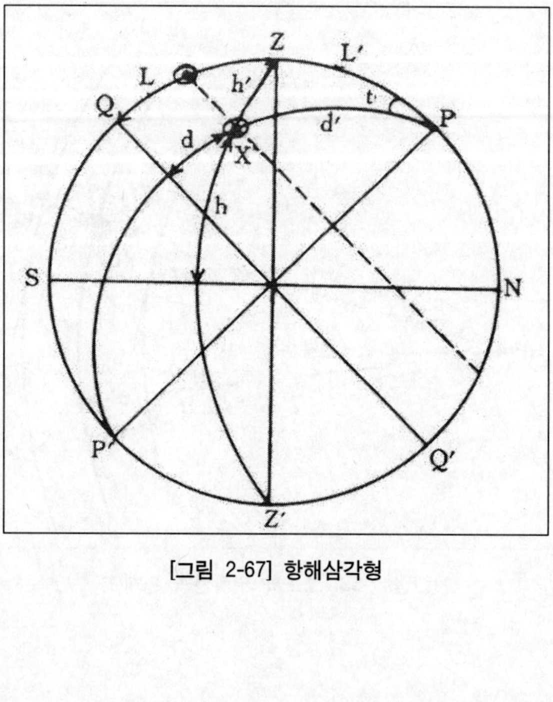

[그림 2-67] 항해삼각형

그러므로, 항해삼각형을 통해서 위도(L), 천체의 적위(d), 자오선각(t)을 알고 있는 관측자는 정해진 시각과 위치에서 그 천체를 볼 수 있는 고도 및 방위각의 계산이 가능하다(삼각형의 2변과 그 사이각을 알고 있다면 나머지 한 변의 길이와 나머지 내각은 삼각함수의 계산을 통해 알 수 있다).

4. 위치선 도출의 원리

어느 지점에서 위치선 도출을 위해 천체의 고도를 관측(관측고도)했다고 가정하자. 그러나 고도 관측을 통한 위치선의 작도는 위에서 설명한 바와 같이 천체의 지위(X)를 알 수 없고, 워낙 반지름이 크기 때문에 해도에 직접 작도하는 것이 불가능하므로 천체의 관측시각에 대한 가정위치(A.P)를 설정하고 항해삼각형을 풀어서 계산고도(Hc) 및 계산방위(Zn)를 구한다.

계산고도 및 계산방위를 구한다는 의미는 만일 관측자가 관측시간에 가정위치에 있다면 천체를 계산고도와 계산방위로 정확하게 볼 수 있다는 것이다. 그러나 현재의 위치를 정확히 알기 위하여 천체를 관측하는 것이므로, 관측시간에 육분의(sextant)를 통해 관측한 관측고도는 계산고도와 차이가 있을 가능성이 크다.

예를 들어 정해진 시간에 가정위치(A.P)에서 계산된 천체의 계산고도가 60°인 경우에 관측자가 육분의를 통해 천체를 60° 10′의 고도로 관측했다고 하자.

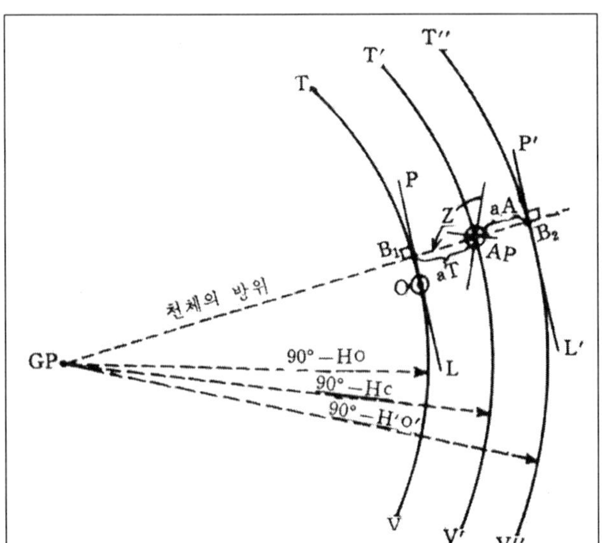

[그림 2-68] 고도와 위치선의 관계(1)

이때 관측자는 [그림 2-68]와 같이 가정위치가 속한 위치권 보다 반지름이 10마일 작은 위치권상에 존재(관측자와 천체의 지위사이의 거리는 90°-고도)하게 되며, [그림 2-69]와 같이 가정위치보다 천체에 가깝게 있는 것이다.

즉 관측자는 가정위치보다 천체를 바라보는 쪽으로 10마일 앞인 T의 위치권상에 있게 된다. 만일 59° 50′으로 천체를 관측한 경우는 천체의 반대편으로 가정위치에서 10마일 뒤쪽에 있는 T″의 위치권상에 있게 된다.

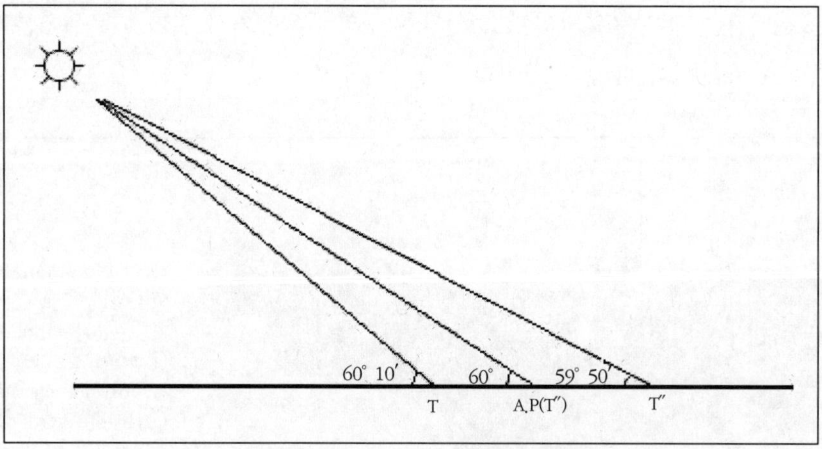

[그림 2-69] 고도와 위치권의 관계(2)

즉, 천문항해에서는 일정한 시각과 위치에서 얻을 수 있는 천체의 고도 및 방위를 천측계산표(통상 229 Table)에 미리 계산해 놓고 관측자가 육분의를 통해 얻은 관측고도와 비교하는 방법을 사용한다. 이러한 방법을 통해 너무 커서 그릴 수 없었던 위치권을 가정위치를 중심으로 천체를 바라보는 방향 혹은 그 반대편으로 조금만 이동하여 그릴 수 있는 것이다.

5. Gyro compass Error 측정의 원리

똑같은 원리로 Gyro compass Error의 측정이 가능하다. 어느 지점에서 Gyro compass Repeater를 통해 천체의 방위를 관측했다고 하자. 위와 같이 그 시각에 대한 가정위치를 설정하고 항해삼각형을 풀어서 천체의 방위를 구한다. 이렇게 얻어진 실측방위와 계산방위를 비교하면 쉽게 Gyro compass Error의 측정이 가능하다.

2.7 천체측정의 도구

2.7.1 육분의

육분의(sextant)는 천문 항법에서 위치선을 구하기 위해 천체의 고도를 측정하거나, 또는 지문 항법에서 두 물표의 수평협각을 측정하여 선위를 결정하는데 사용하는 계기이다. 육분의라는 말은 육분의의 호(arc)가 원주의 1/6과 거의 같기 때문에 6분의 1이라는 라틴어 Sextans에서 유래된 것이다.

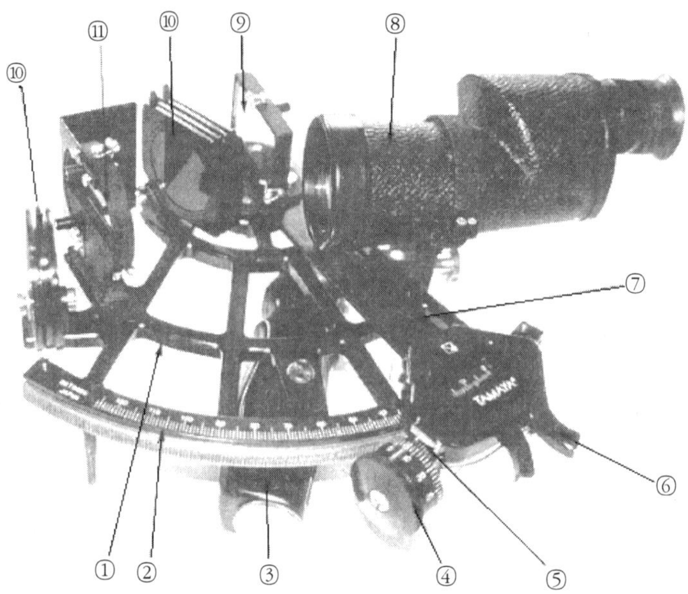

① 기면(frame)
② 호(arc)
③ 손잡이(handle)
④ 마이크로미터 드럼 (micrometer drum)
⑤ 버니어(vernier)
⑥ 고정기(release clutch)
⑦ 인덱스 바(index bar)
⑧ 망원경(telescope)
⑨ 동경(index glass, index mirror)
⑩ 차광유리(shade glass)
⑪ 수평경(horizon glass)

[그림 2-70] 육분의 구조

1. 육분의의 도수 읽는 법

마이크로미터의 1회전이 본 호의 1° 간격이므로 고정기의 화살표의 위치로서 도수를 먼저 읽은 후 분(′)수는 마이크로미터의 눈금과 버니어의 '0' 눈금의 위치로서 읽는다.

즉, [그림 2-71]과 같이 고정기의 화살표는 29°와 30° 사이에 있으므로 본호상에서 29°이다. 마이크로미터와 그 옆에 있는 버니어의 '0' 눈금이 42′과 43′의 사이에 있으므로 42′를 읽는다. 이후 마이크로미터와 버니어의 눈금이 일치되는 점이 '5'이므로 0.′5를 읽는다. 따라서 정확한 판독값은 29° 42.′5이다.

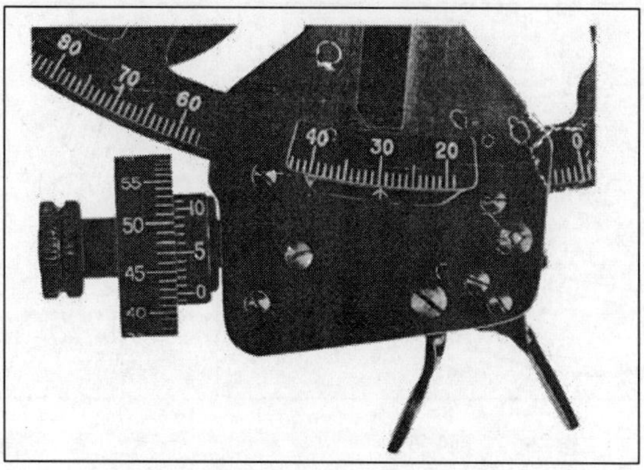

[그림 2-71] 육분의 도수 판독

2. 육분의 오차

1) 수정이 가능한 오차

① Error of perpendicularity(수직차) : 동경이 기면에 수직하지 않아 생기는 오차

② Side Error : 수평경이 기면에 수직하지 않아 생기는 오차

③ Index Error : 고정기의 화살표를 0°의 선에 놓았을 때 수평경의 면이 평행하지 않아 생기는 오차

④ Collimation Error : 망원경의 축이 기면에 평행하지 않아 생기는 오차

2) 수정이 불가능한 오차

① Centering Error(중심차) : 호의 중심과 고정기의 회전중심이 맞지 않아 생기는 오차

② Shade Error : Shade glass의 질과 부착상태가 불량해서 생기는 오차

③ Prismatic Error : 동경이나 수평경의 양면이 평행하지 않아 생기는 오차

④ 기타 제작상의 오차

3) 오차수정 방법

① 제1수정 : 고정기의 화살표를 35° 근처에 맞추고 비스듬히 들여다보아 [그림 2-72]과

같이 동경과 본 호를 비교하여 일직선으로 연속되어 보이면 동경은 기면에 수직하고 오차는 없다. 서로 일치되지 않으면 조정나사로 조정한다.

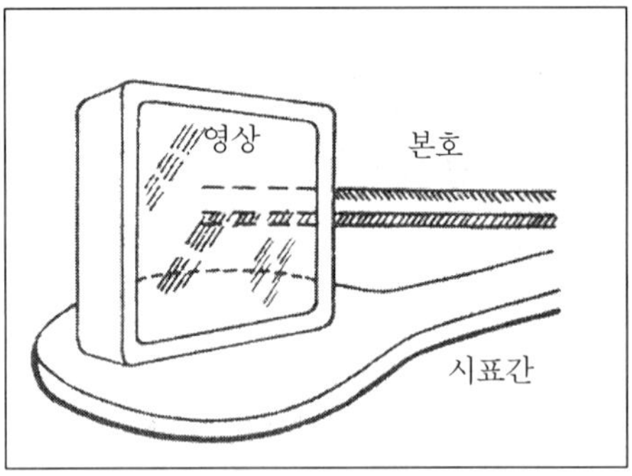

[그림 2-72] 제1수정

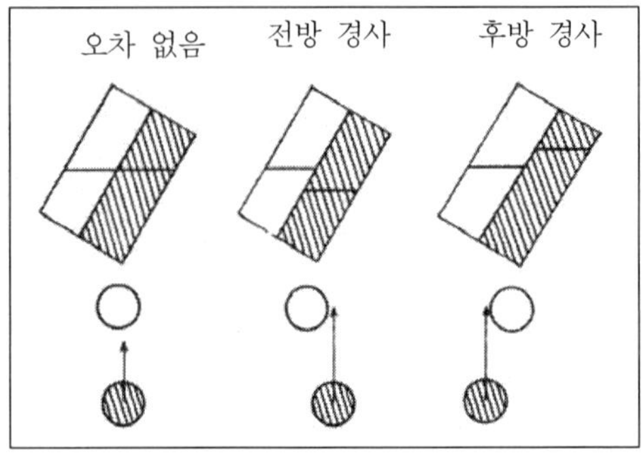

[그림 2-73] 제2수정

② 제2수정 : 고정기의 화살표를 0° 근처에 맞추고 수평선을 보며 좌우로 경사시킨다. [그림 2-73]와 같이 진상과 반사된 상이 일치되면 수평경은 기면에 수직한 것이다. 일치되지 않으면 조정나사로 조정한다.

③ Index Correction 값의 도출 : 1 및 2 수정을 마치고 난 후 고정기의 화살표를 0° 근처에 두고 수평선을 바라본 후 진상과 반사된 상이 정확히 일치되도록 마이크로미터를 조정한다. 만일 화살표가 0°.0′을 나타내면 오차가 없으나 본호상쪽으로 치우치면 그 값에 (−), 여호상에 있으면 (+) 부호를 붙여 육분의 오차값으로 한다.

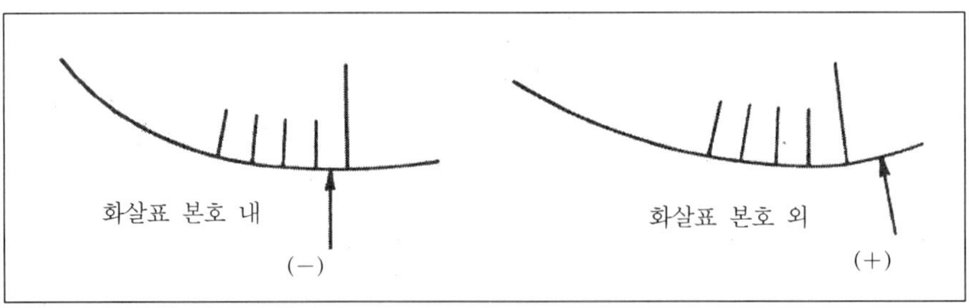

[그림 2-74] Index Correction

2.7.2 시진의 및 초시계

천측을 통해 위치선을 작도하거나 Gyro compass Error를 측정하기 위해서는 정확한 시각이 필요하다. 천체는 항상 움직이고 있으므로 정확한 측정시각을 알아야 그 시각에 해당하는 천체에 관한 자료를 천측력 및 천측계산표(통상 229 Table)에서 찾을 수 있기 때문이다.

시진의(chronometer)는 정밀한 시계로써 선교에 보관하며 주기적으로 무선 시보 신호 등으로 오차수정을 하고 그 내용은 시진의 오차수정 기록부(chronometer Journal)에 기록한다. 초시계(stop watch)는 측정자가 천체를 관측하고 관측시각을 기록하기 위해 움직인 시간을 재는 데 이용되는 도구로써 이 또한 정확한 측정시각을 알기 위해 필요하다.

2.7.3 방위환

목표물의 방위를 측정할 때에는 피프 베인을 통하여 목표물을 보면서 링 위에 있는 핑거 러그(finger lug)를 잡고 링을 돌려서 목표물이 수직 철사에 나타나게 한다. 파 베인 아래쪽에 수직 철사와 일치하는 중심선이 나타나 있는 거울이 있는데, 이 거울을 통하여 컴퍼스 카드를 읽을 수 있게 되어 있다.

또한, 태양 같은 밝고 큰 천체의 방위는 태양 광선을 거울에 반사시켜 광선 반사경에 통과시킨 후 컴퍼스 카드에 비춰진 빛을 통해 측정한다.

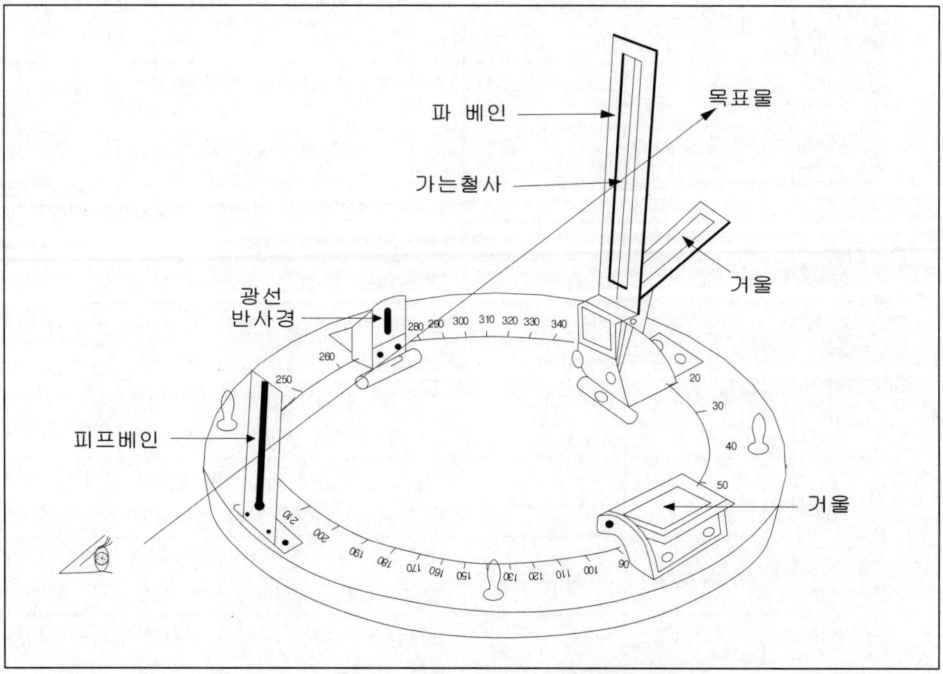

[그림 2-75] 방위환(Azimuth Circle)의 구조

2.8 용어의 정의

2.8.1 천구에 관한 용어

천구(celestial sphere)란 관측자의 눈을 중심으로 하여 무한대의 반지름을 가진 가상의 구로 모든 천체가 구면에 붙어 있는 것이라 가정한다.

〈표 2-4〉

Z	천정	Zenith	관측자와 지구중심을 지나는 직선이 천구와 만난 점 중 관측자의 머리 위쪽에서 만난 점
Na	천저	Nadia	관측자의 발 아래쪽에서 만난 점
	천의 극	Celestial pole	지축을 무한히 연장하여 천구와 만난 점
Pn	천의 북극		천의 극 중 지구의 북극쪽에 있는 것
Ps	천의 남극		천의 극 중 지구의 남극쪽에 있는 것
	동명극	Elevated pole	관측자의 위도와 동명인 극, 수평선 위쪽에 있는 극
	이명극	Dispressed pole	관측자의 위도와 이명인 극, 수평선 아래쪽에 있는 극
PnPs	천의 축	Celestial axis	천의 북극과 남극을 연장한 가상의 선
PnMPs	천의 자오선	Celestial meridian	천의 양극을 지나는 대권
PnZPs	극상반원	Upper branch	천의 자오선을 천의 극에서 양분하여 관측자를 지나는 반원
PsZPn	극하반원	Lower branch	그 반대편 반원
QQ'	천의 적도	Celestial equator	지구의 적도면을 무한히 연장하여 천구와 만나 이루는 대권
LML'	적위의 거등권, 등적위권	Parallel of declination	지구상에서 위도의 거등권에 대응하는 것으로 천의 적도에 평행한 소권
Dec, d	적위	Declination	지구상의 위도에 대응하는 것으로 천체를 지나는 천의 자오선 상에서 천의 적도로부터 천체까지의 호
	극거, 극거리	Polar distance	천체를 지나는 천의 자오선상에서 동명극과 천체사이에 낀 호 ≪ P=90°±d ≫ 동명 : −, 이명 : +
∠♈PnR, ♈R	적경	Right ascension	춘분점을 지나는 시권과 천체를 지나는 시권이 극에서 이루는 각 또는 적도의 호

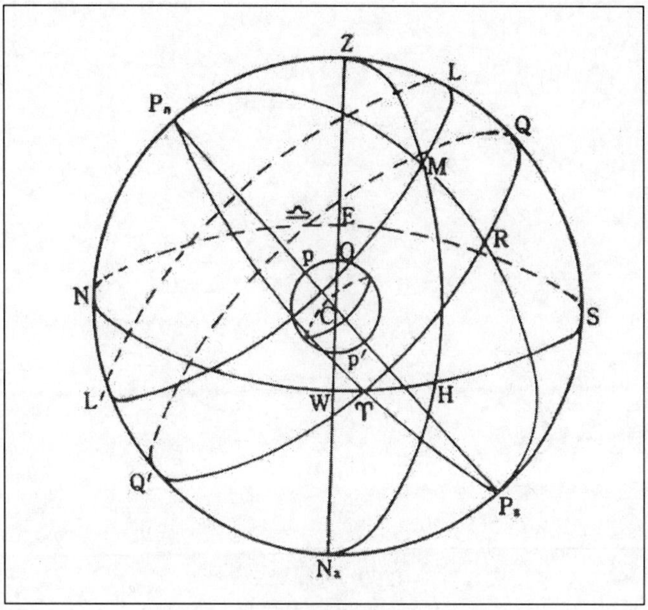

[그림 2-76] 천구

2.8.2 황도 및 천의 적도에 관한 용어

〈표 2-5〉

♈E♎E′	황도	Ecliptic	태양 연주운동(지구공전에 의해 태양이 천구상을 1년에 한번 지구를 중심으로 서에서 동으로 운행하는 것처럼 보이는 것)의 겉보기 궤도
	황도경사		황도와 천의 적도사이의 약 23° 27′의 경사
	분점	Equinoctial Points	황도경사 때문에 생기는 2개의 교점
	지점	Solstial points	황도상에서 천의 적도로부터 가장 멀리 떨어져 있는 2개의 점
♈	춘분점	First point of aries	태양이 황도상을 운행시 적도를 남에서 북으로 가르는 분점
♎	추분점	First point of libra	태양이 황도상을 운행시 적도를 북에서 남으로 가르는 분점
	하지점	Summer solstial point	천의 적도의 북쪽에 있는 지점
	동지점	Winter solstial point	천의 적도의 남쪽에 있는 지점

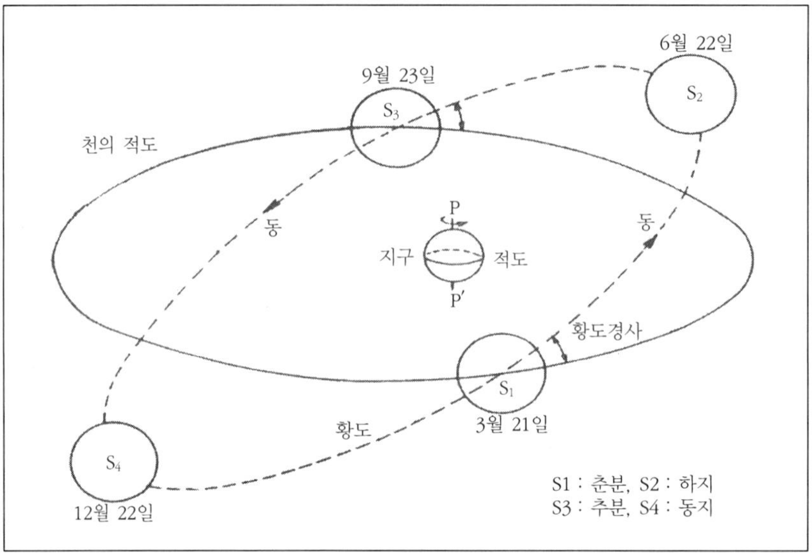

[그림 2-77] 황도

2.8.3 적도면도에 관한 용어

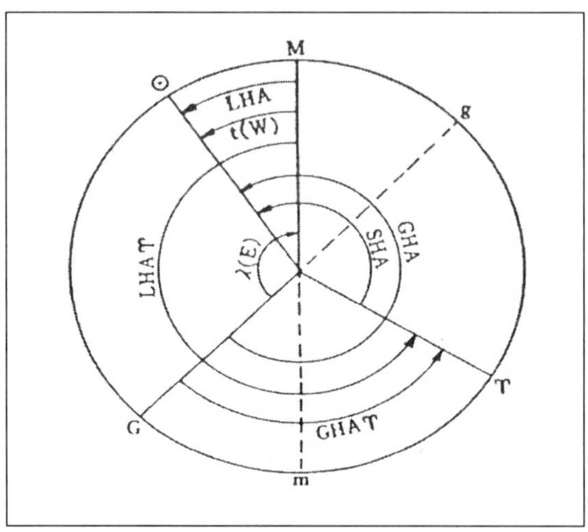

[그림 2-78] 적도면도(남극에서 바라본 그림)

〈표 2-6〉

LHA	지방시각	Local hour angle	관측자의 천의 자오선과 천제 시권이 극에서 이루는 각 또는 그들 사이에 낀 적도의 호
GHA	본초시각	Greenwich hour angle	greenwich 자오선과 천체 시권이 극에서 이루는 각 또는 그들 사이에 낀 적도의 호
t	자오선각	Meridian angle	LHA = t(w) LHA = 360 − t(E)
SHA	항성시각	Sideral hour angle	춘분점과 천체의 시권이 극에서 이루는 각
LHAγ			관측자와 춘분점이 극에서 이루는 각

2.8.4 방위에 관한 용어

① 방위(Azimuth, Zn) : 진북을 기준으로 시계방향으로 360°까지 측정

② 방위각(Azimuth angle, Z) : 가까운 동명극을 0°로 하여 동쪽 또는 서쪽으로 측정 (Zn(방위) 323° = Z(방위각) N 37°W)

③ 출몰방위각(Amplitude) : 동점 혹은 서점을 0°로 하여 북쪽 또는 남쪽으로 측정 (E15°N, W5°S)

2.9 시의 개념 및 계산

2.9.1 시각의 정의

시는 우리들의 일상생활을 규칙적으로 수행하는 데 도움이 될 뿐만 아니라, 천문항해학에 있어서는 천체를 관측하여 천체의 천구상의 위치를 결정하는 데 필요하며, 규칙적으로 반복되는 천구의 일주운동, 즉 지구의 자전과 공전운동을 기준으로 하여 측정한다. 이 일주운동을 측정하는 기준점을 천구상의 어느 점 또는 어느 천체에 두느냐에 따라 여러 종류의 시를 규정할 수 있으며, 각각 목적에 따라서 그에 적합하도록 기준점을 선정

하게 된다. 시는 이 기준점에 의하여 태양시, 항성시 및 태음시의 세 가지로 구별한다.

1. 시 시

여기에서 말하는 태양은 실존하는 태양, 즉 우리들이 직접 눈으로 볼 수 있는 태양이며 다음에 나올 가상의 태양과 구별할 필요가 있을 때에는 시태양이라고 한다. 시태양을 기준하여 측정하는 시를 시태양시 또는 간단히 시시(Apparent Time, AT)라고 하며 0시부터 24시까지로 나타낸다. 즉 시자정은 시시로 0시 또는 24시이고 시정오는 시시로 12시에 해당한다.

그리고 시정오의 지방시각은 $0°(0^h)$이며 시자정의 지방시각은 $180°(12^h)$이므로 시각(H.A.)과 시시는 12시간의 차이가 있으며 이 관계를 이용하면 지방시각을 알고 시시를 구할 수 있다.

$$AT = 12^h \pm t$$

(단 t는 지방시각을 자오선각으로 고친 것이며 t(W)이면 "+", t(E)이면 "−"이다.)

특히 영국의 그리니치에 있어서의 시시를 그리니치 시시(Greenwich Apparent Time, GAT)라고 하며, 임의 지점에 있어서의 시시를 지방시시(Local Apparent Time , LAT)라고 한다.

2. 평 시

시태양이 춘분점을 출발하여 다시 춘분점에 돌아오기까지의 속도를 평균한 속도로써 천의 적도상을 서에서 동으로 운행한다고 상상하는 태양인 것이다. 평균태양을 때로는 천문학적 평균태양이라고도 한다. 평균태양을 기준으로 측정하는 시를 평균태양시 또는 간단히 평시(Mean Time, MT)라고 하며 평자정은 0시(24시), 평정오는 12시에 해당한다.

평균태양을 기준으로 자오선각을 t라고 하면

$$MT = 12^h - t(E), \quad MT = 12^h + t(W)$$

인 관계가 성립한다. 이 평시는 우리들이 일상생활에 사용하는 시이며 그런 이유로 평시를 상용시 또는 평균태양일을 상용일이라고도 한다.

특히 영국의 그리니치에 있어서의 평시를 그리니치 평시(Greenwich Mean Time,

GMT) 또는 세계시(Universal Time, UT 또는 U)라 하며 임의의 지점에 대한 평시를 지방평시(Local Mean Time, LMT)라고 한다.

3. 시 차

시시에서 평시를 뺀 것 즉 평시에 더하면 그 때의 시시가 되는 시간을 평시차 또는 시차(Equation of Time, Eq. T)라고 한다. 이 시차가 생기는 원인은 첫째 황도가 적도와 경사를 이루고 있으며, 둘째 황도가 타원형이기 때문이다. 첫째 원인에 의한 것을 경사시차, 둘째 원인에 의한 것을 타원시차라고 한다.

4. 지방평시 및 지방표준시

일상생활에는 평시가 이용되는데 지방평시는 그 지점의 천의 자오선을 기준으로 하기 때문에 각 지점마다 지방평시(Local Mean Time, LMT)를 사용하면 같은 행정구역 또는 같은 도시 안에서도 각각 위치에 따라, 즉 경도가 다르면 다른 시를 사용하여야 하며 항행 중인 선박에서는 끊임없이 시각을 개정하여야 하는 불편이 있다. 또 세계시를 사용하면 일출몰 현상 등이 일어나는 시각이 지방에 따라 현저한 차이가 생겨 불편하다. 즉, 그린 위치에서 동서로 1° 단위 혹은 그 이상이나 이하로 기준자오선을 설정하였다고 가정할 때 이 기준자오선의 표준시를 지방평시라고 한다.

만일 134° 30′N에 관측자가 있으며 이 자오선이 기준자오선이라면 당연히 12시에 태양이 정중할 것이다. 그러나 실제로 지방평시를 사용하지 않는 이유는 이 경우 같은 도시나 국가 안에서도 경도가 조금 틀리면 다른 시간을 사용해야 하는 문제가 있기 때문이다. 천측력 주표의 오른쪽 페이지에 기록된 시각은 모두 지방평시이다.

그래서 한 나라 또는 한 지방에서는 적당한 구역(그리니치를 기준으로 동서로 각각 15°의 정수배)으로 나누어 어느 적당한 자오선을 표준자오선으로 정하고 이를 기준으로 정한 평시를 사용하고 있다. 이와 같은 평시를 그 나라 또는 한 지방의 표준시(Local Standard Time, LST)라도 하며 보통 이것을 상용하므로 상용시라고도 한다.

5. 시각대 및 대시

지구의 표면을 경도차가 15°되는 자오선들로써 25대역으로 나누고 각 대역을 시각대(time zone)라고 한다. 시각대는 모두 경도차가 15°인 대역이지만 180°자오선을 사이에 두고 있는 2대역만은 폭이 7°30′으로 되어 있다. 그리고 특별한 경우가 아니면 그 시각대의 중앙자오선을 기준자오선으로 정하여 같은 시각대내에서는 경도가 변하더라도 그 기준자오선에 대한 지방평시를 사용하기로 한다. 이와 같은 시를 대시(Zone Time, ZT)라고 하는데 육상에서의 표준시에 대응하는 것이다.

그리고 이들 시각대는 다음과 같이 −12, −11, …, −1, 0, +1, …, +11, +12 등으로 구별하며 이를 시각대명(Zone Description, ZD)이라고 한다. 이 시각대명은 그 수치를 부호에 따라 대시에 가감하면 세계시가 되도록 정한 것이다. 따라서 세계시를 대시로 고치려면 그 부호를 반대로 하여 세계시에 가감하면 된다.

$$UT = ZT + ZD$$

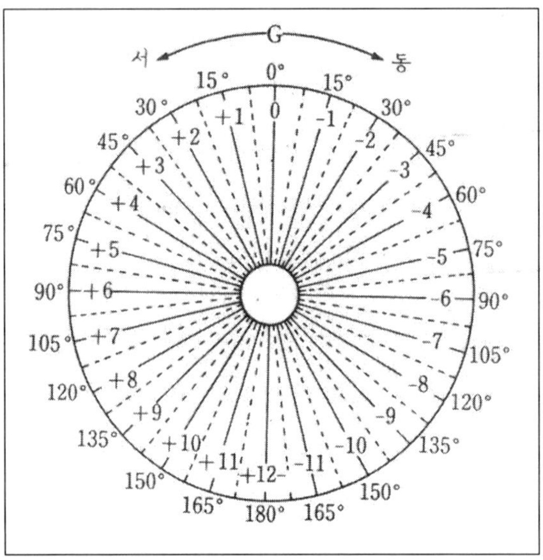

[그림 2-79] 시각대

6. 날짜변경선

180°인 자오선을 출발한 다음 동쪽으로 계속 항해하여 지구를 일주한다면 시각대가 바뀔 때마다 1시간씩 시각을 빨리 하여야 하므로 다시 180° 자오선에 돌아올 때까지에는 24시간, 즉 1일을 빨리 한 셈이 된다. 반대로 180° 자오선을 출발하여 서쪽으로 일주한다면 다시 180° 자오선에 돌아오기까지에는 24시간, 즉 1일을 늦춘 결과가 될 것이다.

즉, 그리니치(경도 000°) 자오선에서의 시각이 10일 12시라고 가정하였을 때 015°E의 대시는 13시, 030°E는 14시, ... , 180°E에서는 10일 24시이며 반대로 015°W의 대시는 11시, 030°W는 10시, ... , 180°W에서는 10일 00시가 된다.

이와 같이 180° 자오선을 기준으로 정확히 1일(24시간)의 시간차가 발생하므로 만일 선박이 항해하면서 이 경계를 지나게 된다면 1일을 중복하거나 생략하여야 할 것이다. 선박이 동쪽으로 진행하면서 통과되는 경우는 1일을 중복(repeated)하고 서쪽으로 진행하면서 통과하는 경우는 1일을 생략(skipped)한다.

그러나 실제로는 선박이 날짜변경선을 통과하였다고 바로 하루를 중복하거나 생략하는 것이 아니고, 자정까지 생활한 다음에 중복일 경우는 똑같은 날짜가 반복되고(예를 들면 9일이 반복), 생략일 경우는 다음 날짜를 그냥 건너뛰고 그 다음 날짜(예를 들면 9일에서 11일로)로 넘어간다.

그리고 Log Book Remark란에 예를 들어 날짜가 중복될 경우, 'Due to passing date line on previous day(9th, May), 9th May repeated.'라고 기재한다.

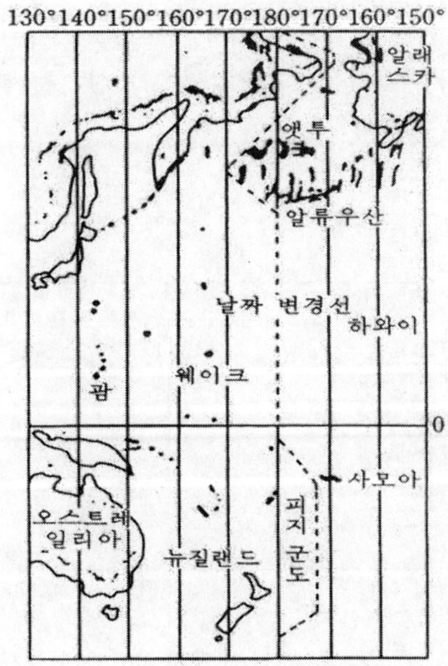

[그림 2-80] 날짜변경선

2.9.2 시의 환산

1. 시간과 호도의 환산

평균태양은 평자정부터 평자정까지 24시간 동안에 지구를 일주하므로 시간과 호도사이에는 다음 관계가 성립한다.

$$1^d = 24^h = 360^\circ$$

$$60^m = 1^h = 15^\circ$$

$$4^m = 1^\circ = 60'$$

$$60^s = 1^m = 15'$$

$$4^s = 1' = 60''$$

$$1^s = 15'' = 0.'25$$

그러므로 위의 관계를 알면 시간을 호도로 또는 호도를 시간으로 환산할 수가 있으나 천측력의 "호도 시간 환산표"를 이용하면 일일이 계산하는 불편을 덜 수 있어 편리하다.

2. 시간과 경도

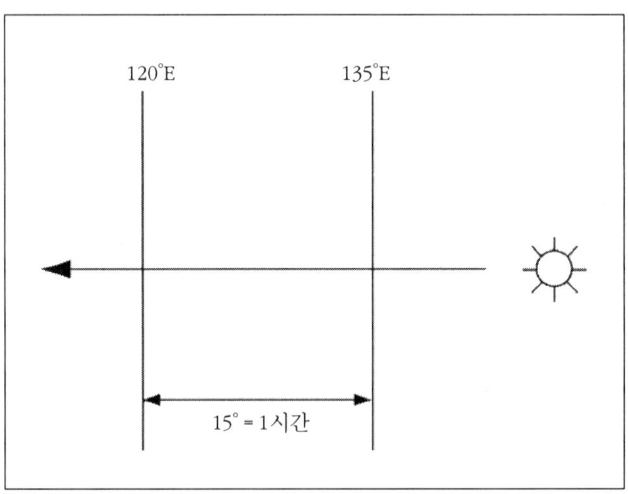

[그림 2-81] 태양의 이동과 시차

평균태양의 운행속도는 1시간에 15°이므로 어느 관측자의 자오선을 기준으로 하여 생각하면 그 자오선의 15° 동쪽인 자오선에는 평균태양이 관측자의 자오선에 정중하기 1시간 전에 정중하고 또 관측자의 자오선의 15° 서쪽에 있는 자오선에는 관측자의 자오선에 정중한 1시간 후에 정중한다. 이와 같은 사실은 두 지점 사이의 경도가 다르면 각 지점에서의 시각에는 그 두 지점 사이의 경도차를 시간으로 환산한 경도시만큼의 차이가 있음을 의미한다.

즉, [그림 2-81]과 같이 태양은 동쪽에서 서쪽으로 이동하므로 135°E 자오선에 태양이 정중해서 12시일 때 120°E는 11시이다. 1시간 후 태양이 120°E 자오선에 정중하게 되면 135°E의 대시는 13시가 되는 것이다. 이와 같이 인접한 두 대시(ZT) 간에는 1시간의 시간차가 있으며 만일 선박이 항해하면서 이 경계를 지나게 된다면 그 지역의 대시에 맞추기 위해 선박사용시를 전진하거나 후진하여야 할 것이다.

"선박이 동쪽으로 진행하게 되면 시간을 전진하고 서쪽으로 항진할 경우는 후진한다."

3. 시간과 시각

시간과 시각은 모두 지구상의 어느 자오선을 기준으로 하고 천구상의 어느 기준점까지의 각을 표현하고 있는 점에서 공통성이 있다. 그러나 시각은 천구상의 어떤 점을 취해도 상관없지만 시간은 보통 시태양, 춘분점 및 때때로 달 등이 천구상의 기준점으로 채택될 뿐이다. 그리고 시각은 보통 각도로 표시하고, 천의 자오선의 극상반원을 기준으로 측정하는데 반하여 시간은 시간단위로 표시하며, 항성시와 태음시는 천의 자오선의 극상반원을 기준으로 측정하지만 태양시는 극하반원을 기준으로 측정한다.

시간과 시각을 모두 호도로 표시하면 이들 사이에는 다음과 같은 관계가 성립한다.

$$\text{LMT} = \text{LHA of Mean sun} \pm 180$$

(단 LHA 〈 180°이면 +, LHA 〉 180°이면 −)

$$\text{LAT} = \text{LHA of Apparent sun} \pm 180^\circ$$

(단 LHA 〈 180°이면 +, LHA 〉 180°이면 −)

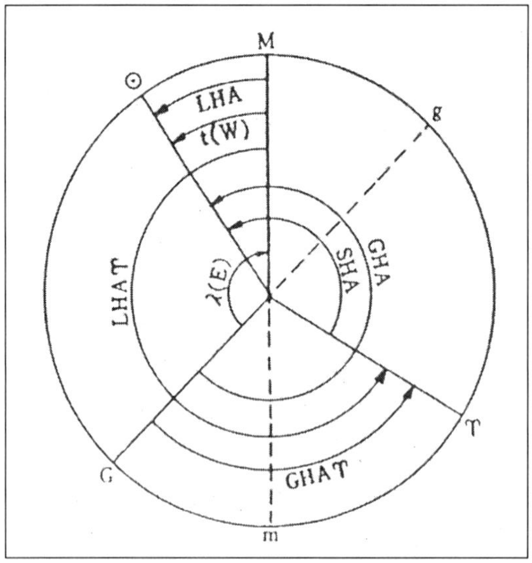

[그림 2-82] 적도면도의 응용

[그림 2-82]에서 LHA은 30° W 이며 이것을 시간으로 환산한다면 천체가 동쪽에서 떠서 관측자(M)에 정중한 후 30°인 경우는(시간으로는 2시간) 간 것이므로 오후 2시가 될 것이다. 앞서의 공식에 대입해보면 30°인 경우 LHA 〈 180°이므로 +180°를 해주어야 하고 30°+180°=220°이므로 이를 시간으로 환산하면 14시(오후 2시)가 된다.

4. 대시와 세계시

대시와 세계시 사이에는 시각대명의 수치에 해당하는 시간만큼 차이가 있으므로 대시를 세계시로 바꾸려면 대시에 시각대명을 그 부호에 따라 가감하고, 세계시를 대시로 고치려면 시각대명의 부호를 반대로 하여 세계시에 가감하면 된다.

5. 세계시와 지방평시

세계시와 지방평시 사이에는 관측자의 경도를 시간으로 환산한 경도시만큼의 차이가 있으므로 다음과 같이 계산하면 된다.

1) 세계시를 알고 지방평시를 구할 때

$$LMT = GMT \pm \lambda T \qquad (\text{단, 동경이면 } +, \text{ 서경이면 } -)$$

2) 지방평시를 알고 세계시를 구할 때

$$GMT = LMT \pm \lambda T$$ (단 동경이면 −, 서경이면 +)

6. 지방평시와 대시

지방평시와 대시 사이에는 관측자의 자오선과 시각대의 기준자오선 사이의 경도차를 시간으로 환산한 양만큼의 시간차가 있다. 관측자의 자오선이 그 시각대의 기준자오선보다 동쪽에 있으면 평균태양이 관측자의 자오선에 먼저 정중하고 나서 기준자오선에 정중하므로 대시보다 지방평시가 빠르고, 관측자의 자오선이 그 시각대의 기준자오선보다 서쪽에 있으면 평균태양이 기준자오선을 통과한 후에 관측자의 자오선을 통과하므로 대시보다 지방평시가 늦다.

따라서 지방평시를 대시로 고치려면 관측자의 자오선과 그것을 포함하는 시각대의 기준자오선 사이의 경도차를 구하고 이를 시간으로 환산하여 관측자가 기준자오선의 서쪽에 있으면 "+", 동쪽에 있으면 "−" 한다.

$$ZT = LMT \pm d\lambda$$

(단 관측자가 기준자오선의 서쪽에 있으면 +, 동쪽에 있으면 −)

천측력에 일출몰시 및 박명시 등으로 기록된 시각은 지방평시이고 우리가 실제 사용하는 시각은 대시이므로 환산이 필요하다.

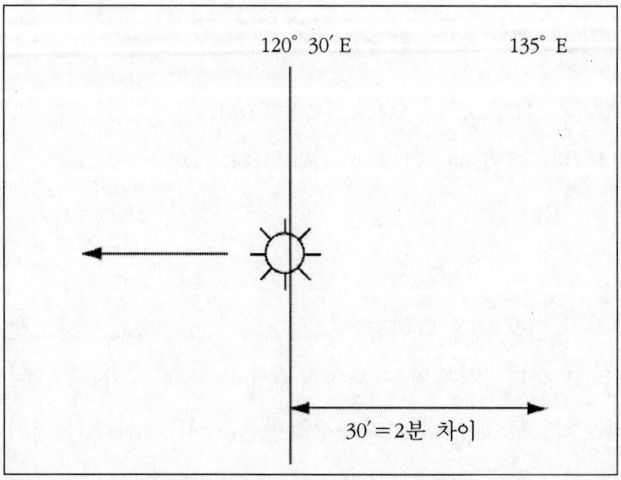

[그림 2-83] 지방평시 및 대시

만일 [그림 2-83]와 같이 134° 30′E에 태양이 정중한 상태일 때 지방평시를 사용한다면 12시가 될 것이다. 그러나 현실적으로 대시를 사용하고 있으므로 기준자오선인 135°E의 시각인 12시 02분에 태양이 정중하는 것이다. 이와 같이 대시와 지방평시에는 차이가 생길 수 있으므로 천측력에 있는 지방평시를 대시로 정확하게 환산하는 능력이 필요하다.

7. 평시와 시시의 환산

Eq. T=AT−MT 이므로 (타원시차 관계식) 평시와 시시의 환산은 다음과 같이 하면 된다.

$$AT = MT + Eq.\ T$$

$$MT = AT - Eq.\ T$$

윗 식에서 Eq. T는 천측력 주표의 오른쪽 페이지에서 구할 수 있는데 천측력에는 GMT 00^h 및 12^h에 대한 Eq. T가 주어져 있으므로 우선 GMT를 알 필요가 있다.

평시를 세계시로 고칠 때에는 경도시만 가감하면 되므로 문제가 없으나 시시를 세계시로 고치는 일은 시차를 구할 수 없어서 사실상 불가능한 것처럼 생각된다. 그러나 실제로는 시차가 미소하여 그 변화량도 작기 때문에 실용상의 계산에서는 세계시와 근사한 그리니치시를 구하고 이것을 세계시로 보아 시차를 구하여도 지장이 없는 것이다.

윗 식은 지방시 또는 그리니치시의 어느 경우에도 성립하지만 일반적으로 그리니치시를 구하고 나서 지방시로 고치는 일이 많다.

또 시시를 구할 때에는 다음을 이용하여 구할 수도 있다.

$$LAT = LHA\ of\ Apparent\ sun\ \pm 180^\circ$$

(단 LHA 〈 180° 이면 +, LHA 〉 180° 이면 −)

8. 정중시

천체가 관측자의 자오선에 합하는 것을 자오선 정중 또는 자오선 통과라고 하며 그 시각을 정중시라고 한다. 천체의 자오선정중시를 구하는 이유는 주로 정중시에 천체의 고도(이것을 자오선 고도라고 함)를 측정하면 간단히 선위를 결정할 수 있으므로 이의 예비계산으로서 필요하기 때문이다(자오선 고도 위도법 참고).

1) 태양의 극상정중시(약산법)

천측력 주표 오른쪽 페이지의 오른편 아래 끝에 있는 정중시를 관측자의 자오선에 대한 지방평시로 보고 선박사용시로 환산하면 된다.

2) 태양의 극상정중시(정밀법)

정중시를 시간의 초 단위까지 구하는 방법을 정밀법이라고 하며 다음 두 가지 방법이 있다.

① GHA method : 태양이 관측자의 자오선에 정중하면 관측자의 경도와 태양의 GHA 사이에는 다음과 같은 관계가 성립한다. 즉, 관측자의 경도가 동경이면 GHA=360° −λ 관계가 성립하며 경도가 서경이면 GHA=λ인 관계가 성립한다.

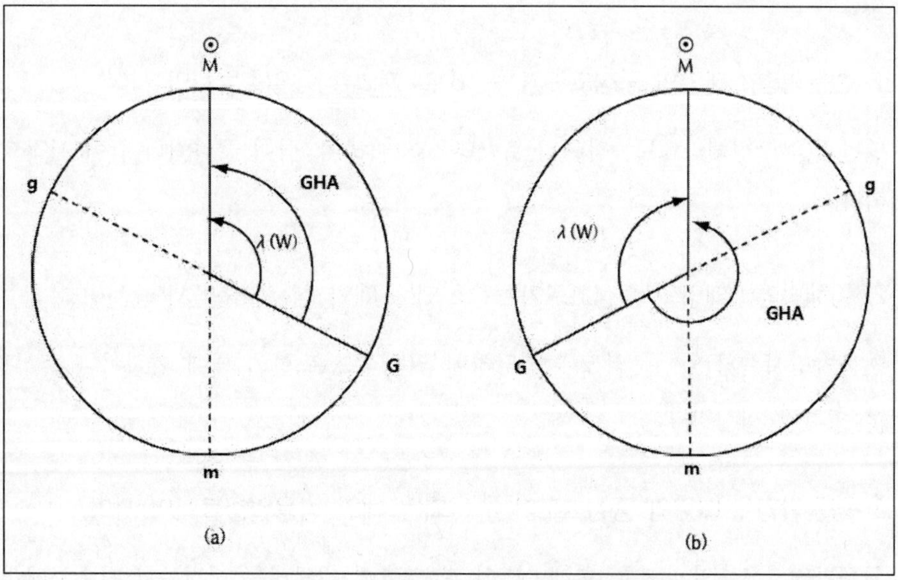

[그림 2-84] 정중시6의 GHA와 경도의 관계

② LAT method : 태양의 극상정중시는 지방시시가 12^h인 때이므로 LAT 12^h에 경도시를 가감하여 GAT를 구하고 이 GAT를 GMT로 보아 천측력에서 Eq. T를 구한다. GAT에 Eq. T를 가감하여 GMT를 구하고 선박사용시로 고치면 된다.

9. 출몰시

천구의 일주운동에 의하여 일반적으로 천체는 '출' 및 '몰'의 현상을 일으킨다. 천체의 중심이 진수평권상에 온 때를 진출몰시라고 하며, 천체의 상변(항성 또는 혹성이 그 중심)이 시수평에 접한 때를 그 천체의 상용출몰시라고 한다. 천측력에서는 안고가 0인 관측자를 기준으로 한 상용일출몰시를 게재하고 있다.

10. 박명시

태양이 수평선의 아래에 있어서 직접 햇볕이 쪼이지 않는 경우라도 일출전 및 일몰후 얼마 동안은 하늘이 밝다. 이것은 공기 중의 작은 먼지나 수증기에 태양광선이 반사되어 빛의 산란을 일으키기 때문이다. 이 현상을 박명이라고 한다. 박명은 태양의 고도에 따라 다음 세 가지로 구별한다.

1) 천문박명 : 태양의 중심고도가 −18°일 때와 상용일출몰시의 사이

일반적인 박명이라 하는 것은 천문박명을 의미하며, 밤이 끝나거나 밤이 시작되는 시기이다.

2) 항해박명 : 태양의 중심고도가 −12°되는 때와 상용일출몰시의 사이

천문박명시간의 약 3분의 2에 해당하며 일반적으로 항해 중에 항성이나 혹성관측에 가장 적합한 시기이다.

3) 상용박명 : 태양의 중심고도가 −6°되는 때와 상용일출몰시의 사이

천문박명의 3분의 1 또는 항해박명의 2분의 1에 해당하는 시간동안 지속하며 태양의 고도가 약 −6°인 때는 저녁이면 1등성이 보이기 시작하거나 또는 아침이면 사라지기 시작하게 된다.

2.9.3 시의 계산

1. 시간의 전후진

부산(대한민국)에서 출항하여 RAS TANURA(사우디아라비아)까지의 시간 전 · 후진을 구한다.

KOREA, BUSAN(Z.D. −9) → SAUDI ARABIA, RAS TANURA(Z.D. −3)

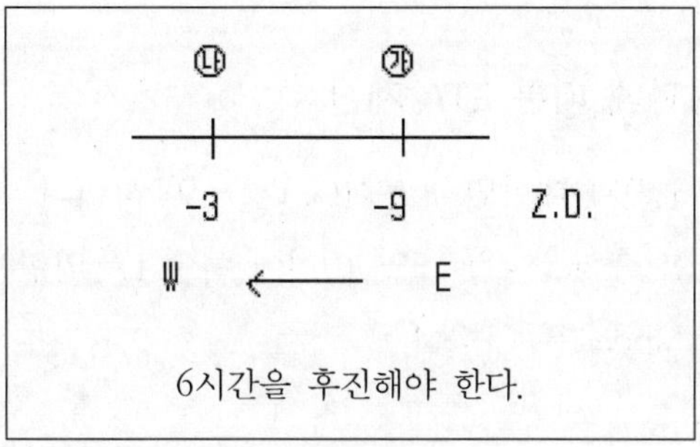

[그림 2-85]

① 상기 두 항구간의 거리를 5760miles, 자선의 속도는 12.0kts라고 가정할 때 부산 출항시간이 3월 5일 10 : 00LT이라면 RAS TANURA의 ETA를 구하는 방법이다.

㉠ 방법-1 : 항해 중 6시간을 후진하므로 6시간 동안 항해시간이 늘어나므로 ETA는 6시간 단축됨

5760′ / 12=480h이므로 3월 25일 10 : 00에 도착하나 6시간 후진하였으므로

ETA : 3월 25일 04 : 00

㉡ 방법-2 : 출발지 및 도착지를 GMT로 계산하고 ZD을 가감하여 ETA 산출함

출항시간 : 3월 5일 01 : 00(GMT)

항해시간 : 20일(480h=5760/12)

도착시간 : 3월 25일 01 : 00(GMT)이나 ZD +3을 더하면

→ 3월 25일 04 : 00(LT)

② 항구간 거리가 5760miles일 때 부산출항이 3월 5일 10 : 00이고 RAS TANURA의 ETA가 3월 21일 10 : 00이면 Target Speed를 구하는 방법이다.

㉠ 방법-1 : 항해 중 6시간을 후진하므로 6시간 동안 항해시간이 늘어나므로
5760′ / (384h + 6h) = 14.76kts

㉡ 방법-2 : 출발지 및 도착지를 GMT로 계산하고 시간 차이를 계산한 후 거리를 나눔
출항시간 : 3월 5일 01 : 00(GMT)
도착시간 : 3월 21일 07 : 00(GMT) 이므로 시간 차이는 390h
5760′ / 390h = 14.76kts

2. 날짜변경선 통과에 따른 ETA 계산

① 미국 LA항(ZD +8)과 대한민국 부산(Z.D. −9) 사이의 항주거리를 3000해리라고 가정하자. 부산항 출항을 4월 15일 10시에 하고 LA 입항은 4월 25일 10시에 해야 할 경우 Target speed를 구한다.

㉠ LT로 구하는 방법

LT 25-10-00
LT 15-10-00

10-00-00

∴ 10 × 24 = 240h, Los Angeles까지 동진하면서 Date line을 통과하면서 하루가 중복되고 7시간 전진을 하니 총 시간에서 하루를 더해 주고 전진한 7시간을 빼준다.

∴ 240h + 24h − 7h = 257h

∴ 3000/257 = 11.67kts

㉡ GMT로 구하는 방법

GMT 25-18-00
GMT 15-01-00

10-17-00

∴ 10일 하고 17시간이 걸린다.

∴ 3000/257 = 11.67kts

② 출항(Seattle, Z.D. : +8) 2015년 4월 15일 1000

입항(Busan, Z.D. : −9) ?

거리 : 4918, Ship's speed : 25kts

㉠ LT로 구하는 방법

```
LT 15-10-00
     8- 4-43
----------------
   23-14-43 → 4918/25=196.72h(8d-4h-43m)
```

∴ 서진을 하면서 하루가 생략되고 7시간 후진을 하여 실제 계산에서는 하루를 더해주고 후진한 7시간만큼 빼주면 된다.

```
∴ LT 23 -  14  - 43
          (+)1 (−)7
   -------------------------
∴ LT 24 -   7  - 43
```

㉡ GMT로 구하는 방법

```
GMT 15-18-00
        8- 4-43
---------------------
       23-22-43
```

∴ Busan은 GMT보다 9시간이 빠르므로 이를 더해 주면 24일 07시 43분이 된다.

3. 태양의 정중시

2006년 8월 8일에 경도 138°16.9′ E인 자오선에 태양이 정중하는 대시(ZT)를 구한다.

1°=4′, 3.3°=13.2′ 차이

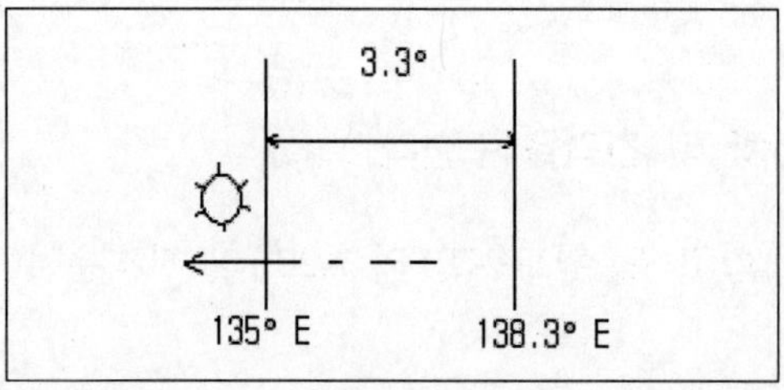

[그림 2-86]

천측력 주표 오른쪽 페이지 하단에 태양과 달의 정중시가 나와 있다. 해당 일의 태양의 정중시(LMT)를 경도 보간하여 ZT로 환산해준다.

Day	SUN					MOON					
	Eqn. of Time				Mer. Pass	Mer. Pass.				Age	Phase
	00h		12h			Upper		Lower			
d	m	s	m	s	h m	h	m	h	m	d %	
8	05	40	05	36	12 06	13	11	00	49	02 03	
9	05	32	05	28	12 05	13	56	01	33	03 07	
10	05	24	05	19	12 05	14	41	02	18	04 13	

[그림 2-87] 태양의 정중시 구하기

① 138°16.9′E ≒ 138.3°E → Mer.pass. time $12^h\ 06^m - 13.2^m = 11^h\ 52.8^m$

$= \underline{11^h}\ \underline{52^m}\ \underline{48^s}$

② 모든 자오선(천측력의 시각은 LMT) → Mer.pass. time $12^h\ 06^m$

③ 138.3°에 태양이 정중할 때 선내시각은 다음과 같다.

태양은 동쪽에서 떠오르고 서쪽으로 지기 때문에 경도 135°E보다 138°16.9′E 에서 해가 더 빨리 뜬다. 즉 자선의 시각은 135°E의 기준자오선을 사용하므로 태양이 138.3°E에 정중할 때 135°E에 있는 사람에게 몇 시인지 물어보는 것과 같다.

4. 일출몰시

2006년 4월 27일에 13°52.0′N, 070°42.1′E인 지점에서의 일몰시를 구하여 본다.

1) 천측력 주표의 오른쪽 페이지의 위도에 따른 Sunset 시간을 구한다

위도 13°52.0′ N은 N10°와 N20°의 사이에 있으므로 이들의 값을 구해준다.

→ N10° $18^h\ 11^m$, N20° $18^h\ 22^m$

2) 두 위도의 시간차 및 간격을 구한다

일몰 시간차는 11^m이고 두 위도(N10°와 N20°)의 간격은 10°이다.

3) 천측력의 일 · 월출몰 개정표 "TABLE Ⅰ - FOR LATITUDE"에서 개정치를 구한다

- Tabular interval : 천측력 주표에서 두 위도 차이
- Difference ～ : 두 위도의 시간차

① Tabular interval 10° 아래에서 원래의 위도(13° 52.0′ N)보다 낮고 가장 가까운 위도 N10°와의 차이만큼인 3° 52′ 인 곳을 찾는다.

→ 3° 30′ 과 4° 00′ 사이에 있다.

② 'Difference ～ '에서 일몰 시간차인 11^m과 만나는 곳을 찾는다.

→ 10^m과 15^m 사이에 있다.

③ 비례식을 이용하여 ①, ②가 교차하는 값을 구한다.

㉠ 3° 30′ 에서 시간차가 10^m일 때 3^m, 15^m일 때 5^m이다.

→ $5^m : 2^m = 1^m : x \quad \therefore x = 0.4^m = 24^s$이므로 11^m일 때는 $3^m + 24^s = 3^m\ 24^s$

㉡ 4° 00′에서 시간차가 10^m일 때 4^m, 15^m일 때 6^m이다.

→ $5^m : 2^m = 1^m : x \quad \therefore x = 0.4^m = 4^s$이므로 11^m일 때는 $4^m + 24^s = 4^m\ 24^s$

㉢ 3° 52′에서 시간차가 11^m일 때의 값은

→ $30' : 1^m = 22' : x \quad \therefore x = 0.7^m = 42^s$이므로 $3^m\ 24^s + 44^s = 4^m\ 08^s$이다.

㉣ 천측력의 일 · 월출몰 개정표를 대신 비례식으로 계산하는 것이 간편하다. 즉, 위도 10°의 간격일 때, 시간차가 11^m일 때 3° 52′에 관한 값을 구하면

$10^\circ : 11^m = 3^\circ 52' : x \rightarrow x = 4^m\ 12^s$

- 비례식과 개정표의 차이는 개정표가 정수 단위로 값을 표시함으로써 생기는 오차이므로 비례식의 수치가 더욱 정확하다고 볼 수 있다.

4) 개정치는 $4^m\ 08^s$이다. 그러므로 13°52.0′ N의 위도보간 한 값은

N10° $18^h\ 11^m + 4^m\ 06^s$(3° 52′ 에 대한 개정치) = <u>18^h</u> <u>15^m</u> <u>08^s</u> 이다(N10°～N20°증가하므로 개정치를 "+" 해준다).

5) 위에서 구한 LMT를 ZT로 환산한다

070° 42.1′ E인 지점의 지방표준시는 경도 075°E이다.

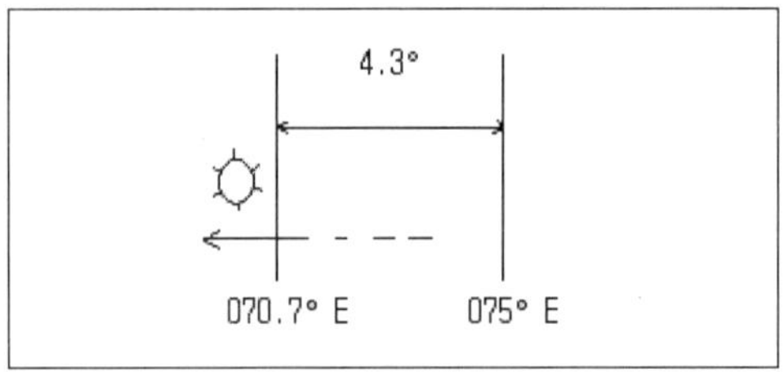

[그림 2-88]

→ $1^\circ = 4'$, $4.3^\circ = 17.2'$ 차이

$\therefore$ LMT + $d\lambda = 18^h\ 15^m\ 08^s + 17^m\ 12^s = \underline{18^h}\ \underline{32^m}\ \underline{20^s}$ (ZT)

〈표 2-7〉 천측력 주표 오른쪽 페이지 위도에 따른 일몰시간차 구하기

Lat.	Sunset	Twilight Civil	Twilight Naut.	Moonset 25	Moonset 26	Moonset 27	Moonset 28
°	h m	h m	h m	h m	h m	h m	h m
N 72	21 32	///	///	02 38	02 17	***	***
N 70	21 01	23 25	///	02 58	02 52	02 45	02 13
68	20 39	22 12	///	03 14	03 18	03 28	03 51
66	20 22	21 36	///	03 27	03 38	03 57	04 31
64	20 07	21 11	23 29	03 37	03 54	04 19	04 58
62	19 56	20 52	22 24	03 46	04 07	04 36	05 19
60	19 46	20 37	21 52	03 54	04 18	04 51	05 37
N 58	19 37	20 24	21 29	04 01	04 28	05 04	05 51
56	19 29	20 13	21 11	04 07	04 37	05 15	06 03
54	19 23	20 03	20 56	04 13	04 45	05 24	06 14
52	19 17	19 55	20 43	04 18	04 52	05 33	06 24
50	19 11	19 47	20 32	04 22	04 58	05 41	06 33
45	18 59	19 31	20 11	04 32	05 11	05 57	06 51
N 40	18 50	19 19	19 54	04 41	05 22	06 11	07 06
35	18 41	19 08	19 40	04 48	05 32	06 22	07 18
30	18 34	18 59	19 29	04 54	05 40	06 32	07 29
20	18 22	18 45	19 12	05 05	05 55	06 50	07 48
10	18 11	18 33	18 58	05 14	06 08	07 05	08 05
0	18 01	18 22	18 47	05 23	06 20	07 19	08 20
S 10	17 51	18 13	18 38	05 32	06 32	07 33	08 36
20	17 41	18 04	18 30	05 42	06 44	07 48	08 52
30	17 29	17 54	18 22	05 53	06 59	08 06	09 11
35	17 23	17 49	18 18	05 59	07 08	08 16	09 22
40	17 15	17 43	18 15	06 06	07 17	08 28	09 34
45	17 06	17 37	18 11	06 15	07 29	08 41	09 49
S 50	16 56	17 29	18 07	06 25	07 43	08 58	10 07
52	16 51	17 26	18 06	06 30	07 49	09 06	10 16
54	16 45	17 22	18 04	06 35	07 56	09 15	10 25
56	16 39	17 19	18 02	06 41	08 04	09 25	10 36
58	16 33	17 14	18 01	06 47	08 14	09 36	10 49
60	16 25	17 10	17 59	06 55	08 24	09 49	11 03

〈표 2-8〉 TABLE I-FOR LATITUDE

일·월출몰 개정표

TABLES FOR INTERPOLATING SUNRISE, MOONRISE, ETC.

TABLE I – FOR LATITUDE

Tabluar Interval			Difference between the times for consecutive latitudes															
10°	5°	2°	5^m	10^m	15^m	20^m	25^m	30^m	35^m	40^m	45^m	50^m	55^m	60^m	$1^h 05^m$	$1^h 10^m$	$1^h 15^m$	$1^h 20^m$
° ′	° ′	° ′	m	m	m	m	m	m	m	m	m	m	m	m	h m	h m	h m	h m
0 30	0 15	0 06	0	0	1	1	1	1	1	2	2	2	2	2	0 02	0 02	0 02	0 02
1 00	0 30	0 12	0	1	1	2	2	3	3	3	4	4	4	5	05	05	05	05
1 30	0 45	0 18	1	1	2	3	3	4	4	5	5	6	7	7	07	07	07	07
2 00	1 00	0 24	1	2	3	4	5	5	6	7	7	8	9	10	10	10	10	10
2 30	1 15	0 30	1	2	4	5	6	7	8	9	9	10	11	12	12	13	13	13
3 00	1 30	0 36	1	3	4	6	7	8	9	10	11	12	13	14	0 15	0 15	0 16	0 16
3 30	1 45	0 42	2	3	5	7	8	10	11	12	13	14	16	17	18	18	19	19
4 00	2 00	0 48	2	4	6	8	9	11	13	14	15	16	18	19	20	21	22	22
4 30	2 15	0 54	2	4	7	9	11	13	15	16	18	19	21	22	23	24	25	26
5 00	2 30	1 00	2	5	7	10	12	14	16	18	20	22	23	25	26	27	28	29
5 30	2 45	1 06	3	5	8	11	13	16	18	20	22	24	26	28	0 29	0 30	0 31	0 32
6 00	3 00	1 12	3	6	9	12	14	17	20	22	24	26	29	31	32	33	34	36
6 30	3 15	1 18	3	6	10	13	16	19	22	24	26	29	31	34	36	37	38	40
7 00	3 30	1 24	3	7	10	14	17	20	23	26	29	31	34	37	39	41	42	44
7 30	3 45	1 30	4	7	11	15	18	22	25	28	31	34	37	40	43	44	46	48
8 00	4 00	1 36	4	8	12	16	20	23	27	30	34	37	41	44	0 47	0 48	0 51	0 53
8 30	4 15	1 42	4	8	13	17	21	25	29	33	36	40	44	48	0 51	0 53	0 56	0 58
9 00	4 30	1 48	4	9	13	18	22	27	31	35	39	43	47	52	0 55	0 58	1 01	1 04
9 30	4 45	1 54	5	9	14	19	24	28	33	38	42	47	51	56	1 00	1 04	1 08	1 12
10 00	5 00	2 00	5	10	15	20	25	30	35	40	45	50	55	60	1 05	1 10	1 15	1 20

표 I은 일출, 월출, 박명시각의 위도에 대한 지방평균시(LMT)의 보정표이다. 왼편 종란의 인수는 진위도에 주표에 수록된 위도에의 차이며, 오른쪽 상란은 주표에서 진위도가 속한 연속된 두 위도간의 시간차이다. 참조할 위도는 진위도보다 낮은 위도를 기준으로 삼는다. 세간극(10°, 5°, 2°, 또는 5^m, 10^m, 15^m)으로 되어 있는 것은 주표에 수록된 위도 간극과 시간차의 크기이다. 이렇게 하여 얻은 개정치의 부호는 상·하 관계를 보아 양(+), 음(−)를 결정하여야 한다.

5. 박명시

2006년 11월 26일 18°27'.6N, 151°52'.0E인 지점에서의 새벽 박명시(ZT)를 구해 본다.

1) 항해박명(Naut. twilight)시

① 천측력 주표의 오른쪽 페이지의 위도에 따른 Naut. twilight 시간을 구한다.

위도 18° 27.6′ N는 N10° 와 N20°의 사이에 있으므로 이들의 값을 구해준다.

→ N10° 05^h 10^m, N20° 05^h 24^m

② 두 위도의 시간차 및 간격을 구한다.

→ 항해박명 시간차는 14^m이고 두 위도(N10°와 N20°)의 간격은 10°이다.

③ 천측력의 일 · 월출몰 개정표 “TABLE I - FOR LATITUDE”에서 개정치를 구한다.

- Tabular interval : 천측력 주표에서 두 위도 차이

- Difference ~ : 두 위도의 시간차

㉠ Tabular interval 10° 아래에서 원래의 위도(18° 27.6′ N) 보다 낮고 가장 가까운 위도 N10°와의 차이 만큼인 8° 27.6′ 인 곳을 찾는다.

→ 8° 00′과 8° 30′ 사이에 있다.

㉡ 'Difference ~'에서 항해박명 시간차인 14^m과 만나는 곳을 찾는다.

→ 10^m과 15^m 사이에 있다.

㉢ 비례식을 이용하여 ㉠, ㉡이 교차하는 값을 구한다.

: 8° 00′에서 시간차가 10^m일 때 8^m, 15^m일 때 12^m이다.

→ $5^m : 4^m = 4^m : x$

∴ $x = 3.2^m = 3^m\ 12^s$ 이므로 14^m일 때는 $8^m + 3^m\ 12^s = 11^m\ 12^s$

: 8° 30′에서 시간차가 10^m일 때 8^m, 15^m일 때 13^m이다.

→ $5^m : 5^m = 4^m : x$ ∴ $x = 4^m$이므로 14^m일 때는 $8^m + 4^m = 12^m$

: 8° 27.6′에서 시간차가 14^m일 때의 값은

→ $30' : 48^s = 27.6' : x$ ∴ $x = 44.2^s$이므로 $11^m\ 12^s + 44.2^s ≒ 11^m\ 56^s$이다.

㉣ 직접 비례식으로 계산한다. 즉, 위도 10°의 간격일 때, 시간차가 14^m일 때, 8.46°에 관한 값을 구하면

$10° : 14^m = 8.46° : x \rightarrow x = 11^m\ \underline{51}^s$

④ 개정치는 $11^m\ 56^s$이다. 그러므로 18°27.6′N의 위도보간 한 값은 N10° $05^h\ 10^m$+8°27.6′에 대한 개정치 $11^m\ 56^s = \underline{05}^h\ \underline{21}^m\ \underline{56}^s$이다. (N10°~N20°증가하므로 개정치를 "+" 해준다.)

⑤ 위에서 구한 LMT를 ZT으로 환산한다.

151° 52.0′ E인 지점의 지방표준시는 경도 150° E이다.

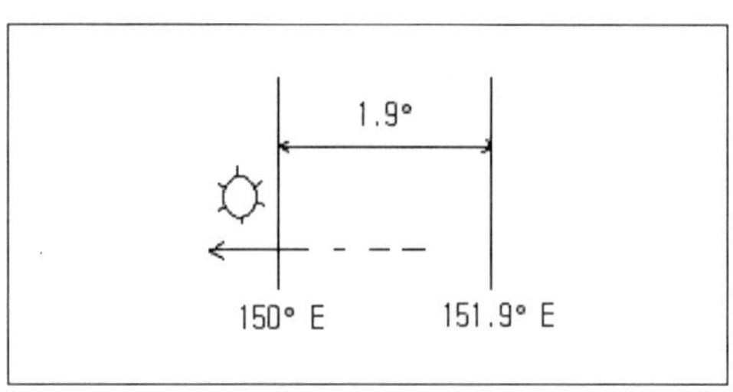

[그림 2-89]

→ $1^\circ = 4'$, $1.9^\circ = 7.6'$ 차이

∴ $LMT + d\lambda = 05^h\ 21^m\ 56^s - 7^m\ 36^s = \underline{05^h}\ \underline{14^m}\ \underline{20^s}$ (ZT)

〈표 2-9〉 천측력 주표의 오른쪽 페이지 위도에 따른 항해박명 시간

Lat.	Twilight Naut.	Twilight Civil	Sunrise	Moonrise 24	Moonrise 25	Moonrise 26	Moonrise 27
°	h m	h m	h m	h m	h m	h m	h m
N 72	07 34	09 22	***	21 18	22 59	24 39	00 39
N 70	07 19	08 50	11 40	21 33	23 06	24 40	00 40
68	07 08	08 27	10 03	21 45	23 12	24 41	00 41
66	06 58	08 08	09 25	21 54	23 17	24 42	00 42
64	06 50	07 53	08 59	22 02	23 21	24 42	00 42
62	06 43	07 41	08 39	22 09	23 25	24 43	00 43
60	06 36	07 30	08 22	22 15	23 28	24 43	00 43
N 58	06 30	07 20	08 08	22 20	23 31	24 44	00 44
56	06 25	07 12	07 56	22 25	23 34	24 44	00 44
54	06 20	07 05	07 46	22 29	23 36	24 44	00 44
52	06 16	06 58	07 37	22 33	23 38	24 45	00 45
50	06 11	06 52	07 28	22 36	23 40	24 45	00 45
45	06 02	06 38	07 11	22 43	23 44	24 45	00 45
N 40	05 53	06 27	06 56	22 50	23 47	24 46	00 46
35	05 46	06 17	06 44	22 55	23 50	24 46	00 46
30	05 38	06 08	06 33	22 59	23 53	24 47	00 47
20	05 24	05 51	06 15	23 07	23 57	24 47	00 47
10	05 10	05 36	05 59	23 14	24 01	00 01	00 48
0	04 56	05 21	05 43	23 21	24 05	00 05	00 49
S 10	04 39	05 05	05 28	23 27	24 08	00 08	00 49
20	04 19	04 47	05 11	23 34	24 12	00 12	00 50
30	03 53	04 25	04 52	23 42	24 16	00 16	00 51
35	03 37	04 12	04 41	23 47	24 19	00 19	00 51
40	03 17	03 56	04 28	23 52	24 22	00 22	00 52
45	02 51	03 37	04 12	23 58	24 25	00 25	00 52
S 50	02 15	03 12	03 53	24 05	00 05	00 29	00 53
52	01 56	03 00	03 44	24 08	00 08	00 31	00 53
54	01 31	02 45	03 34	24 12	00 12	00 33	00 54
56	00 53	02 28	03 22	24 16	00 16	00 35	00 54
58	///	02 08	03 09	00 01	00 20	00 38	00 54
60	///	01 41	02 53	00 08	00 25	00 40	00 55

2) 상용박명(Civil twilight)시

① 천측력 주표의 오른쪽 페이지의 위도에 따른 Civil twilight 시간을 구한다.
위도 18° 27.6′ N는 N10°와 N20°의 사이에 있으므로 이들의 값을 구해준다.

→ N10° $05^h\ 36^m$, N20° $05^h\ 51^m$

② 두 위도의 시간차 및 간격을 구한다.

→ 상용박명 시간차는 15^m이고 두 위도(N10°와 N20°)의 간격은 10°이다.

③ 천측력의 일·월출몰 개정표를 대신 비례식으로 계산해도 된다.

즉, 위도 10°의 간격일 때, 시간차가 15^m일 때, 8° 27′.6에 관한 값을 구하면

$10° : 15^m = 8°.5 : x \rightarrow x = 12.75^m = 12^m\ 45^s$

④ 위에서 구한 LMT을 ZT으로 환산한다.

151° 52.0′ E인 지점의 지방표준시는 경도 150° E이다.

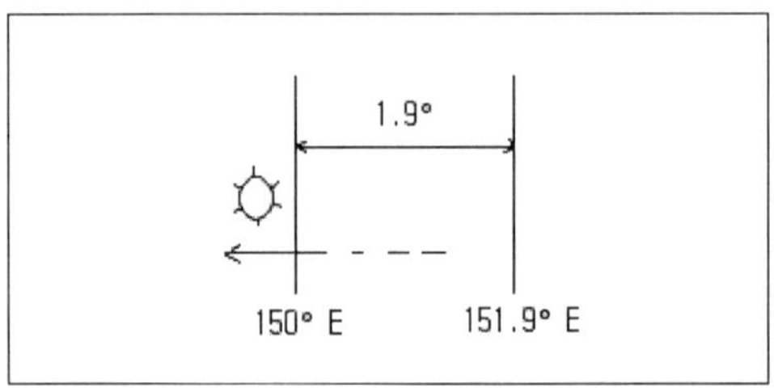

[그림 2-90]

→ 1°=4′, 1.9°=7.6′ 차이

$\therefore$ LMT+dλ = $05^h\ 48^m\ 55^s - 7^m\ 36^s = 05^h\ 41^m\ 19^s$ (ZT)

2.10 태양의 고도 관측을 통한 위치선 작도

2.10.1 태양의 고도 측정법

① 육분의에 망원경(telescope)을 끼우고 적당한 Shade Glass를 내린다.

② 오른손에 육분의를 수직되게 가지고 천체직하의 수평선을 향한다.

③ 수평경을 통하여 투시하면서 왼손으로 Release를 잡고 Index Arm을 서서히 앞으로 움직인다.

④ 그러면 천체의 영상이 시계안에 들어오면서 점차 하강한다.

⑤ 기면을 좌우로 흔들면서 태양이나 달의 하변 또는 상변이 수평선에 접하도록 Tangent Screw를 돌린다.

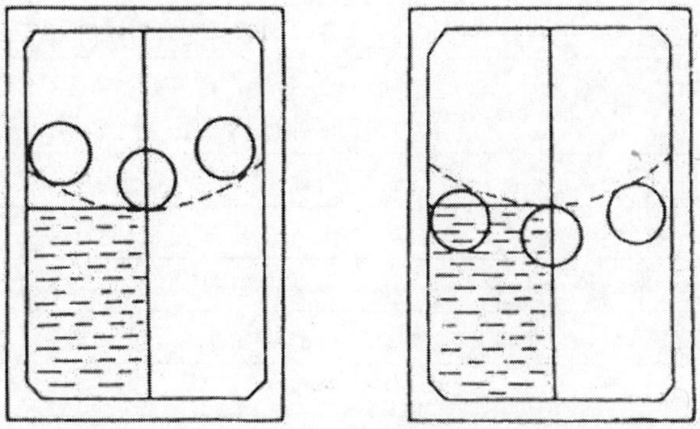

(a) 하변고도를 관측한 것　　(b) 상변고도를 관측한 것

[그림 2-91] 하변고도 및 상변고도

⑥ 정확하게 수평선에 접촉하였을 때 시각과 고도를 기록해둔다(이 때의 고도가 관측 시각에 대한 육분의 고도이다).

2.10.2 고도의 환산

1. 고도에 관한 용어

① 육분의 고도(Sextant Altitude ; Hs) : 천체를 측정하여 얻은 고도이다.

② 관측고도(Observered Altitude ; Ho) : SEXTANT 고도에 모든 개정을 실시하여 얻어진 고도이다.

③ 계산고도(Computed Altitude ; Hc) : 항해삼각형의 계산으로 풀어서 얻은 고도이다.

2. 육분의 고도를 관측고도로 환산

천측계산표에서 얻은 계산고도는 진수평상으로부터 천체의 중심까지를 측정한 고도이고, 우리가 측정한 육분의 고도는 시수평상으로부터 천체의 하변 혹은 상변까지를 측정한 고도이다. 즉, 육분의 고도(Hs)를 계산고도(Hc)와 동일한 기준으로 개정하여 관측고도(Ho)를 구한다.

〈표 2-10〉

	+	★	−	내 용
I.C.				1)의 내용. 육분의 기차를 확인하여 기록
Dip				2)의 내용. 천측력에서 안고차 개정값을 확인
◎				3), 4), 5), 6), 7)의 내용. 천측력에는 고도에 따른 관련값을 모두 계산하여 제공
A4 Table				기압 및 기온 개정치
sum				+, − 값을 따로 계산
corr.				수정값 표시
Hs				육분의 고도
Ho				관측고도

① 육분의 기차(Index Correction, I.C.) : 육분의 자체가 가지는 오차이다.

② 안고차(Dip) : 눈의 높이가 선박의 높이에 따라 변화하여 진수평과 시수평이 이루는 각이다. 안고가 높아지면 시수평이 낮아져서 결과적으로 고도를 높게 측정하게 되므로 개정값은 항상 (−)값을 가진다.

③ 시반경(Semidiameter, S) : 천체의 반경으로 하변고도를 관측한 경우 반경을 (+)하고 상변고도를 관측한 경우는 반경을 (−)한다.

④ 기차(Refrection, R) : 천체가 대기권을 통과하면서 굴절하는 값으로 이로 인해 실제보다 천체의 고도가 높게 측정되므로 항상 (−)개정을 한다.

⑤ 시차(Parallax, P) : 천체를 지구표면상의 관측자에서 본 고도와 지구 중심에서의 고도의 차를 말하며, 진수평은 지구 중심에 위치하므로 항상 지표면에 있는 관측자는 시차만큼 천체의 고도를 측정하므로 항상 (+)개정을 한다.

⑥ 광참(Irradiation, I) : 밝은 물체가 어두운 배경을 갖고 있을 때에 실제보다 약간 크게 보이는 현상으로 태양의 경우에만 해당된다. 그러나 태양의 하변고도 관측시는 태양과 수평선의 광참치가 서로 상쇄되어 개정 필요가 없다.

⑦ 위상(Phase, P) : 태양과의 상대위치에 따라 관측자가 보는 천체의 모양이 달라지는 현상을 말하며, 태양과 달의 경우 상변이나 하변의 고도를 관측하므로 개정의 필요가 없다. 그러나 금성, 화성의 경우에는 개정이 필요하다.

⑧ 겉보기 고도가 50 이하이고 대기가 평균과 다를 때는 A4 페이지의 기온, 기압개정치(TB)를 구하거나 항해표 제23표, 제24표를 이용하여 기온개정치(T), 기압개정치(B)

를 구한다.

⑨ 위에서 구한 개정치를 육분의 고도에 가감하여 관측고도(Ho)로 한다.

3. 천측계산의 방법

〈표 2-11〉

계산방법			내 용
GMT	GHA	Dec.(적위)	고도 관측시간을 GMT로 환산
h			시각에 대한 GHA, d값을 천측력에서 구함
m s			분, 초에 대한 GHA, d값을 천측력에서 구함
GHA		d	시각, 분, 초에 대한 GHA, d값 합산
aλ			가정경도. LHA값이 정수가 되도록 조정 (동경이면 +, 서경이면 −)
LHA			천체의 지방시각
t			자오선각, 관측자와 천체의 사이각을 가까운 쪽으로 측정(E or W)
aL			

① 육분의(sextant)를 통해 태양의 하변고도를 관측(육분의 고도, Hs)하고 초시계를 이용하여 관측시점의 시각을 기록한다.

② 관측시(LMT)를 GMT로 고쳐준다.

③ 천측력 주표에서 해당 시각(GMT)에 대한 GHA를 구하고 증분 및 개정표에서 분과 초에 대한 GHA를 구한 후 두 값을 더해준다.

④ 천측력 주표에서 해당 시각(GMT)에 대한 d(적위)를 구하고 증분 및 개정표에서 분과 초에 대한 d(적위)를 구한 후 두 값을 더해준다.

⑤ ③에서 구한 GHA에 가정경도(aλ)를 가감하여(동경일 때는 +, 서경일 때는 −) 지방시각(LHA)을 구한다. 이때 계산이 편리하도록 지방시각(LHA)이 정수로 되도록 가정경도의 분 단위를 조정한다.

⑥ 가정위도를 정한다. 이때 계산이 편리하도록 가정위치에 가까운 위도를 정수로 선택한다.

⑦ LHA, 가정위도, dec.을 이용하여 천측계산표(PUB No.229)에서 Hc와 Zn을 구한다.

㉠ dec.과 가정위도의 부호가 같으면 Same name을, 다르면 Contrary name을 사용한다.

㉡ 적위(dec.)값에 대한 보간을 한다.

㉢ 방위각(Z)은 dec.를 이용하여 Z값을 구한다.

방위(Zn) = 동명극(N or S) / 천측계산표의 방위각(Z) / 자오선각(E or W)

4. 천측위치선의 작도

태양을 관측해서 얻은 관측고도(Ho)와 그 시간에 추측위치(aL, aλ)에서 관측할 수 있는 계산고도(Hc)를 비교해 보면 그 차이만큼 관측자가 추측위치로부터 태양쪽으로 혹은 그 반대쪽으로 떨어져 있음을 확인할 수 있다.

[그림 2-92]에서와 같이 관측고도가 계산고도보다 크다면 관측자는 태양 가까이 위치할 것이므로 고도의 차이만큼 태양쪽으로 위치선을 작도하여야 할 것이다. 만일 관측고도가 계산고도보다 작다면 태양을 바라보는 반대쪽으로 이동해서 위치선을 표시한다. 이 위치선은 태양을 중심으로 하고 여고도를 반지름으로 하는 커다란 원이므로, 곡선으로 표시되어야 하지만 그 원이 너무 크기 때문에 해도상에는 직선으로 표시해도 무방하다.

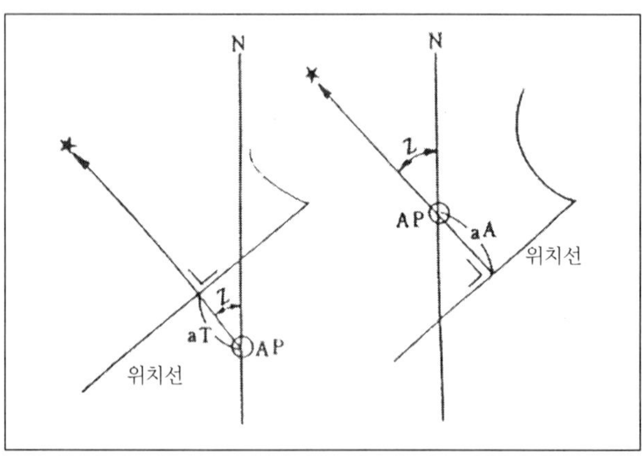

[그림 2-92] 천측위치선의 작도

① 육분의 고도(Hs)에 DIP, App. Alt, Sextant 오차, 온도기압보정(50° 이하로 관측시 -A4 Table)의 개정을 통해 Ho를 구한다.

② Hc와 Ho의 차 (a)를 구한다.

Hc 〉 Ho이면 A, Hc 〈 Ho이면 T부호를 취해준다(HoMoTo).

③ 가정위도, 가정경도의 위치에서 방위선을 기입하고 a 값 만큼 떨어진 곳에 수선을 긋는다. 부호가 T이면 방위쪽으로, A이면 반대쪽으로 이동한다.

[예제] 2013년 5월 22일 D.R. 13° 10'N, 50° 10'E에서 09시 05분 55초에 태양의 고도를 52° 35'.3로 관측했다. 이때의 태양의 방위각 및 계산고도를 구하고, 태양의 위치선을 작도하시오. (ZD −3 / Index Correction 값 : −0.3 / 안고 26m)

〈표 2-12〉

	계산방법	
GMT	GHA	Dec.(적위)
06h	270° 50′.8	N 20°21′.9
05m 55s	1° 28′.8	0′.0
GHA	272° 19′.6	d N 20°21'.9
aλ	49° 40′.4	
LHA	322° 00′.0	
t	38°(E)	
aL	13 N	

〈표 2-13〉

	+	★	−
I.C.			0′.3
Dip			9′.0
◎	15′.2		
A4 Table	−		−
sum	15′.2		9′.3
corr.	5′.9		
Hs	52° 35′.3		
Ho	52° 41′.2		

1) 천측계산표에서 dec. 20° → Hc 52° 58′.9, Z 73°.9

dec. 21° → Hc 52° 53′.1, Z 72°.3

구하는 dec.(적위)값은 20° 21′.9에 대한 값이므로 비례식을 세우면

60′ : −5′.8 : 21′.9 : x → x = −2′.1

52° 58′.9 − 2′.1 = 52° 56′.8(계산고도(Hc))

를 얻는다. 또한, 구하는 Z(방위각)값은 20° 21′. 9에 대한 값이므로 비례식을 세우면

60′ : −1′.6 : 21′.9 : x → x = −0′.6

73°.9 − 0′.6 = 73°.3(방위각, Z)

를 얻는다.

2) 계산고도(Hc)와 관측고도(Ho)의 차이는 15′.6이다.

3) 방위(Zn) = 동명극(N or S) / 천측계산표의 방위각(Z) / 자오선각(E or W)

73°.7 = N 73°.7 E

4) 가정위도, 가정경도의 위치에서 천체의 방위선을 긋고 천체를 바라보는 반대쪽으로 (A, away) 15′.6 떨어진 곳에 방위선에 대하여 수선을 그으면 위치선이 작도된다.

〈표 2-14〉 2006. 5/22～5/24(WED, THU, FRI)

UT	SUN		MOON				
	GHA	Dec	GHA	v	Dec	d	HP
d h	° ′	° ′	° ′	′	° ′	′	′
22 00	180 50.7	N20 22.3	44 31.2	9.6	S09 18.4	10.8	58.9
01	195 50.7	N20 22.8	58 59.9	9.5	S09 29.2	10.8	59.0
02	210 50.6	N20 23.3	73 28.4	9.5	S09 39.9	10.7	59.0
03	225 50.6	N20 23.8	87 56.9	9.4	S09 50.7	10.7	59.0
04	240 50.6	N20 24.3	102 25.3	9.3	S10 1.3	10.6	59.1
05	255 50.5	N20 24.8	116 53.6	9.2	S10 12.0	10.6	59.1
WEDNESDAY 06	270 50.5	N20 25.2	131 21.8	9.1	S10 22.6	10.5	59.2
07	285 50.4	N20 25.7	145 49.9	9.0	S10 33.1	10.5	59.2
08	300 50.4	N20 26.2	160 17.9	9.0	S10 43.6	10.5	59.2
09	315 50.3	N20 26.7	174 45.9	8.9	S10 54.1	10.4	59.3
10	330 50.3	N20 27.2	189 13.8	8.8	S11 4.5	10.3	59.3
11	345 50.2	N20 27.7	203 41.5	8.7	S11 14.8	10.3	59.3
12	0 50.2	N20 28.2	218 9.2	8.6	S11 25.1	10.2	59.4
13	15 50.1	N20 28.7	232 36.8	8.5	S11 35.4	10.2	59.4
14	30 50.1	N20 29.1	247 4.4	8.4	S11 45.5	10.1	59.4
15	45 50.0	N20 29.6	261 31.8	8.3	S11 55.7	10.1	59.5
16	60 50.0	N20 30.1	275 59.1	8.3	S12 5.7	10.0	59.5
17	75 49.9	N20 30.6	290 26.4	8.2	S12 15.7	9.9	59.6
18	90 49.9	N20 31.1	304 53.5	8.1	S12 25.7	9.9	59.6
19	105 49.8	N20 31.6	319 20.6	8.0	S12 35.6	9.8	59.6
20	120 49.8	N20 32.0	333 47.6	7.9	S12 45.4	9.7	59.7
21	135 49.7	N20 32.5	348 14.5	7.8	S12 55.1	9.7	59.7
22	150 49.7	N20 33.0	2 41.3	7.7	S13 4.8	9.6	59.7
23	165 49.6	N20 33.5	17 8.0	7.6	S13 14.4	9.5	59.8
23 00	180 49.6	N20 34.0	31 34.6	7.5	S13 23.9	9.5	59.8
01	195 49.5	N20 34.4	46 1.2	7.4	S13 33.4	9.4	59.8
02	210 49.5	N20 34.9	60 27.6	7.4	S13 42.8	9.3	59.9
03	225 49.4	N20 35.4	74 54.0	7.3	S13 52.1	9.2	59.9
04	240 49.4	N20 35.9	89 20.2	7.2	S14 1.3	9.2	59.9
05	255 49.3	N20 36.3	103 46.4	7.1	S14 10.5	9.1	60.0
THURSDAY 06	270 49.3	N20 36.8	118 12.5	7.0	S14 19.6	9.0	60.0
07	285 49.2	N20 37.3	132 38.5	6.9	S14 28.6	8.9	60.0
08	300 49.2	N20 37.8	147 4.4	6.8	S14 37.5	8.8	60.1
09	315 49.1	N20 38.2	161 30.2	6.7	S14 46.3	8.7	60.1
10	330 49.1	N20 38.7	175 55.9	6.6	S14 55.1	8.7	60.1
11	345 49.0	N20 39.2	190 21.6	6.5	S15 3.7	8.6	60.2
12	0 49.0	N20 39.7	204 47.1	6.5	S15 12.3	8.5	60.2
13	15 48.9	N20 40.1	219 12.6	6.4	S15 20.7	8.4	60.2
14	30 48.9	N20 40.6	233 37.9	6.3	S15 29.1	8.3	60.2
15	45 48.8	N20 41.1	248 3.2	6.2	S15 37.4	8.2	60.3
16	60 48.7	N20 41.5	262 28.4	6.1	S15 45.6	8.1	60.3
17	75 48.7	N20 42.0	276 53.5	6.0	S15 53.7	8.0	60.3
18	90 48.6	N20 42.5	291 18.5	5.9	S16 1.6	7.9	60.4
19	105 48.6	N20 42.9	305 43.5	5.8	S16 9.5	7.8	60.4
20	120 48.5	N20 43.4	320 8.3	5.8	S16 17.3	7.7	60.4
21	135 48.5	N20 43.9	334 33.1	5.7	S16 25.0	7.6	60.4
22	150 48.4	N20 44.3	348 57.8	5.6	S16 32.6	7.5	60.5
23	165 48.4	N20 44.8	3 22.4	5.5	S16 40.0	7.4	60.5
24 00	180 48.3	N20 45.3	17 46.9	5.4	S16 47.4	7.3	60.5
01	195 48.2	N20 45.7	32 11.3	5.3	S16 54.7	7.1	60.5
02	210 48.2	N20 46.2	46 35.7	5.3	S17 1.8	7.0	60.6
03	225 48.1	N20 46.7	60 59.9	5.2	S17 8.8	6.9	60.6
04	240 48.1	N20 47.1	75 24.1	5.1	S17 15.7	6.8	60.6
05	255 48.0	N20 47.6	89 48.2	5.0	S17 22.5	6.7	60.6
FRIDAY 06	270 48.0	N20 48.0	104 12.2	5.0	S17 29.2	6.6	60.7
07	285 47.9	N20 48.5	118 36.2	4.9	S17 35.8	6.4	60.7
08	300 47.9	N20 49.0	133 0.1	4.8	S17 42.2	6.3	60.7
09	315 47.8	N20 49.4	147 23.9	4.7	S17 48.6	6.2	60.7
10	330 47.7	N20 49.9	161 47.6	4.7	S17 54.8	6.1	60.8
11	345 47.7	N20 50.3	176 11.3	4.6	S18 0.8	6.0	60.8
12	0 47.6	N20 50.8	190 34.9	4.5	S18 6.8	5.8	60.8
13	15 47.6	N20 51.2	204 58.4	4.4	S18 12.6	5.7	60.8
14	30 47.5	N20 51.7	219 21.8	4.4	S18 18.3	5.6	60.8
15	45 47.4	N20 52.2	233 45.2	4.3	S18 23.9	5.4	60.9
16	60 47.4	N20 52.6	248 8.5	4.2	S18 29.3	5.3	60.9
17	75 47.3	N20 53.1	262 31.8	4.2	S18 34.7	5.2	60.9
18	90 47.3	N20 53.5	276 54.9	4.1	S18 39.8	5.0	60.9
19	105 47.2	N20 54.0	291 18.1	4.1	S18 44.9	4.9	60.9
20	120 47.1	N20 54.4	305 41.1	4.0	S18 49.8	4.8	60.9
21	135 47.1	N20 54.9	320 4.1	3.9	S18 54.6	4.6	61.0
22	150 47.0	N20 55.3	334 27.1	3.9	S18 59.2	4.5	61.0
23	165 47.0	N20 55.8	348 50.0	3.8	S19 3.7	4.4	61.0
	SD 15.8	d 0.5	SD	16.2	16.4		16.6

〈표 2-15〉 증분 및 개정표

증 분 및 개 정 표

INCREMENTS AND CORRECTIONS

4^m

4 m	SUN PLANETS	ARIES	MOON	v or d	Corr^n	v or d	Corr^n	v or d	Corr^n
s	° ′	° ′	° ′	′	′	′	′	′	′
00	01 00.0	01 00.2	00 57.3	0.0	0.0	6.0	0.5	12.0	0.9
01	01 00.3	01 00.4	00 57.5	0.1	0.0	6.1	0.5	12.1	0.9
02	01 00.5	01 00.7	00 57.7	0.2	0.0	6.2	0.5	12.2	0.9
03	01 00.8	01 00.9	00 58.0	0.3	0.0	6.3	0.5	12.3	0.9
04	01 01.0	01 01.2	00 58.2	0.4	0.0	6.4	0.5	12.4	0.9
05	01 01.3	01 01.4	00 58.5	0.5	0.0	6.5	0.5	12.5	0.9
06	01 01.5	01 01.7	00 58.7	0.6	0.0	6.6	0.5	12.6	0.9
07	01 01.8	01 01.9	00 58.9	0.7	0.1	6.7	0.5	12.7	1.0
08	01 02.0	01 02.2	00 59.2	0.8	0.1	6.8	0.5	12.8	1.0
09	01 02.3	01 02.4	00 59.4	0.9	0.1	6.9	0.5	12.9	1.0
10	01 02.5	01 02.7	00 59.7	1.0	0.1	7.0	0.5	13.0	1.0
11	01 02.8	01 02.9	00 59.9	1.1	0.1	7.1	0.5	13.1	1.0
12	01 03.0	01 03.2	01 00.1	1.2	0.1	7.2	0.5	13.2	1.0
13	01 03.3	01 03.4	01 00.4	1.3	0.1	7.3	0.5	13.3	1.0
14	01 03.5	01 03.7	01 00.6	1.4	0.1	7.4	0.6	13.4	1.0
15	01 03.8	01 03.9	01 00.8	1.5	0.1	7.5	0.6	13.5	1.0
16	01 04.0	01 04.2	01 01.1	1.6	0.1	7.6	0.6	13.6	1.0
17	01 04.3	01 04.4	01 01.3	1.7	0.1	7.7	0.6	13.7	1.0
18	01 04.5	01 04.7	01 01.6	1.8	0.1	7.8	0.6	13.8	1.0
19	01 04.8	01 04.9	01 01.8	1.9	0.1	7.9	0.6	13.9	1.0
20	01 05.0	01 05.2	01 02.0	2.0	0.2	8.0	0.6	14.0	1.1
21	01 05.3	01 05.4	01 02.3	2.1	0.2	8.1	0.6	14.1	1.1
22	01 05.5	01 05.7	01 02.5	2.2	0.2	8.2	0.6	14.2	1.1
23	01 05.8	01 05.9	01 02.8	2.3	0.2	8.3	0.6	14.3	1.1
24	01 06.0	01 06.2	01 03.0	2.4	0.2	8.4	0.6	14.4	1.1
25	01 06.3	01 06.4	01 03.2	2.5	0.2	8.5	0.6	14.5	1.1
26	01 06.5	01 06.7	01 03.5	2.6	0.2	8.6	0.6	14.6	1.1
27	01 06.8	01 06.9	01 03.7	2.7	0.2	8.7	0.7	14.7	1.1
28	01 07.0	01 07.2	01 03.9	2.8	0.2	8.8	0.7	14.8	1.1
29	01 07.3	01 07.4	01 04.2	2.9	0.2	8.9	0.7	14.9	1.1
30	01 07.5	01 07.7	01 04.4	3.0	0.2	9.0	0.7	15.0	1.1
31	01 07.8	01 07.9	01 04.7	3.1	0.2	9.1	0.7	15.1	1.1
32	01 08.0	01 08.2	01 04.9	3.2	0.2	9.2	0.7	15.2	1.1
33	01 08.3	01 08.4	01 05.1	3.3	0.2	9.3	0.7	15.3	1.1
34	01 08.5	01 08.7	01 05.4	3.4	0.3	9.4	0.7	15.4	1.2
35	01 08.8	01 08.9	01 05.6	3.5	0.3	9.5	0.7	15.5	1.2
36	01 09.0	01 09.2	01 05.9	3.6	0.3	9.6	0.7	15.6	1.2
37	01 09.3	01 09.4	01 06.1	3.7	0.3	9.7	0.7	15.7	1.2
38	01 09.5	01 09.7	01 06.3	3.8	0.3	9.8	0.7	15.8	1.2
39	01 09.8	01 09.9	01 06.6	3.9	0.3	9.9	0.7	15.9	1.2
40	01 10.0	01 10.2	01 06.8	4.0	0.3	10.0	0.8	16.0	1.2
41	01 10.3	01 10.4	01 07.0	4.1	0.3	10.1	0.8	16.1	1.2
42	01 10.5	01 10.7	01 07.3	4.2	0.3	10.2	0.8	16.2	1.2
43	01 10.8	01 10.9	01 07.5	4.3	0.3	10.3	0.8	16.3	1.2
44	01 11.0	01 11.2	01 07.8	4.4	0.3	10.4	0.8	16.4	1.2
45	01 11.3	01 11.4	01 08.0	4.5	0.3	10.5	0.8	16.5	1.2
46	01 11.5	01 11.7	01 08.2	4.6	0.3	10.6	0.8	16.6	1.2
47	01 11.8	01 11.9	01 08.5	4.7	0.4	10.7	0.8	16.7	1.3
48	01 12.0	01 12.2	01 08.7	4.8	0.4	10.8	0.8	16.8	1.3
49	01 12.3	01 12.4	01 09.0	4.9	0.4	10.9	0.8	16.9	1.3
50	01 12.5	01 12.7	01 09.2	5.0	0.4	11.0	0.8	17.0	1.3
51	01 12.8	01 12.9	01 09.4	5.1	0.4	11.1	0.8	17.1	1.3
52	01 13.0	01 13.2	01 09.7	5.2	0.4	11.2	0.8	17.2	1.3
53	01 13.3	01 13.5	01 09.9	5.3	0.4	11.3	0.8	17.3	1.3
54	01 13.5	01 13.7	01 10.2	5.4	0.4	11.4	0.9	17.4	1.3
55	01 13.8	01 14.0	01 10.4	5.5	0.4	11.5	0.9	17.5	1.3
56	01 14.0	01 14.2	01 10.6	5.6	0.4	11.6	0.9	17.6	1.3
57	01 14.3	01 14.5	01 10.9	5.7	0.4	11.7	0.9	17.7	1.3
58	01 14.5	01 14.7	01 11.1	5.8	0.4	11.8	0.9	17.8	1.3
59	01 14.8	01 15.0	01 11.3	5.9	0.4	11.9	0.9	17.9	1.3
60	01 15.0	01 15.2	01 11.6	6.0	0.5	12.0	0.9	18.0	1.4

5^m

5 m	SUN PLANETS	ARIES	MOON	v or d	Corr^n	v or d	Corr^n	v or d	Corr^n
s	° ′	° ′	° ′	′	′	′	′	′	′
00	01 15.0	01 15.2	01 11.6	0.0	0.0	6.0	0.6	12.0	1.1
01	01 15.3	01 15.5	01 11.8	0.1	0.0	6.1	0.6	12.1	1.1
02	01 15.5	01 15.7	01 12.1	0.2	0.0	6.2	0.6	12.2	1.1
03	01 15.8	01 16.0	01 12.3	0.3	0.0	6.3	0.6	12.3	1.1
04	01 16.0	01 16.2	01 12.5	0.4	0.0	6.4	0.6	12.4	1.1
05	01 16.3	01 16.5	01 12.8	0.5	0.0	6.5	0.6	12.5	1.1
06	01 16.5	01 16.7	01 13.0	[illegible]	[illegible]	6.6	0.6	12.6	1.2
07	01 16.8	01 17.0	01 13.3	0.7	0.1	6.7	0.6	12.7	1.2
08	01 17.0	01 17.2	01 13.5	0.8	0.1	6.8	0.6	12.8	1.2
09	01 17.3	01 17.5	01 13.7	0.9	0.1	6.9	0.6	12.9	1.2
10	01 17.5	01 17.7	01 14.0	1.0	0.1	7.0	0.6	13.0	1.2
11	01 17.8	01 18.0	01 14.2	1.1	0.1	7.1	0.7	13.1	1.2
12	01 18.0	01 18.2	01 14.4	1.2	0.1	7.2	0.7	13.2	1.2
13	01 18.3	01 18.5	01 14.7	1.3	0.1	7.3	0.7	13.3	1.2
14	01 18.5	01 18.7	01 14.9	1.4	0.1	7.4	0.7	13.4	1.2
15	01 18.8	01 19.0	01 15.2	1.5	0.1	7.5	0.7	13.5	1.2
16	01 19.0	01 19.2	01 15.4	1.6	0.1	7.6	0.7	13.6	1.2
17	01 19.3	01 19.5	01 15.6	1.7	0.2	7.7	0.7	13.7	1.3
18	01 19.5	01 19.7	01 15.9	1.8	0.2	7.8	0.7	13.8	1.3
19	01 19.8	01 20.0	01 16.1	1.9	0.2	7.9	0.7	13.9	1.3
20	01 20.0	01 20.2	01 16.4	2.0	0.2	8.0	0.7	14.0	1.3
21	01 20.3	01 20.5	01 16.6	2.1	0.2	8.1	0.7	14.1	1.3
22	01 20.5	01 20.7	01 16.8	2.2	0.2	8.2	0.8	14.2	1.3
23	01 20.8	01 21.0	01 17.1	2.3	0.2	8.3	0.8	14.3	1.3
24	01 21.0	01 21.2	01 17.3	2.4	0.2	8.4	0.8	14.4	1.3
25	01 21.3	01 21.5	01 17.5	2.5	0.2	8.5	0.8	14.5	1.3
26	01 21.5	01 21.7	01 17.8	2.6	0.2	8.6	0.8	14.6	1.3
27	01 21.8	01 22.0	01 18.0	2.7	0.2	8.7	0.8	14.7	1.3
28	01 22.0	01 22.2	01 18.3	2.8	0.3	8.8	0.8	14.8	1.4
29	01 22.3	01 22.5	01 18.5	2.9	0.3	8.9	0.8	14.9	1.4
30	01 22.5	01 22.7	01 18.7	3.0	0.3	9.0	0.8	15.0	1.4
31	01 22.8	01 23.0	01 19.0	3.1	0.3	9.1	0.8	15.1	1.4
32	01 23.0	01 23.2	01 19.2	3.2	0.3	9.2	0.8	15.2	1.4
33	01 23.3	01 23.5	01 19.5	3.3	0.3	9.3	0.9	15.3	1.4
34	01 23.5	01 23.7	01 19.7	3.4	0.3	9.4	0.9	15.4	1.4
35	01 23.8	01 24.0	01 19.9	3.5	0.3	9.5	0.9	15.5	1.4
36	01 24.0	01 24.2	01 20.2	3.6	0.3	9.6	0.9	15.6	1.4
37	01 24.3	01 24.5	01 20.4	3.7	0.3	9.7	0.9	15.7	1.4
38	01 24.5	01 24.7	01 20.7	3.8	0.3	9.8	0.9	15.8	1.4
39	01 24.8	01 25.0	01 20.9	3.9	0.4	9.9	0.9	15.9	1.5
40	01 25.0	01 25.2	01 21.1	4.0	0.4	10.0	0.9	16.0	1.5
41	01 25.3	01 25.5	01 21.4	4.1	0.4	10.1	0.9	16.1	1.5
42	01 25.5	01 25.7	01 21.6	4.2	0.4	10.2	0.9	16.2	1.5
43	01 25.8	01 26.0	01 21.8	4.3	0.4	10.3	0.9	16.3	1.5
44	01 26.0	01 26.2	01 22.1	4.4	0.4	10.4	1.0	16.4	1.5
45	01 26.3	01 26.5	01 22.3	4.5	0.4	10.5	1.0	16.5	1.5
46	01 26.5	01 26.7	01 22.6	4.6	0.4	10.6	1.0	16.6	1.5
47	01 26.8	01 27.0	01 22.8	4.7	0.4	10.7	1.0	16.7	1.5
48	01 27.0	01 27.2	01 23.0	4.8	0.4	10.8	1.0	16.8	1.5
49	01 27.3	01 27.5	01 23.3	4.9	0.4	10.9	1.0	16.9	1.5
50	01 27.5	01 27.7	01 23.5	5.0	0.5	11.0	1.0	17.0	1.6
51	01 27.8	01 28.0	01 23.8	5.1	0.5	11.1	1.0	17.1	1.6
52	01 28.0	01 28.2	01 24.0	5.2	0.5	11.2	1.0	17.2	1.6
53	01 28.3	01 28.5	01 24.2	5.3	0.5	11.3	1.0	17.3	1.6
54	01 28.5	01 28.7	01 24.5	5.4	0.5	11.4	1.0	17.4	1.6
55	01 28.8	01 29.0	01 24.7	5.5	0.5	11.5	1.1	17.5	1.6
56	01 29.0	01 29.2	01 24.9	5.6	0.5	11.6	1.1	17.6	1.6
57	01 29.3	01 29.5	01 25.2	5.7	0.5	11.7	1.1	17.7	1.6
58	01 29.5	01 29.7	01 25.4	5.8	0.5	11.8	1.1	17.8	1.6
59	01 29.8	01 30.0	01 25.7	5.9	0.5	11.9	1.1	17.9	1.6
60	01 30.0	01 30.2	01 25.9	6.0	0.6	12.0	1.1	18.0	1.7

〈표 2-16〉 229 Table

38°, 322° L.H.A. LATITUDE **SAME** NAME AS DECLINATION N. Lat. {L.H.A. greater than 180° ....Zn=Z; L.H.A. less than 180°............Zn=360°−Z}

Dec.	8°			9°			10°			11°			12°			13°			14°			15°			Dec.
	Hc	d	Z	Hc	d	Z	Hc	d	Z	Hc	d	Z	Hc	d	Z	Hc	d	Z	Hc	d	Z	Hc	d	Z	
°	° ′	′	°	° ′	′	°	° ′	′	°	° ′	′	°	° ′	′	°	° ′	′	°	° ′	′	°	° ′	′	°	°
0	51 17.5	+12.7	100.1	51 06.4	+14.3	101.3	50 54.0	+15.9	102.5	50 40.3	+17.5	103.7	50 25.5	+19.0	104.9	50 09.5	+20.5	106.1	49 52.3	+22.0	107.2	49 34.0	+23.4	108.3	0
1	51 30.2	+11.5	98.5	51 20.7	+13.1	99.8	51 09.9	+14.7	101.0	50 57.8	+16.3	102.2	50 44.5	+17.9	103.4	50 30.0	+19.4	104.6	50 14.3	+20.9	105.8	49 57.4	+22.4	106.9	1
2	51 41.7	+10.2	97.0	51 33.8	+11.8	98.2	51 24.6	+13.4	99.4	51 14.1	+15.1	100.7	51 02.4	+16.6	101.9	50 49.4	+18.2	103.1	50 35.2	+19.7	104.3	50 19.8	+21.2	105.5	2
3	51 51.9	+8.9	95.4	51 45.6	+10.6	96.6	51 38.0	+12.3	97.9	51 29.2	+13.8	99.1	51 19.0	+15.5	100.4	51 07.6	+17.0	101.6	50 54.9	+18.6	102.8	50 41.0	+20.2	104.0	3
4	52 00.8	+7.5	93.8	51 56.2	+9.2	95.0	51 50.3	+10.9	96.3	51 43.0	+12.6	97.6	51 34.5	+14.2	98.8	51 24.6	+15.9	100.1	51 13.5	+17.4	101.3	51 01.2	+18.9	102.5	4
5	52 08.3	+6.3	92.1	52 05.4	+8.0	93.4	52 01.2	+9.6	94.7	51 55.6	+11.3	96.0	51 48.7	+12.9	97.2	51 40.5	+14.5	98.5	51 30.9	+16.2	99.7	51 20.1	+17.8	101.0	5
6	52 14.6	+4.9	90.5	52 13.4	+6.6	91.8	52 10.8	+8.3	93.1	52 06.9	+10.0	94.4	52 01.6	+11.7	95.7	51 55.0	+13.4	96.9	51 47.1	+15.0	98.2	51 37.9	+16.6	99.4	6
7	52 19.5	+3.5	88.9	52 20.0	+5.2	90.2	52 19.1	+7.0	91.5	52 16.9	+8.6	92.8	52 13.3	+10.3	94.1	52 08.4	+12.0	95.3	52 02.1	+13.7	96.6	51 54.5	+15.4	97.9	7
8	52 23.0	+2.2	87.3	52 25.2	+3.9	88.6	52 26.1	+5.6	89.9	52 25.5	+7.4	91.2	52 23.6	+9.1	92.5	52 20.4	+10.7	93.7	52 15.8	+12.4	95.0	52 09.9	+14.0	96.3	8
9	52 25.2	+0.9	85.6	52 29.1	+2.6	86.9	52 31.7	+4.2	88.2	52 32.9	+5.9	89.5	52 32.7	+7.7	90.8	52 31.1	+9.4	92.1	52 28.2	+11.1	93.4	52 23.9	+12.8	94.7	9
10	52 26.1	−0.8	84.0	52 31.7	+1.2	85.3	52 35.9	+2.9	86.6	52 38.8	+4.6	87.9	52 40.4	+6.3	89.2	52 40.5	+8.1	90.5	52 39.3	+9.8	91.8	52 36.7	+11.5	93.1	10
11	52 25.5	−1.9	82.3	52 32.9	−0.2	83.6	52 38.8	+1.6	84.9	52 43.4	+3.3	86.2	52 46.7	+5.0	87.6	52 48.6	+6.7	88.9	52 49.1	+8.4	90.2	52 48.2	+10.1	91.5	11
12	52 23.6	−3.2	80.7	52 32.7	−1.6	82.0	52 40.4	+0.1	83.3	52 46.7	+1.9	84.6	52 51.7	+3.8	85.9	52 55.3	+5.3	87.2	52 57.5	+7.0	88.5	52 58.3	+8.8	89.9	12
13	52 20.4	−4.6	79.1	52 31.1	−2.9	80.3	52 40.5	−1.2	81.6	52 48.6	+0.5	82.9	52 55.3	+2.2	84.3	53 00.6	+3.9	85.6	53 04.5	+5.7	86.9	53 07.1	+7.4	88.2	13
14	52 15.8	−5.9	77.4	52 28.2	−4.3	78.7	52 39.3	−2.6	80.0	52 49.1	−0.9	81.3	52 57.5	+0.8	82.6	53 04.5	+2.8	83.9	53 10.2	+4.3	85.2	53 14.5	+6.0	86.6	14
15	52 09.9	−7.3	75.8	52 23.9	−5.6	77.1	52 36.7	−3.9	78.3	52 48.2	−2.3	79.6	52 58.3	−0.6	80.9	53 07.1	+1.1	82.2	53 14.5	+2.9	83.6	53 20.5	+4.6	84.9	15
16	52 02.6	−8.6	74.2	52 18.3	−7.0	75.4	52 32.8	−5.4	76.7	52 45.9	−3.6	78.0	52 57.7	−1.9	79.3	53 08.2	−0.2	80.6	53 17.4	+1.5	81.9	53 25.1	+3.3	83.2	16
17	51 54.0	−9.9	72.6	52 11.3	−8.2	73.8	52 27.4	−6.6	75.1	52 42.3	−5.0	76.3	52 55.8	−3.3	77.6	53 08.0	−1.6	78.9	53 18.9	0.0	80.2	53 28.4	+1.8	81.6	17
18	51 44.1	−11.2	71.0	52 03.1	−9.7	72.2	52 20.8	−8.0	73.4	52 37.3	−6.4	74.7	52 52.5	−4.7	76.0	53 06.4	−3.1	77.2	53 18.9	−1.3	78.6	53 30.2	+0.4	79.9	18
19	51 32.9	−12.4	69.4	51 53.4	−10.9	70.6	52 12.8	−9.3	71.8	52 30.9	−7.7	73.0	52 47.8	−6.1	74.3	53 03.3	−4.4	75.6	53 17.6	−2.7	76.9	53 30.6	−1.1	78.2	19
20	51 20.5	−13.7	67.8	51 42.5	−12.1	69.0	52 03.5	−10.7	70.2	52 23.2	−9.1	71.4	52 41.7	−7.5	72.7	52 58.9	−5.8	73.9	53 14.9	−4.1	75.2	53 29.5	−2.4	76.5	20
21	51 06.8	−14.9	66.3	51 30.4	−13.5	67.4	51 52.8	−11.9	68.6	52 14.1	−10.3	69.8	52 34.2	−8.7	71.0	52 53.1	−7.1	72.3	53 10.8	−5.6	73.5	53 27.1	−3.8	74.8	21
22	50 51.9	−16.1	64.7	51 16.9	−14.6	65.9	51 40.9	−13.2	67.0	52 03.8	−11.7	68.2	52 25.5	−10.1	69.4	52 46.0	−8.5	70.6	53 05.2	−6.8	71.9	53 23.3	−5.3	73.2	22
23	50 35.8	−17.3	63.2	51 02.3	−15.9	64.3	51 27.7	−14.4	65.5	51 52.1	−12.9	66.6	52 15.4	−11.4	67.8	52 37.5	−9.9	69.0	52 58.4	−8.3	70.2	53 18.0	−6.6	71.5	23
24	50 18.5	−18.5	61.7	50 46.4	−17.1	62.8	51 13.3	−15.6	63.9	51 39.2	−14.2	65.0	52 04.0	−12.8	66.2	52 27.6	−11.2	67.4	52 50.1	−9.6	68.6	53 11.4	−8.0	69.8	24
25	50 00.0	−19.6	60.2	50 29.3	−18.2	61.3	50 57.7	−16.9	62.4	51 25.0	−15.4	63.5	51 51.2	−13.9	64.6	52 16.4	−12.4	65.8	52 40.5	−11.0	67.0	53 03.4	−9.4	68.2	25
26	49 40.4	−20.6	58.8	50 11.1	−19.4	59.8	50 40.8	−18.0	60.8	51 09.6	−16.7	61.9	51 37.3	−15.2	63.0	52 04.0	−13.8	64.2	52 29.5	−12.2	65.3	52 54.0	−10.7	66.5	26
27	49 19.8	−21.7	57.3	49 51.7	−20.4	58.3	50 22.8	−19.2	59.3	50 52.9	−17.8	60.4	51 22.1	−16.5	61.5	51 50.2	−15.0	62.6	52 17.3	−13.6	63.7	52 43.3	−12.1	64.9	27
28	48 58.1	−22.8	55.9	49 31.3	−21.6	56.9	50 03.6	−20.2	57.9	50 35.1	−19.0	58.9	51 05.6	−17.6	59.9	51 35.2	−16.2	61.0	52 03.7	−14.8	62.2	52 31.2	−13.3	63.3	28
29	48 35.3	−23.8	54.5	49 09.7	−22.5	55.4	49 43.4	−21.4	56.4	50 16.1	−20.1	57.4	50 48.0	−18.8	58.4	51 19.0	−17.5	59.5	51 48.9	−16.0	60.6	52 17.9	−14.6	61.7	29
30	48 11.5	−24.7	53.1	48 47.2	−23.7	54.0	49 22.0	−22.4	55.0	49 56.0	−21.2	55.9	50 29.2	−19.9	56.9	51 01.5	−18.6	58.0	51 32.9	−17.3	59.0	52 03.3	−15.9	60.1	30
31	47 46.8	−25.7	51.8	48 23.5	−24.5	52.6	48 59.6	−23.5	53.5	49 34.8	−22.2	54.5	50 09.3	−21.0	55.5	50 42.9	−19.8	56.5	51 15.6	−18.5	57.5	51 47.4	−17.1	58.6	31
32	47 21.1	−26.7	50.4	47 59.0	−25.6	51.3	48 36.1	−24.4	52.1	49 12.6	−23.3	53.1	49 48.3	−22.2	54.0	50 23.1	−20.9	55.0	50 57.1	−19.6	56.0	51 30.3	−18.4	57.0	32
33	46 54.4	−27.5	49.1	47 33.4	−26.5	49.9	48 11.7	−25.4	50.8	48 49.3	−24.3	51.6	49 26.1	−23.1	52.6	50 02.2	−22.0	53.5	50 37.5	−20.8	54.5	51 11.9	−19.5	55.5	33
34	46 26.9	−28.4	47.8	47 06.9	−27.4	48.6	47 46.3	−26.4	49.4	48 25.0	−25.3	50.3	49 03.0	−24.2	51.1	49 40.2	−23.0	52.1	50 16.7	−21.8	53.0	50 52.4	−20.6	54.0	34
35	45 58.5	−29.2	46.5	46 39.5	−28.2	47.3	47 19.9	−27.2	48.1	47 59.7	−26.2	48.9	48 38.8	−25.2	49.8	49 17.2	−24.1	50.6	49 54.9	−22.9	51.6	50 31.8	−21.7	52.5	35
36	45 29.3	−30.0	45.3	46 11.3	−29.1	46.0	46 52.7	−28.2	46.8	47 33.5	−27.2	47.6	48 13.6	−26.1	48.4	48 53.1	−25.0	49.2	49 32.0	−24.0	50.1	50 10.1	−22.9	51.0	36
37	44 59.3	−30.8	44.0	45 42.2	−29.9	44.8	46 24.5	−29.0	45.5	47 06.3	−28.0	46.3	47 47.5	−27.1	47.0	48 28.1	−26.1	47.9	49 08.0	−25.0	48.7	49 47.2	−23.9	49.6	37
38	44 28.5	−31.5	42.8	45 12.3	−30.7	43.5	45 55.5	−29.8	44.2	46 38.3	−28.9	45.0	47 20.4	−27.9	45.7	48 02.0	−26.9	46.5	48 43.0	−25.9	47.3	49 23.3	−24.9	48.2	38
39	43 57.0	−32.3	41.6	44 41.6	−31.5	42.3	45 25.7	−30.6	43.0	46 09.4	−29.8	43.7	46 52.5	−28.8	44.4	47 35.1	−27.9	45.2	48 17.1	−26.9	46.0	48 58.4	−25.8	46.8	39
40	43 24.7	−33.0	40.5	44 10.1	−32.2	41.1	44 55.1	−31.4	41.8	45 39.6	−30.5	42.4	46 23.7	−29.7	43.1	47 07.2	−28.8	43.9	47 50.2	−27.9	44.6	48 32.6	−26.9	45.4	40
41	42 51.7	−33.6	39.3	43 37.9	−32.9	39.9	44 23.7	−32.1	40.6	45 09.1	−31.3	41.2	45 54.0	−30.5	41.9	46 38.4	−29.6	42.6	47 22.3	−28.7	43.3	48 05.7	−27.8	44.1	41
42	42 18.1	−34.3	38.2	43 05.0	−33.5	38.8	43 51.6	−32.8	39.4	44 37.8	−32.1	40.0	45 23.5	−31.2	40.7	46 08.8	−30.4	41.3	46 53.6	−29.5	42.0	47 37.9	−28.6	42.8	42
43	41 43.8	−34.9	37.1	42 31.5	−34.2	37.7	43 18.8	−33.5	38.2	44 05.7	−32.7	38.8	44 52.3	−32.0	39.4	45 38.4	−31.2	40.1	46 24.1	−30.4	40.8	47 09.3	−29.6	41.5	43

〈표 2-17〉 태양, 항성 및 행성의 고도개정표

A2 ALTITUDE CORRECTION TABLES 10° - 90° - SUN, STARS, PLANETS

OCT.-MAR. SUN App. Alt.	Lower Limb / Upper Limb	APR.-SEPT. App. Alt.	Lower Limb / Upper Limb	STARS AND PLANETS App. Alt.	Corrⁿ	DIP Ht. of Eye (m)	Corrⁿ	Ht. of Eye (ft.)
° ′	′ ′	° ′	′ ′	° ′	′	m	′	ft.
9 33	+10.8-21.5	9 39	+10.6-21.2	9 55	-5.3	2.4	-2.8	8.0
9 45	+10.9-21.4	9 50	+10.7-21.1	10 07	-5.2	2.6	-2.9	8.6
9 56	+11.0-21.3	10 02	+10.8-21.0	10 20	-5.1	2.8	-3.0	9.2
10 08	+11.1-21.2	10 14	+10.9-20.9	10 32	-5.0	3.0	-3.1	9.8
10 20	+11.2-21.1	10 27	+11.0-20.8	10 46	-4.9	3.2	-3.2	10.5
10 33	+11.3-21.0	10 40	+11.1-20.7	10 59	-4.8	3.4	-3.3	11.2
10 46	+11.4-20.9	10 53	+11.2-20.6	11 14	-4.7	3.6	-3.4	11.9
11 00	+11.5-20.8	11 07	+11.3-20.5	11 29	-4.6	3.8	-3.5	12.6
11 15	+11.6-20.7	11 22	+11.4-20.4	11 44	-4.5	4.0	-3.6	13.3
11 30	+11.7-20.6	11 37	+11.5-20.3	12 00	-4.4	4.3	-3.7	14.1
11 45	+11.8-20.5	11 53	+11.6-20.2	12 17	-4.3	4.5	-3.8	14.9
12 01	+11.9-20.4	12 10	+11.7-20.1	12 35	-4.2	4.7	-3.9	15.7
12 18	+12.0-20.3	12 27	+11.8-20.0	12 53	-4.1	5.0	-4.0	16.5
12 36	+12.1-20.2	12 45	+11.9-19.9	13 12	-4.0	5.2	-4.1	17.4
12 54	+12.2-20.1	13 04	+12.0-19.8	13 32	-3.9	5.5	-4.2	18.3
13 14	+12.3-20.0	13 24	+12.1-19.7	13 53	-3.8	5.8	-4.3	19.1
13 34	+12.4-19.9	13 44	+12.2-19.6	14 16	-3.7	6.1	-4.4	20.1
13 55	+12.5-19.8	14 06	+12.3-19.5	14 39	-3.6	6.3	-4.5	21.0
14 17	+12.6-19.7	14 29	+12.4-19.4	15 03	-3.5	6.6	-4.6	22.0
14 41	+12.7-19.6	14 53	+12.5-19.3	15 29	-3.4	6.9	-4.7	22.9
15 05	+12.8-19.5	15 18	+12.6-19.2	15 56	-3.3	7.2	-4.8	23.9
15 31	+12.9-19.4	15 45	+12.7-19.1	16 25	-3.2	7.5	-4.9	24.9
15 59	+13.0-19.3	16 13	+12.8-19.0	16 55	-3.1	7.9	-5.0	26.0
16 27	+13.1-19.2	16 43	+12.9-18.9	17 27	-3.0	8.2	-5.1	27.1
16 58	+13.2-19.1	17 14	+13.0-18.8	18 01	-2.9	8.5	-5.2	28.1
17 30	+13.3-19.0	17 47	+13.1-18.7	18 37	-2.8	8.8	-5.3	29.2
18 05	+13.4-18.9	18 23	+13.2-18.6	19 16	-2.7	9.2	-5.4	30.4
18 41	+13.5-18.8	19 00	+13.3-18.5	19 56	-2.6	9.5	-5.5	31.5
19 20	+13.6-18.7	19 41	+13.4-18.4	20 40	-2.5	9.9	-5.6	32.7
20 02	+13.7-18.6	20 24	+13.5-18.3	21 27	-2.4	10.3	-5.7	33.9
20 46	+13.8-18.5	21 10	+13.6-18.2	22 17	-2.3	10.6	-5.8	35.1
21 34	+13.9-18.4	21 59	+13.7-18.1	23 11	-2.2	11.0	-5.9	36.3
22 25	+14.0-18.3	22 52	+13.8-18.0	24 09	-2.1	11.4	-6.0	37.6
23 20	+14.1-18.2	23 49	+13.9-17.9	25 12	-2.0	11.8	-6.1	38.9
24 20	+14.2-18.1	24 51	+14.0-17.8	26 20	-1.9	12.2	-6.2	40.1
25 24	+14.3-18.0	25 58	+14.1-17.7	27 34	-1.8	12.6	-6.3	41.5
26 34	+14.4-17.9	27 11	+14.2-17.6	28 54	-1.7	13.0	-6.4	42.8
27 50	+14.5-17.8	28 31	+14.3-17.5	30 22	-1.6	13.4	-6.5	44.2
29 13	+14.6-17.7	29 58	+14.4-17.4	31 58	-1.5	13.8	-6.6	45.5
30 44	+14.7-17.6	31 33	+14.5-17.3	33 43	-1.4	14.2	-6.7	46.9
32 24	+14.8-17.5	33 18	+14.6-17.2	35 38	-1.3	14.7	-6.8	48.4
34 15	+14.9-17.4	35 15	+14.7-17.1	37 45	-1.2	15.1	-6.9	49.8
36 17	+15.0-17.3	37 24	+14.8-17.0	40 06	-1.1	15.5	-7.0	51.3
38 34	+15.1-17.2	39 48	+14.9-16.9	42 42	-1.0	16.0	-7.1	52.8
41 06	+15.2-17.1	42 28	+15.0-16.8	45 34	-0.9	16.5	-7.2	54.3
43 56	+15.3-17.0	45 29	+15.1-16.7	48 45	-0.8	16.9	-7.3	55.8
47 07	+15.4-16.9	48 52	+15.2-16.6	52 16	-0.7	17.4	-7.4	57.4
50 43	+15.5-16.8	52 41	+15.3-16.5	56 09	-0.6	17.9	-7.5	58.9
54 46	+15.6-16.7	56 59	+15.4-16.4	60 26	-0.5	18.4	-7.6	60.5
59 21	+15.7-16.6	61 50	+15.5-16.3	65 06	-0.4	18.8	-7.7	62.1
64 28	+15.8-16.5	67 15	+15.6-16.2	70 09	-0.3	19.3	-7.8	63.8
70 10	+15.9-16.4	73 14	+15.7-16.1	75 32	-0.2	19.8	-7.9	65.4
76 24	+16.0-16.3	79 42	+15.8-16.0	81 12	-0.1	20.4	-8.0	67.1
83 05	+16.1-16.2	86 31	+15.9-15.9	87 03	-0.0	20.9	-8.1	68.8
90 00		90 00		90 00		21.4		70.5

STARS AND PLANETS — App. Alt. / Additional Corrⁿ (2006)

	App. Alt. (°)	Additional Corrⁿ (′)
VENUS Jan. 1 - Feb. 3	0	+0.5
	26	+0.4
	46	+0.3
	60	+0.2
	73	+0.1
	84	
Feb. 4 - Feb. 18	0	+0.4
	29	+0.3
	51	+0.2
	68	+0.1
	83	
Feb. 19 - Mar. 13	0	+0.3
	34	+0.2
	60	+0.1
	80	
Mar. 14 - May 02	0	+0.2
	41	+0.1
	76	
May 03 - Dec. 31	0	+0.1
	60	
MARS Jan. 01 - Jan. 23	0	+0.2
	41	+0.1
	76	
Jan. 24 - Dec. 31	0	+0.1
	60	

DIP — Ht. of Eye / Corrⁿ

Ht. of Eye	Corrⁿ
m	′
1.0	- 1.8
1.5	- 2.2
2.0	- 2.5
2.5	- 2.8
3.0	- 3.0
See Table ←	
m	′
20	- 7.9
22	- 8.3
24	- 8.6
26	- 9.0
28	- 9.3
30	- 9.6
32	-10.0
34	-10.3
36	-10.6
38	-10.8
40	-11.1
42	-11.4
44	-11.7
46	-11.9
48	-12.2
ft.	′
2	- 1.4
4	- 1.9
6	- 2.4
8	- 2.7
10	- 3.1
See Table ←	
ft.	′
70	- 8.1
75	- 8.4
80	- 8.7
85	- 8.9
90	- 9.2
95	- 9.5
100	- 9.7
105	- 9.9
110	-10.2
115	-10.4
120	-10.6
125	-10.8
130	-11.1
135	-11.3
140	-11.5
145	-11.7
150	-11.9
155	-12.1

App. Alt. = Apparent Altitude = 굴절 및 안고차를 개정한 육분의 고도

〈표 2-18〉 태양, 항성 및 행성의 추가고도개정표

A4 ALTITUDE CORRECTION — ADDITIONAL CORRECTIONS

ADDITIONAL REFRACTION CORRECTIONS FOR NO-STANDARD CONDITIONS

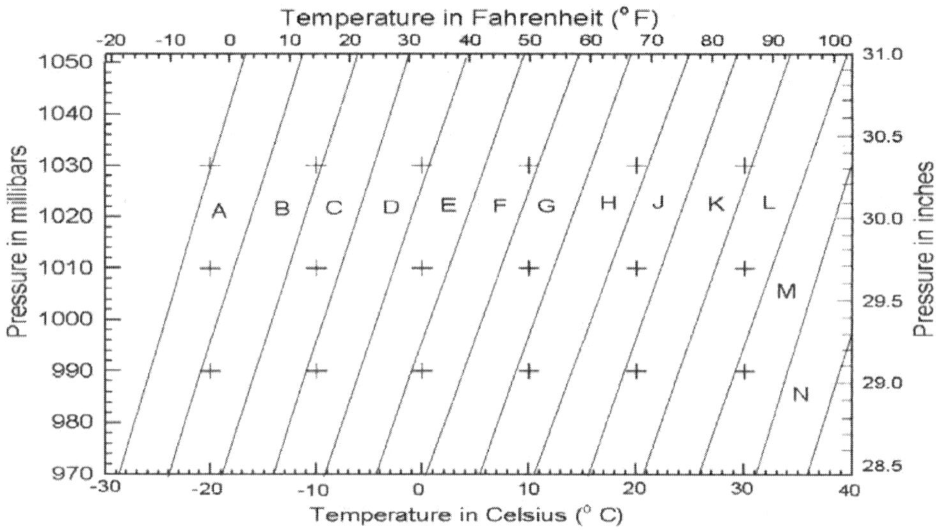

App. Alt.	A	B	C	D	E	F	G	H	J	K	L	M	N	P	App. Alt.
° ′	′	′	′	′	′	′	′	′	′	′	′	′	′	′	° ′
00 00	-7.3	-5.9	-4.6	-3.4	-2.2	-1.1	0.0	+1.0	+2.0	+3.0	+4.0	+4.9	+5.9	+6.9	00 00
00 30	5.5	4.5	3.5	2.6	1.7	0.8	0.0	0.8	1.6	2.3	3.1	3.8	4.5	5.3	00 30
01 00	4.4	3.5	2.8	2.0	1.3	0.7	0.0	0.6	1.2	1.8	2.4	3.0	3.6	4.2	01 00
01 30	3.5	2.9	2.2	1.7	1.1	0.5	0.0	0.5	1.0	1.5	2.0	2.5	2.9	3.4	01 30
02 00	2.9	2.4	1.9	1.4	0.9	0.4	0.0	0.4	0.8	1.3	1.7	2.0	2.4	2.8	02 00
02 30	-2.5	-2.0	-1.6	-1.2	-0.8	-0.4	0.0	+0.4	+0.7	+1.1	+1.4	+1.7	+2.1	+2.4	02 30
03 00	2.1	1.7	1.4	1.0	0.7	0.3	0.0	0.3	0.6	0.9	1.2	1.5	1.8	2.1	03 00
03 30	1.9	1.5	1.2	0.9	0.6	0.3	0.0	0.3	0.5	0.8	1.1	1.3	1.6	1.8	03 30
04 00	1.6	1.3	1.1	0.8	0.5	0.3	0.0	0.2	0.5	0.7	0.9	1.2	1.4	1.6	04 00
04 30	1.5	1.2	0.9	0.7	0.5	0.2	0.0	0.2	0.4	0.6	0.8	1.0	1.3	1.5	04 30
05 00	-1.3	-1.1	-0.9	-0.6	-0.4	-0.2	0.0	+0.2	+0.4	+0.6	+0.8	+0.9	+1.1	+1.3	05 00
06	1.1	0.9	0.7	0.5	0.3	0.2	0.0	0.2	0.3	0.5	0.6	0.8	0.9	1.1	06
07	1.0	0.8	0.6	0.5	0.3	0.1	0.0	0.1	0.3	0.4	0.5	0.7	0.8	0.9	07
08	0.8	0.7	0.5	0.4	0.3	0.1	0.0	0.1	0.2	0.4	0.5	0.6	0.7	0.8	08
09	0.7	0.6	0.5	0.4	0.2	0.1	0.0	0.1	0.2	0.3	0.4	0.5	0.6	0.7	09
10 00	-0.7	-0.5	-0.4	-0.3	-0.2	-0.1	0.0	+0.1	+0.2	+0.3	+0.4	+0.5	+0.6	+0.7	10 00
12	0.6	0.5	0.4	0.3	0.2	0.1	0.0	0.1	0.2	0.2	0.3	0.4	0.5	0.5	12
14	0.5	0.4	0.3	0.2	0.1	0.1	0.0	0.1	0.1	0.2	0.3	0.3	0.4	0.5	14
16	0.4	0.3	0.3	0.2	0.1	0.1	0.0	0.1	0.1	0.2	0.2	0.3	0.3	0.4	16
18	0.4	0.3	0.2	0.2	0.1	-0.1	0.0	+0.1	0.1	0.2	+0.2	0.3	0.3	0.4	18
20 00	-0.3	-0.3	-0.2	-0.2	-0.1	0.0	0.0	0.0	+0.1	+0.1	+0.2	+0.2	+0.3	+0.3	20 00
25	0.3	0.2	0.2	0.1	0.1	0.0	0.0	0.0	0.1	0.1	0.1	0.2	0.2	0.2	25
30	0.2	0.2	0.1	0.1	0.1	0.0	0.0	0.0	+0.1	0.1	0.1	0.1	0.2	0.2	30
35	0.2	0.1	0.1	0.1	-0.1	0.0	0.0	0.0	0.0	0.1	0.1	0.1	0.1	0.2	35
40	0.1	0.1	0.1	-0.1	0.0	0.0	0.0	0.0	0.0	+0.1	0.1	0.1	0.1	0.1	40
50 00	-0.1	-0.1	-0.1	0.0	0.0	0.0	0.0	0.0	0.0	0.0	+0.1	+0.1	+0.1	+0.1	50 00

위의 도표에서 온도와 기압을 인수로 하여 사선대의 문자(A, B, C---N)을 얻은 후에, 이 문자표와 시고도(App. Alt.)를 또다시 인수로 하여 표에서 개정치를 얻는다. 그리고 구해진 개정치를 표준 상태에 대한 개정에 추가하여서 육분의 고도에 적용한다. (태양, 항성과 행성에 대하여는 A2, A3페이지, 태음에 대해서는 310-311 페이지의 자료를 참조).

2.11 항성의 고도 관측을 통한 위치선 작도

2.11.1 항성의 색성

〈표 2-19〉

	계산방법		내 용
GMT	GHA♈	Dec.(적위)	고도 관측시간을 GMT로 환산
h			시각에 대한 GHA♈, d값을 천측력에서 구함
m, s			분, 초에 대한 GHA♈, d값을 천측력에서 구함
GHA♈		d	시각, 분, 초에 대한 GHA, d값 합산
aλ			가정경도. LHA값이 정수가 되도록 조정 (동경이면 +, 서경이면 −)
LHA♈			관측자와 춘분점의 사이각을 서쪽으로 측정
aL			

① 항성을 관측할 수 있는 시기는 수평선도 보이고 항성도 보이는 박명시이므로 천측력을 통해 항해 박명시를 구한다(시의 계산참조).

② 항해박명시를 GMT로 환산하고 해당시간에 대한 춘분점의 지방시각(LHA♈) 및 가정위도(aL)를 천측력을 통해 구한다.

③ LHA과 가정위도(aL)를 이용하여 천측계산표(249 Table) 혹은 색성판(star finder)을 이용하여 박명시간 동안 볼 수 있는 항성들의 계산고도 및 방위를 확인한다.

[예제] 2013년 11월 13일에 D.R. Lat. 35° 10′, Long. 129° 10′에서의 저녁 박명시가 ZT으로 18시 16분 20초라고 가정한다.

〈표 2-20〉

	계산방법			계산방법	
GMT	GHA♈	Dec.(적위)	aλ	129° 34′.5	
09h	187° 20′.5		LHA♈	321° 00′.0	
16m 20s	04° 05′.0		aL	35°N	
GHA♈	191° 25′.5	d			

① 대시(ZT)를 GMT로 환산한다.

② 해당 GMT에 대하여 천측력 주표의 왼쪽 페이지 및 증분 및 개정표에서 GHAγ를 찾는다.

③ GHAγ에 가정경도를 가감하여(동경은 +, 서경은 −) LHAγ를 구한다.

④ 가정경도를 정한다.

⑤ LHAγ 및 가정경도를 이용하여 색성판 혹은 249 Table을 이용하여 색성한다.

---

1. Star Finder로 3개의 항성에 대한 고도와 방위 구하기

① Starbase에 가장 가까운 위도인 35°N Template를 끼운다.

② 위에서 구한 LHAγ는 321°이므로 0°~180° 방위선의 연장선 끝에 표시되어 있는 화살표를 321°에 맞춘다.

③ 항성의 교각이 60°~90° 정도 되도록 선택하도록 한다. 우선 교각이 90° 정도가 생기도록 항성을 정하여 보면, Kochab, Mirfak, Hamal을 선택할 수 있다.

- Kochab의 대략적인 고도 및 방위는 32°, 342°

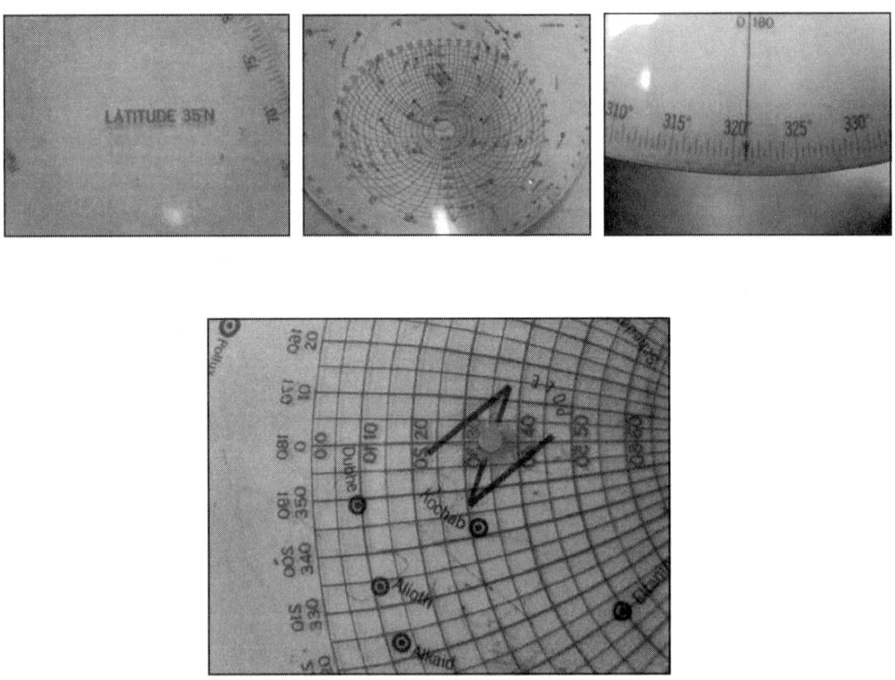

[그림 2-93] Star Finder

• Mirfak의 대략적인 고도 및 방위는 27°, 046°

• Hamal의 대략적인 고도 및 방위는 29°, 080°

④ 박명시(18시 16분 20초)에 Sextant를 32°에 맞추고 방위 342°를 보았을 때 수평선 가까이 위치하는 별이 Kochab이다.

2. 249 Table로 3개의 항성에 대한 고도와 방위 구하기

249 Table VOL.1에서 가정위도(aL) 35°인 페이지에서 LHAγ 321°를 찾으면 Mirfak, Hamal, Diphda, FOMALHAUT, ALTAIR, VEGA, Kochab이 있다. 이중에 고도 및 상호교각이 양호한 항성에는 ◆ 표시가 되어 있으며 Mirfak, FOMALHAUT, VEGA가 그러한 항성이다.

• Mirfak의 계산고도(Hc) 및 방위(Zn)는 25° 55′, 046°

• FOMALHAUT의 계산고도(Hc) 및 방위(Zn)는 21° 43′, 158°

• VEGA의 계산고도(Hc) 및 방위(Zn)는 56° 42′, 289°

〈표 2-21〉

	◆ Mirfak		Hamal		Diphda		◆ FOMALHAUT		ALTAIR		◆ VEGA		Kochab	
315	22 27	044	23 32	077	15 02	125	19 41	153	59 29	215	61 22	288	32 52	341
316	23 01	044	24 20	077	15 42	126	20 03	154	59 00	217	60 35	288	32 36	341
317	23 35	045	25 08	078	16 21	127	20 24	155	58 30	219	59 49	288	32 20	341
318	24 10	045	25 56	078	17 01	127	20 45	156	57 59	220	59 02	289	32 04	341
319	24 45	045	26 44	079	17 39	128	21 05	156	57 27	222	58 15	289	31 48	341
320	25 20	045	27 33	079	18 18	129	21 24	157	56 54	223	57 29	289	31 33	341
321	25 55	046	28 21	080	18 56	130	21 43	158	56 20	225	56 42	289	31 17	342
322	26 30	046	29 09	080	19 34	130	22 01	159	55 45	226	55 56	289	31 02	342
323	27 06	046	29 58	081	20 11	131	22 18	160	55 09	227	55 10	289	30 46	342
324	27 41	047	30 47	081	20 48	132	22 35	161	54 32	229	54 23	290	30 31	342

2.11.2 항성의 고도 관측을 통한 위치선 작도

① Sextant 및 Azimuth Circle 등을 이용하여 색성한 항성들을 찾아 정확하게 고도 및 관측시간을 기록한다.

② 229 Table을 이용할 경우 관측시간에 대한 가정위도(aL), 지방시각(LHA), 적위(d)를 이용하여 항성 관측시점의 가정위치에 대한 계산고도(Hc) 및 방위(Zn)를 구한다.

③ 249 Table을 이용할 경우 관측시간에 대한 시춘분점의 지방시각(LHAγ) 및 가정위도(aL)에 대한 계산고도(Hc) 및 방위(Zn)를 구한다.

[예제] 2013년 11월 13일에 D.R. Lat. 35° 10′, Long. 129° 18′.3에서 Hamal 항성을 Hamal 관측시각 18시 17분 07초(기록하기까지 10초 경과) 같이 관측하였다.

---

1. 229 Table을 이용하여 작도

① Hamal의 관측시각 18시 16분 57초(보정된 시각)를 GMT로 환산하면 09시 16분 57초가 된다.

② GHA♈를 천측력을 통해 구한다(ARIES 기준).

③ 가정경도를 가감하여(동경이면 +, 서경이면 −) LHA♈를 구한다.

④ Hamal의 SHA을 천측력 주표에서 찾아 LHA★을 구한다.

⑤ LHA★, dec, aL의 값을 이용하여 229 Table을 이용하여 작도한다(태양을 이용한 위치선 작도와 동일).

〈표 2-22〉

	계산방법		내 용
GMT	GHA♈	Dec.(적위)	고도 관측시간을 GMT로 환산함
09h	187° 20′.5	N23° 29′.8	시각에 대한 GHA♈, d값을 천측력에서 구함
16m 57s	4° 14′.9	−	분, 초에 대한 GHA♈, d값을 천측력에서 구함
GHA♈	191° 35′.4	d N23° 29'.8	시각, 분, 초에 대한 GHA, d값 합산함
aλ	129° 18.′3		가정경도. LHA값이 정수가 되도록 조정함 (동경이면 +, 서경이면 −)
LHA♈	320° 53′.7		관측자와 춘분점의 사이각을 서쪽으로 측정함
SHA	328° 06′.3		춘분점과 천체의 사이각을 서쪽으로 측정함
LHA★	289° 00′.0		관측자와 천체의 사이각을 서쪽으로 측정함
t	071°(E)		자오선각, 관측자와 천체의 사이각을 가까운 쪽으로 측정(E or W)함
aL	35°		

2. 249 Table을 이용하여 작도

① 위에서 구한 LHAΥ(320° 53′.7) 및 가정위도(35°) 값으로 계산고도(Hc)와 방위(Zn)를 구한다. 항성은 춘분점을 기준으로 거의 같이 움직이고 적위의 변화량도 미소하므로 249 Table에서는 SHA, d 값을 평균값을 대입하여 LHA, d, aL을 계산하므로 사용자는 LHAΥ, aL까지만 구해도 계산고도 및 방위를 쉽게 얻을 수 있다.

② 249 Table은 5년마다 개정되며 SHA, d 값은 개정년도의 평균값을 사용하므로 오차값의 개정은 관측고도(Ho)와 계산고도(Hc)를 비교하여 해도에 위치선을 작도한 이후에 부록에 나와 있는 개정값만큼 변위시킴으로써 제거한다.

55	4 13·8	4 14·4	4 02·2	5·5	1·5	11·5	3·2	17·5	4·8
56	4 14·0	4 14·7	4 02·4	5·6	1·5	11·6	3·2	17·6	4·8
57	4 14·3	4 14·9	4 02·7	5·7	1·6	11·7	3·2	17·7	4·9
58	4 14·5	4 15·2	4 02·9	5·8	1·6	11·8	3·2	17·8	4·9
59	4 14·8	4 15·4	4 03·1	5·9	1·6	11·9	3·3	17·9	4·9

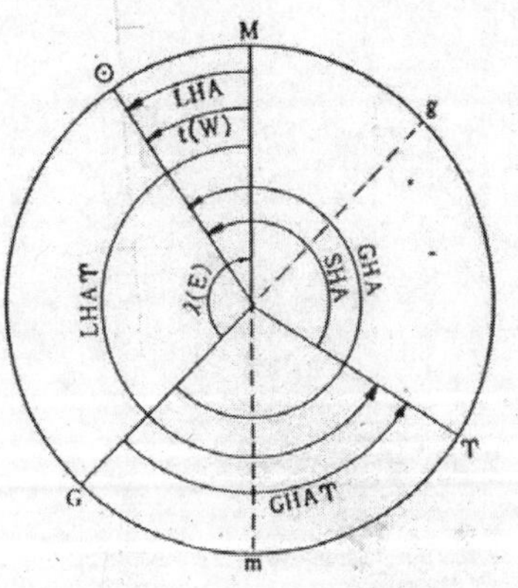

Gacrux	172	7.4	S57	8.9
Gienah	175	57.9	S17	34.7
Hadar	148	56.1	S60	24.3
Hamal	328	6.3	N23	29.9
Kaus Aust.	83	50.8	S34	23.1
Kochab	137	20.4	N74	7.5
Markab	13	43.3	N15	14.7
Menkar	314	20.1	N 4	7.2
Menkent	148	14.1	S36	24.2
Miaplacidus	221	41.1	S69	44.3
Mirfak	308	47.3	N49	53.3
Nunki	76	4.8	S26	17.4
Peacock	53	27.2	S56	43.1
Pollux	243	33.8	N28	0.6
Procyon	245	4.9	N 5	12.6

[그림 2-94] 적도면도의 활용

〈표 2-23〉 2013. 11/12-11/14(TUE, WED, THU)

UT		ARIES	VENUS −4.7		MARS +1.4		JUPITER −2.5		SATURN +0.5		STARS		
d	h	GHA ° ′	GHA ° ′	Dec ° ′	GHA ° ′	Dec ° ′	GHA ° ′	Dec ° ′	GHA ° ′	Dec ° ′	NAME	SHA ° ′	Dec ° ′
12	00	51 17.3	134 15.3	S27 1.5	243 30.9	N07 2.8	299 8.8	N21 52.9	188 12.8	S14 16.8	Acamar	315 17.7	S40 15.0
	01	66 19.8	149 15.2	S27 1.4	258 32.1	N07 2.2	314 11.3	N21 52.9	203 14.9	S14 16.9	Achernar	335 26.1	S57 10.1
	02	81 22.2	164 15.1	S27 1.3	273 33.2	N07 1.7	329 13.8	N21 52.9	218 17.1	S14 17.0	Acrux	173 9.4	S63 10.3
	03	96 24.7	179 15.0	S27 1.1	288 34.4	N07 1.2	344 16.3	N21 52.9	233 19.3	S14 17.1	Adhara	255 12.1	S28 59.4
	04	111 27.2	194 14.9	S27 1.0	303 35.5	N07 0.7	359 18.8	N21 52.9	248 21.5	S14 17.2	Aldebaran	290 48.7	N16 32.1
	05	126 29.6	209 14.8	S27 0.9	318 36.7	N07 0.1	14 21.3	N21 53.0	263 23.6	S14 17.2			
TUESDAY	06	141 32.1	224 14.7	S27 0.7	333 37.8	N06 59.6	29 23.8	N21 53.0	278 25.8	S14 17.3	Alioth	166 21.0	N55 52.9
	07	156 34.6	239 14.6	S27 0.6	348 39.0	N06 59.1	44 26.3	N21 53.0	293 28.0	S14 17.4	Alkaid	152 59.2	N49 14.6
	08	171 37.0	254 14.5	S27 0.5	3 40.1	N06 58.6	59 28.8	N21 53.0	308 30.1	S14 17.5	Al Na'ir	27 43.4	S46 53.7
	09	186 39.5	269 14.5	S27 0.3	18 41.3	N06 58.0	74 31.3	N21 53.0	323 32.3	S14 17.6	Alnilam	275 45.8	S01 11.7
	10	201 41.9	284 14.4	S27 0.2	33 42.4	N06 57.5	89 33.8	N21 53.0	338 34.5	S14 17.7	Alphard	217 55.8	S08 43.1
	11	216 44.4	299 14.3	S27 0.1	48 43.6	N06 57.0	104 36.3	N21 53.0	353 36.6	S14 17.8			
	12	231 46.9	314 14.2	S26 59.9	63 44.7	N06 56.5	119 38.9	N21 53.0	8 38.8	S14 17.9	Alphecca	126 11.2	N26 40.3
	13	246 49.3	329 14.1	S26 59.8	78 45.9	N06 55.9	134 41.4	N21 53.0	23 41.0	S14 17.9	Alpheratz	357 42.9	N29 10.3
	14	261 51.8	344 14.0	S26 59.7	93 47.0	N06 55.4	149 43.9	N21 53.1	38 43.1	S14 18.0	Altair	62 8.1	N08 54.6
	15	276 54.3	359 13.9	S26 59.5	108 48.2	N06 54.9	164 46.4	N21 53.1	53 45.3	S14 18.1	Ankaa	353 15.2	S42 13.9
	16	291 56.7	14 13.9	S26 59.4	123 49.3	N06 54.4	179 48.9	N21 53.1	68 47.5	S14 18.2	Antares	112 26.3	S26 27.6
	17	306 59.2	29 13.8	S26 59.2	138 50.5	N06 53.8	194 51.4	N21 53.1	83 49.6	S14 18.3			
	18	322 1.7	44 13.7	S26 59.1	153 51.6	N06 53.3	209 53.9	N21 53.1	98 51.8	S14 18.4	Arcturus	145 55.8	N19 6.7
	19	337 4.1	59 13.6	S26 58.9	168 52.8	N06 52.8	224 56.4	N21 53.1	113 54.0	S14 18.5	Atria	107 28.3	S69 3.0
	20	352 6.6	74 13.6	S26 58.8	183 53.9	N06 52.3	239 58.9	N21 53.1	128 56.2	S14 18.5	Avior	234 17.7	S59 33.1
	21	7 9.0	89 13.5	S26 58.7	198 55.1	N06 51.7	255 1.5	N21 53.1	143 58.3	S14 18.6	Bellatrix	278 31.4	N06 21.6
	22	22 11.5	104 13.4	S26 58.5	213 56.2	N06 51.2	270 4.0	N21 53.2	159 0.5	S14 18.7	Betelgeuse	271 0.7	N07 24.5
	23	37 14.0	119 13.3	S26 58.4	228 57.4	N06 50.7	285 6.5	N21 53.2	174 2.7	S14 18.8			
13	00	52 16.4	134 13.3	S26 58.2	243 58.5	N06 50.2	300 9.0	N21 53.2	189 4.8	S14 18.9	Canopus	263 55.6	S52 42.1
	01	67 18.9	149 13.2	S26 58.1	258 59.7	N06 49.6	315 11.5	N21 53.2	204 7.0	S14 19.0	Capella	280 33.6	N46 0.5
	02	82 21.4	164 13.1	S26 57.9	274 0.8	N06 49.1	330 14.0	N21 53.2	219 9.2	S14 19.1	Deneb	49 31.4	N45 20.3
	03	97 23.8	179 13.1	S26 57.8	289 2.0	N06 48.6	345 16.5	N21 53.2	234 11.3	S14 19.2	Denebola	182 33.6	N14 29.6
	04	112 26.3	194 13.0	S26 57.6	304 3.2	N06 48.1	0 19.1	N21 53.2	249 13.5	S14 19.2	Diphda	348 55.4	S17 54.6
	05	127 28.8	209 12.9	S26 57.5	319 4.3	N06 47.5	15 21.6	N21 53.2	264 15.7	S14 19.3			
WEDNESDAY	06	142 31.2	224 12.9	S26 57.3	334 5.5	N06 47.0	30 24.1	N21 53.3	279 17.8	S14 19.4	Dubhe	193 51.7	N61 40.2
	07	157 33.7	239 12.8	S26 57.2	349 6.6	N06 46.5	45 26.6	N21 53.3	294 20.0	S14 19.5	Elnath	278 11.9	N28 37.0
	08	172 36.2	254 12.8	S26 57.0	4 7.8	N06 46.0	60 29.1	N21 53.3	309 22.2	S14 19.6	Eltanin	90 46.4	N51 29.6
	09	187 38.6	269 12.7	S26 56.8	19 8.9	N06 45.4	75 31.6	N21 53.3	324 24.4	S14 19.7	Enif	33 46.8	N09 56.6
	10	202 41.1	284 12.6	S26 56.7	34 10.1	N06 44.9	90 34.2	N21 53.3	339 26.5	S14 19.8	Fomalhaut	15 23.6	S29 32.9
	11	217 43.5	299 12.6	S26 56.5	49 11.2	N06 44.4	105 36.7	N21 53.3	354 28.7	S14 19.9			
	12	232 46.0	314 12.5	S26 56.4	64 12.4	N06 43.9	120 39.2	N21 53.3	9 30.9	S14 19.9	Gacrux	172 1.0	S57 11.2
	13	247 48.5	329 12.5	S26 56.2	79 13.6	N06 43.3	135 41.7	N21 53.4	24 33.0	S14 20.0	Gienah	175 52.2	S17 37.0
	14	262 50.9	344 12.4	S26 56.1	94 14.7	N06 42.8	150 44.2	N21 53.4	39 35.2	S14 20.1	Hadar	148 48.0	S60 26.2
	15	277 53.4	359 12.4	S26 55.9	109 15.9	N06 42.3	165 46.8	N21 53.4	54 37.4	S14 20.2	Hamal	328 0.1	N23 31.8
	16	292 55.9	14 12.3	S26 55.7	124 17.0	N06 41.8	180 49.3	N21 53.4	69 39.5	S14 20.3	Kaus Aust.	83 43.7	S34 22.5
	17	307 58.3	29 12.3	S26 55.6	139 18.2	N06 41.2	195 51.8	N21 53.4	84 41.7	S14 20.4			
	18	323 0.8	44 12.2	S26 55.4	154 19.3	N06 40.7	210 54.3	N21 53.4	99 43.9	S14 20.5	Kochab	137 21.3	N74 5.9
	19	338 3.3	59 12.2	S26 55.2	169 20.5	N06 40.2	225 56.8	N21 53.4	114 46.0	S14 20.5	Markab	13 37.9	N15 17.1
	20	353 5.7	74 12.2	S26 55.1	184 21.6	N06 39.7	240 59.4	N21 53.4	129 48.2	S14 20.6	Menkar	314 14.4	N04 8.6
	21	8 8.2	89 12.1	S26 54.9	199 22.8	N06 39.1	256 1.9	N21 53.5	144 50.4	S14 20.7	Menkent	148 7.6	S36 26.1
	22	23 10.7	104 12.1	S26 54.7	214 24.0	N06 38.6	271 4.4	N21 53.5	159 52.6	S14 20.8	Miaplacidus	221 39.4	S69 46.3
	23	38 13.1	119 12.0	S26 54.6	229 25.1	N06 38.1	286 6.9	N21 53.5	174 54.7	S14 20.9			
14	00	53 15.6	134 12.0	S26 54.4	244 26.3	N06 37.6	301 9.5	N21 53.5	189 56.9	S14 21.0	Mirfak	308 39.4	N49 54.6
	01	68 18.0	149 12.0	S26 54.2	259 27.4	N06 37.0	316 12.0	N21 53.5	204 59.1	S14 21.1	Nunki	75 58.2	S26 16.6
	02	83 20.5	164 11.9	S26 54.1	274 28.6	N06 36.5	331 14.5	N21 53.5	220 1.2	S14 21.1	Peacock	53 19.0	S56 41.5
	03	98 23.0	179 11.9	S26 53.9	289 29.8	N06 36.0	346 17.0	N21 53.5	235 3.4	S14 21.2	Pollux	243 27.3	N27 59.3
	04	113 25.4	194 11.9	S26 53.7	304 30.9	N06 35.5	1 19.6	N21 53.6	250 5.6	S14 21.3	Procyon	244 59.3	N05 11.2
	05	128 27.9	209 11.8	S26 53.5	319 32.1	N06 34.9	16 22.1	N21 53.6	265 7.7	S14 21.4			
THURSDAY	06	143 30.4	224 11.8	S26 53.4	334 33.2	N06 34.4	31 24.6	N21 53.6	280 9.9	S14 21.5	Rasalhague	96 6.5	N12 33.3
	07	158 32.8	239 11.8	S26 53.2	349 34.4	N06 33.9	46 27.1	N21 53.6	295 12.1	S14 21.6	Regulus	207 43.3	N11 53.8
	08	173 35.3	254 11.7	S26 53.0	4 35.5	N06 33.4	61 29.7	N21 53.6	310 14.2	S14 21.7	Rigel	281 11.5	S08 11.2
	09	188 37.8	269 11.7	S26 52.8	19 36.7	N06 32.8	76 32.2	N21 53.6	325 16.4	S14 21.8	Rigil Kent.	139 52.0	S60 53.4
	10	203 40.2	284 11.7	S26 52.6	34 37.9	N06 32.3	91 34.7	N21 53.6	340 18.6	S14 21.8	Sabik	102 12.5	S15 44.3
	11	218 42.7	299 11.7	S26 52.5	49 39.0	N06 31.8	106 37.2	N21 53.7	355 20.8	S14 21.9			
	12	233 45.1	314 11.6	S26 52.3	64 40.2	N06 31.3	121 39.8	N21 53.7	10 22.9	S14 22.0	Schedar	349 39.7	N56 37.1
	13	248 47.6	329 11.6	S26 52.1	79 41.3	N06 30.8	136 42.3	N21 53.7	25 25.1	S14 22.1	Shaula	96 21.9	S37 6.7
	14	263 50.1	344 11.6	S26 51.9	94 42.5	N06 30.2	151 44.8	N21 53.7	40 27.3	S14 22.2	Sirius	258 33.3	S16 44.1
	15	278 52.5	359 11.6	S26 51.7	109 43.7	N06 29.7	166 47.4	N21 53.7	55 29.4	S14 22.3	Spica	158 31.2	S11 13.9
	16	293 55.0	14 11.6	S26 51.6	124 44.8	N06 29.2	181 49.9	N21 53.7	70 31.6	S14 22.4	Suhail	222 52.2	S43 29.2
	17	308 57.5	29 11.6	S26 51.4	139 46.0	N06 28.7	196 52.4	N21 53.7	85 33.8	S14 22.4			
	18	323 59.9	44 11.5	S26 51.2	154 47.1	N06 28.1	211 54.9	N21 53.8	100 35.9	S14 22.5	Vega	80 39.1	N38 48.2
	19	339 2.4	59 11.5	S26 51.0	169 48.3	N06 27.6	226 57.5	N21 53.8	115 38.1	S14 22.6	Zuben'ubi	137 5.4	S16 5.8
	20	354 4.9	74 11.5	S26 50.8	184 49.5	N06 27.1	242 0.0	N21 53.8	130 40.3	S14 22.7		SHA	Mer.Pass.
	21	9 7.3	89 11.5	S26 50.6	199 50.6	N06 26.6	257 2.5	N21 53.8	145 42.4	S14 22.8		° ′	h m
	22	24 9.8	104 11.5	S26 50.4	214 51.8	N06 26.0	272 5.1	N21 53.8	160 44.6	S14 22.9	Venus	81 56.8	15 3
	23	39 12.3	119 11.5	S26 50.2	229 53.0	N06 25.5	287 7.6	N21 53.8	175 46.8	S14 23.0	Mars	191 42.1	7 44
		h m									Jupiter	247 52.6	3 59
Mer.Pass.		20 27.5	v −0.1	d 0.2	v 1.2	d 0.5	v 2.5	d 0.0	v 2.2	d 0.1	Saturn	136 48.4	11 22

2.12 방위각의 측정

2.12.1 출몰방위각법

천체의 진출몰시에 방위각을 산출하는 방법으로 방위측정이 쉽고 오차가 작으나 관측시기에 제한을 받으며 이용할 수 있는 천체가 태양에 한정된다.

1. 태양의 출몰방위각 측정시기

시수평상 태양의 하변고도가 그 겉보기 반지름과 같은 높이이거나 또는 그 보다 약간 높은 시기에 관측한다.

2. 측정방법(태양방위각표의 이용)

관측자의 위도에 해당하는 페이지를 열고 그 날짜가 기재되어 있는 쪽의 맨 아래쪽 "Azimuth" 표에서 출몰시의 방위각을 구한다.

① 위도와 적위가 동명 : 제2부

② 위도와 적위가 이명 : 제3부

③ 방위각 앞에는 위도에 따라 N · S, 출시 E · 몰시 W 붙임

[예제] 2013년 2월 15일 02시 09분 19초(UTC)에 본선의 위치가 05° 25′.1N, 107° 00′.5E일 때 AZIMUTH CIRCLE을 이용하여 태양의 출몰시 방위측정 결과 104° 였다. 이때의 Gyro Error를 구하시오.

① 먼저 천측력을 이용하여 2013년 2월 15일 02시 09분 19초에 대한 GHA 및 Dec를 구한다.

천측력 참고표

GHA → 206° 27.7′   Dec → S 12° 46.3′
동명극과 적위값이 이명이므로 3부의 표를 찾는다.

〈표 3-24〉 2013. 2/15~2/17(FRI, SAT, SUN)

Day	UT (d h)	SUN GHA (° ′)	SUN Dec (° ′)	MOON GHA (° ′)	v (′)	MOON Dec (° ′)	d (′)	HP (′)
FRIDAY	15 00	176 28.1	S12 41.5	123 3.8	12.9	N11 38.4	9.2	55.8
	01	191 28.1	S12 40.7	137 35.6	12.9	N11 47.6	9.1	55.7
	02	206 28.1	S12 39.8	152 7.5	12.9	N11 56.7	9.1	55.7
	03	221 28.2	S12 39.0	166 39.4	12.9	N12 5.8	9.0	55.7
	04	236 28.2	S12 38.1	181 11.3	12.9	N12 14.7	8.9	55.6
	05	251 28.2	S12 37.2	195 43.1	12.9	N12 23.7	8.9	55.6
	06	266 28.2	S12 36.4	210 15.0	12.9	N12 32.5	8.8	55.6
	07	281 28.3	S12 35.5	224 46.9	12.9	N12 41.3	8.7	55.6
	08	296 28.3	S12 34.7	239 18.7	12.9	N12 50.0	8.7	55.5
	09	311 28.3	S12 33.8	253 50.6	12.9	N12 58.7	8.6	55.5
	10	326 28.4	S12 32.9	268 22.5	12.9	N13 7.3	8.5	55.5
	11	341 28.4	S12 32.1	282 54.3	12.8	N13 15.8	8.4	55.5
	12	356 28.4	S12 31.2	297 26.2	12.8	N13 24.3	8.4	55.4
	13	11 28.5	S12 30.3	311 58.0	12.8	N13 32.6	8.3	55.4
	14	26 28.5	S12 29.5	326 29.8	12.8	N13 40.9	8.2	55.4
	15	41 28.5	S12 28.6	341 1.7	12.8	N13 49.2	8.2	55.4
	16	56 28.6	S12 27.8	355 33.5	12.8	N13 57.3	8.1	55.3
	17	71 28.6	S12 26.9	10 5.3	12.8	N14 5.4	8.0	55.3
	18	86 28.6	S12 26.0	24 37.1	12.8	N14 13.5	7.9	55.3
	19	101 28.7	S12 25.2	39 8.9	12.8	N14 21.4	7.9	55.3
	20	116 28.7	S12 24.3	53 40.7	12.8	N14 29.3	7.8	55.2
	21	131 28.7	S12 23.4	68 12.5	12.8	N14 37.1	7.7	55.2
	22	146 28.8	S12 22.6	82 44.3	12.8	N14 44.8	7.7	55.2
	23	161 28.8	S12 21.7	97 16.1	12.8	N14 52.5	7.6	55.2
SATURDAY	16 00	176 28.9	S12 20.8	111 47.8	12.8	N15 0.0	7.5	55.1
	01	191 28.9	S12 20.0	126 19.6	12.7	N15 7.5	7.4	55.1
	02	206 28.9	S12 19.1	140 51.3	12.7	N15 15.0	7.3	55.1
	03	221 29.0	S12 18.2	155 23.1	12.7	N15 22.3	7.3	55.1
	04	236 29.0	S12 17.4	169 54.8	12.7	N15 29.6	7.2	55.0
	05	251 29.0	S12 16.5	184 26.5	12.7	N15 36.8	7.1	55.0
	06	266 29.1	S12 15.6	198 58.2	12.7	N15 43.9	7.0	55.0
	07	281 29.1	S12 14.8	213 29.9	12.7	N15 50.9	7.0	55.0
	08	296 29.2	S12 13.9	228 1.6	12.7	N15 57.9	6.9	55.0
	09	311 29.2	S12 13.0	242 33.3	12.7	N16 4.8	6.8	54.9
	10	326 29.2	S12 12.2	257 5.0	12.7	N16 11.6	6.7	54.9
	11	341 29.3	S12 11.3	271 36.6	12.6	N16 18.3	6.6	54.9
	12	356 29.3	S12 10.4	286 8.3	12.6	N16 25.0	6.6	54.9
	13	11 29.4	S12 9.5	300 39.9	12.6	N16 31.5	6.5	54.9
	14	26 29.4	S12 8.7	315 11.5	12.6	N16 38.0	6.4	54.8
	15	41 29.4	S12 7.8	329 43.1	12.6	N16 44.4	6.3	54.8
	16	56 29.5	S12 6.9	344 14.7	12.6	N16 50.7	6.2	54.8
	17	71 29.5	S12 6.1	358 46.3	12.6	N16 57.0	6.2	54.8
	18	86 29.6	S12 5.2	13 17.9	12.6	N17 3.1	6.1	54.8
	19	101 29.6	S12 4.3	27 49.5	12.6	N17 9.2	6.0	54.7
	20	116 29.7	S12 3.4	42 21.0	12.5	N17 15.2	5.9	54.7
	21	131 29.7	S12 2.6	56 52.5	12.5	N17 21.1	5.8	54.7
	22	146 29.7	S12 1.7	71 24.1	12.5	N17 27.0	5.7	54.7
	23	161 29.8	S12 0.8	85 55.6	12.5	N17 32.7	5.7	54.7
SUNDAY	17 00	176 29.8	S11 59.9	100 27.1	12.5	N17 38.4	5.6	54.7
	01	191 29.9	S11 59.1	114 58.6	12.5	N17 43.9	5.5	54.6
	02	206 29.9	S11 58.2	129 30.0	12.5	N17 49.4	5.4	54.6
	03	221 30.0	S11 57.3	144 1.5	12.4	N17 54.8	5.3	54.6
	04	236 30.0	S11 56.4	158 33.0	12.4	N18 0.2	5.2	54.6
	05	251 30.1	S11 55.6	173 4.4	12.4	N18 5.4	5.2	54.6
	06	266 30.1	S11 54.7	187 35.8	12.4	N18 10.6	5.1	54.6
	07	281 30.1	S11 53.8	202 7.2	12.4	N18 15.6	5.0	54.6
	08	296 30.2	S11 52.9	216 38.6	12.4	N18 20.6	4.9	54.5
	09	311 30.2	S11 52.1	231 10.0	12.4	N18 25.5	4.8	54.5
	10	326 30.3	S11 51.2	245 41.4	12.4	N18 30.3	4.7	54.5
	11	341 30.3	S11 50.3	260 12.7	12.3	N18 35.0	4.6	54.5
	12	356 30.4	S11 49.4	274 44.1	12.3	N18 39.6	4.5	54.5
	13	11 30.4	S11 48.5	289 15.4	12.3	N18 44.2	4.5	54.5
	14	26 30.5	S11 47.7	303 46.7	12.3	N18 48.6	4.4	54.5
	15	41 30.5	S11 46.8	318 18.0	12.3	N18 53.0	4.3	54.4
	16	56 30.6	S11 45.9	332 49.3	12.3	N18 57.3	4.2	54.4
	17	71 30.6	S11 45.0	347 20.6	12.3	N19 1.5	4.1	54.4
	18	86 30.7	S11 44.1	1 51.9	12.3	N19 5.6	4.0	54.4
	19	101 30.7	S11 43.3	16 23.1	12.2	N19 9.6	3.9	54.4
	20	116 30.8	S11 42.4	30 54.4	12.2	N19 13.5	3.8	54.4
	21	131 30.8	S11 41.5	45 25.6	12.2	N19 17.3	3.7	54.4
	22	146 30.9	S11 40.6	59 56.8	12.2	N19 21.1	3.7	54.4
	23	161 30.9	S11 39.7	74 28.0	12.2	N19 24.7	3.6	54.4
		SD 16.2	d 0.9	SD	15.1	15.0		14.8

Lat. (°)	Twilight Naut. (h m)	Twilight Civil (h m)	Sunrise (h m)	Moonrise 15 (h m)	Moonrise 16 (h m)	Moonrise 17 (h m)	Moonrise 18 (h m)
N 72	06 10	07 29	08 47	07 07	06 45	ooo	ooo
N 70	06 08	07 20	08 28	07 30	07 23	07 14	06 55
68	06 07	07 12	08 13	07 47	07 49	07 55	08 08
66	06 06	07 06	08 01	08 01	08 09	08 23	08 44
64	06 05	07 00	07 51	08 13	08 26	08 44	09 11
62	06 04	06 56	07 42	08 23	08 39	09 01	09 31
60	06 03	06 51	07 35	08 31	08 51	09 16	09 48
N 58	06 02	06 48	07 28	08 39	09 01	09 28	10 02
56	06 01	06 44	07 23	08 45	09 10	09 39	10 14
54	06 00	06 41	07 18	08 51	09 17	09 48	10 25
52	05 59	06 38	07 13	08 57	09 24	09 56	10 34
50	05 58	06 35	07 09	09 02	09 31	10 04	10 42
45	05 55	06 29	06 59	09 12	09 44	10 20	11 00
N 40	05 52	06 24	06 52	09 21	09 56	10 33	11 15
35	05 49	06 19	06 45	09 29	10 05	10 45	11 27
30	05 46	06 14	06 39	09 35	10 14	10 55	11 38
20	05 40	06 06	06 29	09 47	10 29	11 12	11 57
10	05 33	05 58	06 19	09 57	10 42	11 27	12 13
0	05 25	05 49	06 11	10 07	10 54	11 41	12 28
S 10	05 15	05 40	06 02	10 17	11 06	11 55	12 44
20	05 03	05 30	05 52	10 27	11 19	12 10	13 00
30	04 47	05 16	05 41	10 39	11 34	12 27	13 19
35	04 37	05 08	05 35	10 46	11 43	12 38	13 30
40	04 24	04 59	05 27	10 54	11 53	12 49	13 43
45	04 09	04 47	05 19	11 04	12 05	13 03	13 58
S 50	03 49	04 33	05 08	11 15	12 19	13 20	14 16
52	03 39	04 26	05 04	11 20	12 26	13 28	14 25
54	03 28	04 19	04 58	11 26	12 33	13 37	14 34
56	03 15	04 10	04 52	11 33	12 42	13 46	14 45
58	02 59	04 00	04 46	11 40	12 51	13 58	14 58
60	02 40	03 49	04 38	11 48	13 02	14 11	15 13

Lat. (°)	Sunset (h m)	Twilight Civil (h m)	Twilight Naut. (h m)	Moonset 15 (h m)	Moonset 16 (h m)	Moonset 17 (h m)	Moonset 18 (h m)
N 72	15 43	17 01	18 21	00 25	02 24	ooo	ooo
N 70	16 02	17 10	18 22	00 04	01 48	03 34	05 32
68	16 16	17 17	18 23	25 23	01 23	02 54	04 20
66	16 28	17 24	18 24	25 03	01 03	02 27	03 43
64	16 38	17 29	18 25	24 47	00 47	02 06	03 17
62	16 47	17 34	18 26	24 35	00 35	01 49	02 57
60	16 54	17 38	18 27	24 24	00 24	01 35	02 41
N 58	17 01	17 42	18 27	24 14	00 14	01 23	02 27
56	17 06	17 45	18 28	24 06	00 06	01 13	02 15
54	17 11	17 48	18 29	23 59	25 04	01 04	02 05
52	17 16	17 51	18 30	23 52	24 56	00 56	01 55
50	17 20	17 54	18 31	23 46	24 49	00 49	01 47
45	17 29	18 00	18 34	23 33	24 33	00 33	01 30
N 40	17 37	18 05	18 37	23 23	24 20	00 20	01 15
35	17 44	18 10	18 39	23 14	24 09	00 09	01 03
30	17 50	18 14	18 42	23 06	24 00	00 00	00 53
20	18 00	18 22	18 48	22 52	23 44	24 34	00 34
10	18 09	18 30	18 55	22 41	23 30	24 19	00 19
0	18 17	18 39	19 03	22 30	23 16	24 04	00 04
S 10	18 26	18 48	19 13	22 19	23 03	23 49	24 36
20	18 36	18 58	19 25	22 07	22 49	23 33	24 19
30	18 46	19 11	19 41	21 53	22 33	23 15	24 00
35	18 53	19 19	19 51	21 46	22 24	23 05	23 49
40	19 00	19 28	20 03	21 37	22 13	22 53	23 36
45	19 08	19 40	20 18	21 27	22 01	22 38	23 21
S 50	19 19	19 54	20 37	21 14	21 45	22 21	23 03
52	19 23	20 00	20 47	21 09	21 38	22 13	22 54
54	19 29	20 08	20 58	21 02	21 30	22 04	22 44
56	19 34	20 16	21 11	20 55	21 22	21 54	22 33
58	19 41	20 26	21 26	20 47	21 12	21 42	22 20
60	19 48	20 37	21 44	20 39	21 00	21 29	22 05

Day (d)	SUN Eqn. of Time 00^h (m s)	SUN Eqn. of Time 12^h (m s)	SUN Mer. Pass. (h m)	MOON Mer. Pass. Upper (h m)	MOON Mer. Pass. Lower (h m)	MOON Age (d)	MOON Age (%)	Phase
15	14 08	14 06	12 14	16 18	03 55	05	28	
16	14 04	14 03	12 14	17 05	04 42	06	37	
17	14 00	13 58	12 14	17 52	05 29	07	47	

〈표 2-25〉 TRUE BEARING OR AZIMUTH

LATITUDE 5°.

DECLINATION—CONTRARY NAME TO—LATITUDE.

Dec.	0°	1°	2°	3°	4°	5°	6°	7°	8°	9°	10°	11°	12°	13°	14°	15°	16°	17°	18°	19°
Apparent Time, A. M.	March.					April.										May.				
	21	23	26	28	31	3	5	8	11	13	16	19	22	25	28	1	5	8	12	16
	September.									August.										
	23	21	18	16	13	10	8	5	2	30	28	25	22	19	16	12	9	5	2	28
	September.			October.												November.				
	23	26	28	1	4	6	9	11	14	17	19	22	25	28	31	3	6	10	14	17
	March.									February.										
	21	18	16	13	11	8	6	3	1	26	23	20	18	15	12	9	5	2	29	25
h. m.	° ′	° ′	° ′	° ′	° ′	° ′	° ′	° ′	° ′	° ′	° ′	° ′	° ′	° ′	° ′	° ′	° ′	° ′	° ′	° ′
VI 00	90 00	91 00	......	......	......	......	......	......	......	......	......	......	......	......	......	......	......	......	......	......
10	90 13	91 13	92 13	93 12	94 12	95 12	96 12	97 12	98 11	99 11	100 11	101 11	102 10	103 10	104 10	105 10	106 10	107 09	108 09	109 09
20	90 26	91 26	92 26	93 26	94 26	95 26	96 26	97 26	98 26	99 26	100 25	101 25	102 25	103 25	104 25	105 24	106 24	107 24	108 24	109 23
30	90 39	91 40	92 40	93 40	94 41	95 41	96 41	97 41	98 41	99 41	100 41	101 41	102 41	103 41	104 41	105 41	106 41	107 41	108 40	109 40
40	90 53	91 54	92 54	93 55	94 55	95 56	96 56	97 57	98 57	99 58	100 58	101 58	102 58	103 59	104 59	105 59	106 59	107 59	108 59	109 59
50	91 06	92 07	93 09	94 10	95 11	96 12	97 13	98 14	99 15	100 15	101 16	102 17	103 18	104 18	105 19	106 19	107 19	108 20	109 20	110 20
VII 00	91 20	92 22	93 24	94 26	95 27	96 29	97 30	98 32	99 33	100 34	101 36	102 37	103 38	104 39	105 40	106 41	107 42	108 43	109 43	110 44
10	91 34	92 37	93 40	94 42	95 44	96 47	97 49	98 51	99 53	100 55	101 57	102 59	104 01	105 02	106 04	107 05	108 06	109 08	110 09	111 10
20	91 49	92 53	93 56	94 59	96 03	97 06	98 09	99 12	100 15	101 17	102 20	103 23	104 25	105 27	106 29	107 31	108 33	109 35	110 37	111 38
30	92 04	93 09	94 13	95 17	96 22	97 26	98 30	99 34	100 38	101 41	102 44	103 48	104 51	105 55	106 58	108 00	109 03	110 05	111 08	112 10
40	92 20	93 25	94 31	95 37	96 42	97 47	98 52	99 58	101 02	102 07	103 12	104 16	105 20	106 24	107 28	108 32	109 35	110 38	111 41	112 44
50	92 35	93 43	94 50	95 57	97 04	98 10	99 17	100 23	101 29	102 35	103 41	104 46	105 52	106 57	108 01	109 06	110 10	111 15	112 18	113 22
VIII 00	92 53	94 02	95 10	96 19	97 27	98 35	99 43	100 51	101 58	103 05	104 12	105 19	106 26	107 32	108 38	109 44	110 49	111 54	112 59	114 03
10	93 11	94 21	95 32	96 42	97 52	99 02	100 11	101 20	102 29	103 38	104 47	105 55	107 03	108 10	109 17	110 24	111 31	112 37	113 43	114 48
20	93 30	94 42	95 54	97 07	98 19	99 30	100 42	101 53	103 04	104 14	105 24	106 34	107 43	108 52	110 01	111 09	112 17	113 24	114 31	115 38
30	93 50	95 04	96 19	97 33	98 47	100 01	101 15	102 28	103 41	104 53	106 05	107 16	108 27	109 38	110 48	111 58	113 07	114 16	115 25	116 32
40	94 11	95 28	96 45	98 02	99 19	100 35	101 51	103 06	104 21	105 36	106 50	108 03	109 16	110 28	111 40	112 51	114 02	115 12	116 22	117 31
50	94 34	95 54	97 14	98 34	99 53	101 12	102 31	103 48	105 05	106 23	107 39	108 54	110 09	111 24	112 37	113 51	115 03	116 15	117 25	118 36
IX 00	94 59	96 23	97 46	99 09	100 31	101 53	103 15	104 35	105 55	107 14	108 33	109 51	111 08	112 25	113 40	114 55	116 09	117 23	118 35	119 47
10	95 26	96 53	98 20	99 47	101 13	102 38	104 02	105 26	106 49	108 12	109 33	110 54	112 13	113 32	114 50	116 07	117 23	118 38	119 52	121 05
20	95 56	97 28	98 59	100 29	101 59	103 28	104 56	106 23	107 50	109 15	110 40	112 03	113 26	114 47	116 07	117 26	118 44	120 01	121 17	122 31
30	96 29	98 05	99 41	101 16	102 51	104 24	105 56	107 28	108 58	110 27	111 54	113 21	114 46	116 10	117 33	118 54	120 14	121 33	122 50	124 06
40	97 06	98 48	100 29	102 10	103 49	105 27	107 04	108 40	110 14	111 47	113 18	114 48	116 17	117 43	119 08	120 32	121 54	123 14	124 34	125 51
50	97 48	99 36	101 24	103 10	104 55	106 39	108 21	110 01	111 40	113 17	114 53	116 26	117 58	119 28	120 56	122 22	123 45	125 08	126 28	127 47
X 00	98 35	100 31	102 26	104 19	106 11	108 01	109 49	111 35	113 19	115 01	116 40	118 17	119 53	121 25	122 56	124 24	125 50	127 14	128 36	129 56
10	99 30	101 35	103 39	105 40	107 39	109 36	111 31	113 23	115 12	116 59	118 43	120 24	122 03	123 39	125 12	126 42	128 10	129 35	130 58	132 19
20	100 35	102 51	105 03	107 15	109 24	111 28	113 30	115 28	117 23	119 15	121 05	122 50	124 32	126 10	127 46	129 18	130 47	132 14	133 37	134 58
30	101 52	104 21	106 46	109 08	111 26	113 40	115 50	117 56	119 57	121 55	123 48	125 37	127 22	129 03	130 41	132 14	133 44	135 11	136 34	137 54
40	103 27	106 11	108 50	111 25	113 54	116 18	118 37	120 50	122 57	125 01	126 59	128 51	130 38	132 22	134 00	135 34	137 03	138 30	139 52	141 10
50	105 27	108 29	111 25	114 14	116 56	119 31	121 59	124 20	126 35	128 42	130 43	132 38	134 26	136 09	137 47	139 20	140 48	142 12	143 32	144 47
XI 00	108 00	111 26	114 41	117 47	120 43	123 30	126 06	128 35	130 54	133 04	135 07	137 02	138 49	140 31	142 06	143 36	145 01	146 21	147 36	148 48
10	111 27	115 20	118 59	122 23	125 34	128 31	131 15	133 47	136 07	138 18	140 17	142 10	143 54	145 31	147 01	148 25	149 43	150 57	152 06	153 11
20	116 18	120 44	124 46	128 29	131 50	134 53	137 39	140 10	142 27	144 31	146 25	148 09	149 45	151 12	152 32	153 47	154 57	156 02	157 02	157 57
30	123 30	128 30	132 53	136 43	140 06	143 03	145 40	147 59	150 02	151 52	153 31	155 01	156 21	157 35	158 42	159 44	160 40	161 32	162 21	163 06
40	134 53	140 06	144 22	147 55	150 53	153 24	155 32	157 22	158 59	160 23	161 38	162 44	163 43	164 37	165 26	166 10	166 50	167 27	168 01	168 33
XI 50	153 23	157 20	160 18	162 37	164 27	165 57	167 11	168 14	169 08	169 54	170 34	171 10	171 42	172 10	172 35	172 59	173 19	173 39	173 56	174 13
	h. m.	*h. m.*	*h. m.*	*h. m.*	*h. m.*	*h. m.*	*h. m.*	*h. m.*	*h. m.*	*h. m.*	*h. m.*	*h. m.*	*h. m.*	*h. m.*	*h. m.*	*h. m.*	*h. m.*	*h. m.*	*h. m.*	*h. m.*
Sun rises. .	6 00	6 00	6 01	6 01	6 01	6 02	6 02	6 02	6 03	6 03	6 04	6 04	6 04	6 05	6 05	6 05	6 06	6 06	6 07	6 07
Sun sets. .	6 00	6 00	5 59	5 59	5 59	5 58	5 58	5 58	5 57	5 57	5 56	5 56	5 56	5 55	5 55	5 55	5 54	5 54	5 53	5 53
	° ′	° ′	° ′	° ′	° ′	° ′	° ′	° ′	° ′	° ′	° ′	° ′	° ′	° ′	° ′	° ′	° ′	° ′	° ′	° ′
Azimuth. .	90 00	91 00	92 00	93 01	94 01	95 01	96 01	97 02	98 02	99 02	100 02	101 03	102 03	103 03	104 03	105 04	106 04	107 04	108 04	109 05

② 위도가 05° 25′.1N이므로 2월 15일의 5°와 6°의 출몰방위각(amplitued)을 읽어 보간하면 103°03'.5 가 된다.

③ 103° 03′.5는 방위각이므로 (N, 동명극) 103°03'.5 (E, 출시)를 붙혀 방위로 환산하면 방위는 103° 03′.5가 된다.

④ 따라서 자이로 에러는 103° 03′.5−104°=−56′.5 즉 0.94°(W)가 된다.

〈표 2-26〉 TRUE BEARING OR AZIMUTH

LATITUDE 6°.

DECLINATION—CONTRARY NAME TO—LATITUDE.

Dec.	0°	1°	2°	3°	4°	5°	6°	7°	8°	9°	10°	11°	12	13°	14°	15°	16°	17°	18°	19°	20°
Apparent Time. A. M.	March.					April.										May.					
	21	23	26	28	31	3	5	8	11	13	16	19	22	25	28	1	5	8	12	16	21
	September.									August.											
	23	21	18	16	13	10	8	5	2	30	28	25	22	19	16	12	9	5	2	28	24
	September.			October.												November.					
	23	26	28	1	4	6	9	11	14	17	19	22	25	28	31	3	6	10	14	17	22
	March.									February.									Janu		
	21	18	16	13	11	8	6	3	1	26	23	20	18	15	12	9	5	2	29	25	21
h. m.	° ′	° ′	° ′	° ′	° ′	° ′	° ′	° ′	° ′	° ′	° ′	° ′	° ′	° ′	° ′	° ′	° ′	° ′	° ′	° ′	° ′
VI 00	90 00	91 00																			
10	90 16	91 15	92 15	93 15	94 15	95 14	96 14	97 14	98 13	99 13	100 13	101 12	102 12	103 11	104 11	105 11	106 10	107 10	108 10	109 09	110 09
20	90 31	91 31	92 31	93 31	94 31	95 31	96 30	97 30	98 30	99 30	100 29	101 29	102 29	103 29	104 28	105 28	106 27	107 27	108 27	109 26	110 26
30	90 47	91 47	92 48	93 48	94 48	95 48	96 48	97 48	98 48	99 48	100 48	101 47	102 47	103 47	104 47	105 47	106 46	107 46	108 46	109 45	110 45
40	91 03	92 04	93 04	94 05	95 05	96 06	97 06	98 06	99 07	100 07	101 07	102 07	103 07	104 07	105 07	106 07	107 07	108 07	109 07	110 07	111 06
50	91 20	92 21	93 22	94 23	95 24	96 24	97 25	98 26	99 27	100 27	101 28	102 28	103 29	104 29	105 29	106 30	107 30	108 30	109 30	110 30	111 30
VII 00	91 36	92 38	93 40	94 41	95 43	96 44	97 45	98 47	99 48	100 49	101 50	102 51	103 52	104 53	105 54	106 54	107 55	108 55	109 56	110 56	111 56
10	91 53	92 56	93 58	95 00	96 03	97 05	98 07	99 09	100 11	101 12	102 14	103 16	104 17	105 18	106 20	107 21	108 22	109 23	110 24	111 25	112 25
20	92 11	93 14	94 17	95 21	96 24	97 27	98 29	99 32	100 35	101 37	102 40	103 42	104 44	105 46	106 48	107 50	108 51	109 53	110 54	111 55	112 56
30	92 29	93 33	94 37	95 42	96 46	97 50	98 53	99 57	101 01	102 04	103 07	104 10	105 13	106 16	107 19	108 21	109 23	110 26	111 27	112 29	113 31
40	92 47	93 53	94 58	96 04	97 09	98 14	99 19	100 24	101 28	102 33	103 37	104 41	105 45	106 48	107 52	108 55	109 58	111 01	112 04	113 06	114 08
50	93 07	94 14	95 21	96 27	97 34	98 40	99 46	100 52	101 58	103 04	104 09	105 14	106 19	107 24	108 28	109 32	110 36	111 40	112 43	113 46	114 49
VIII 00	93 27	94 36	95 44	96 52	98 00	99 08	100 16	101 23	102 30	103 37	104 44	105 50	106 56	108 02	109 07	110 12	111 17	112 22	113 26	114 30	115 34
10	93 49	94 59	96 09	97 19	98 29	99 38	100 47	101 56	103 05	104 13	105 21	106 29	107 36	108 43	109 50	110 56	112 02	113 08	114 13	115 18	116 22
20	94 11	95 24	96 36	97 48	98 59	100 10	101 21	102 32	103 42	104 52	106 02	107 11	108 20	109 28	110 36	111 44	112 51	113 58	115 04	116 10	117 15
30	94 35	95 50	97 04	98 18	99 32	100 45	101 58	103 11	104 23	105 35	106 46	107 57	109 07	110 17	111 27	112 35	113 44	114 52	115 59	117 06	118 13
40	95 01	96 18	97 35	98 51	100 07	101 23	102 38	103 53	105 07	106 21	107 34	108 47	109 59	111 11	112 22	113 32	114 42	115 51	117 00	118 08	119 16
50	95 28	96 48	98 08	99 27	100 46	102 04	103 22	104 39	105 56	107 12	108 27	109 42	110 56	112 09	113 22	114 34	115 46	116 56	118 06	119 16	120 24
IX 00	95 58	97 21	98 44	100 06	101 28	102 49	104 10	105 30	106 49	108 07	109 25	110 42	111 59	113 14	114 29	115 42	116 55	118 08	119 19	120 29	121 39
10	96 30	97 57	99 24	100 50	102 15	103 39	105 03	106 26	107 48	109 09	110 29	111 49	113 07	114 25	115 42	116 57	118 12	119 26	120 39	121 50	123 01
20	97 06	98 37	100 08	101 37	103 06	104 34	106 01	107 28	108 53	110 17	111 40	113 02	114 24	115 43	117 02	118 20	119 36	120 51	122 06	123 19	124 31
30	97 45	99 21	100 56	102 30	104 04	105 36	107 07	108 37	110 06	111 33	112 59	114 24	115 48	117 11	118 32	119 51	121 10	122 27	123 42	124 57	126 10
40	98 29	100 11	101 51	103 30	105 08	106 45	108 20	109 55	111 27	112 58	114 28	115 56	117 22	118 47	120 10	121 32	122 53	124 12	125 29	126 44	127 58
50	99 19	101 07	102 53	104 38	106 21	108 04	109 44	111 22	112 59	114 34	116 08	117 39	119 08	120 36	122 02	123 26	124 48	126 08	127 26	128 43	129 58
X 00	100 16	102 11	104 04	105 55	107 45	109 33	111 19	113 02	114 44	116 23	118 00	119 35	121 07	122 38	124 06	125 31	126 55	128 16	129 36	130 53	132 09
10	101 21	103 25	105 26	107 25	109 22	111 16	113 08	114 57	116 43	118 27	120 08	121 47	123 22	124 55	126 25	127 53	129 18	130 40	132 00	133 18	134 34
20	102 38	104 51	107 02	109 10	111 15	113 16	115 15	117 10	119 02	120 50	122 35	124 17	125 55	127 30	129 02	130 31	131 57	133 20	134 41	135 58	137 14
30	104 09	106 35	108 56	111 14	113 28	115 38	117 43	119 43	121 42	123 35	125 24	127 09	128 50	130 27	132 00	133 30	134 56	136 19	137 39	138 55	140 09
40	106 01	108 41	111 15	113 43	116 08	118 28	120 40	122 47	124 50	126 46	128 39	130 26	132 09	133 47	135 20	136 50	138 16	139 38	140 56	142 11	143 23
50	108 21	111 16	114 02	116 47	119 20	121 50	124 11	126 25	128 32	130 33	132 27	134 15	135 58	137 36	139 08	140 36	142 00	143 19	144 35	145 47	146 55
XI 00	111 18	114 34	117 39	120 36	123 23	125 59	128 27	130 46	132 55	134 58	136 52	138 40	140 22	141 56	143 26	144 50	146 10	147 25	148 36	149 44	150 48
10	115 13	118 53	122 17	125 28	128 24	131 08	133 40	136 00	138 10	140 11	142 02	143 46	145 22	146 52	148 16	149 34	150 48	151 57	153 01	154 02	155 00
20	120 40	124 43	128 24	131 46	134 48	137 34	140 04	142 21	144 25	146 19	148 02	149 37	151 05	152 26	153 40	154 49	155 54	156 54	157 50	158 42	159 31
30	128 27	132 50	136 40	140 02	143 00	145 36	147 55	149 58	151 48	153 27	154 56	156 16	157 30	158 37	159 38	160 34	161 27	162 15	163 00	163 41	164 20
40	140 05	144 21	147 53	150 51	153 22	155 29	157 20	158 56	160 20	161 35	162 41	163 40	164 34	165 22	166 06	166 46	167 23	167 57	168 28	168 57	169 24
XI 50	157 20	160 18	162 36	164 26	165 55	167 10	168 13	169 06	169 52	170 33	171 08	171 40	172 08	172 34	172 57	173 18	173 37	173 54	174 10	174 26	174 39
	h. m.	h. m.	h. m.	h. m.	h. m.	h. m.	h. m.	h. m.	h. m.	h. m.	h. m.	h. m.	h. m.	h. m.	h. m.	h. m.	h. m.	h. m.	h. m.	h. m.	h. m.
Sun rises. .	6 00	6 00	6 01	6 01	6 02	6 02	6 03	6 03	6 03	6 04	6 04	6 05	6 05	6 06	6 06	6 06	6 07	6 07	6 08	6 08	6 09
Sun sets . .	6 00	6 00	5 59	5 59	5 58	5 58	5 57	5 57	5 57	5 56	5 56	5 55	5 55	5 54	5 54	5 54	5 53	5 53	5 52	5 52	5 51
Azimuth. .	90 00	91 00	92 01	93 01	94 01	95 02	96 02	97 02	98 03	99 03	100 03	101 04	102 04	103 04	104 05	105 05	106 05	107 06	108 06	109 07	110 07

2.12.2 시진방위각법

1. 229 Table의 이용

방위환을 통해 천체의 방위를 측정하고 측정시각에 대하여 천측력과 계산을 통해 항해삼각형을 풀기 위한 3가지 값(LHA, d, aL)을 ①, ②, ③에서 구한다. 이후 천측계산표를 통해 방위(Zn)를 구하고 실측방위와 비교하여 자이로 오차를 구한다.

① 천체를 관측한 시각을 GMT로 환산하여 천측력 주표에서 해당 천체의 GHA(본초시각, 그린 위치와 천체까지의 사이각를 서쪽 방향으로 측정한 값)와 d(적위, 천체와 적도사이의 각)를 확인한다.

② GHA에 가정경도를 가감하여(동경일 때는 +, 서경일 때는 −) LHA(지방시각, 관측자와 천체까지의 사이각을 서쪽 방향으로 측정한 값)를 구한다. 또한 위에서 1시간 단위로 구한 d(적위)값을 분단위까지 천측력의 부록을 통해 정확하게 구한다.

③ 가정위도(dL)를 설정한다.

④ 천측계산표에서 해당되는 LHA, d, aL 값으로 계산고도(Hc)와 방위각(Z)를 구하고 정확히 보간한다.

⑤ 천측계산표에서 구한 방위각(Z)를 방위(Zn)로 환산한다.
방위(Zn) = 동명극(N or S) / 천측계산표의 방위각(Z) / 자오선각(E or W)

⑥ 계산방위와 실측방위를 비교하여 자이로 오차를 구한다.

〈표 2-27〉

계산방법			내 용
GMT	GHA	Dec.(적위)	고도 관측시간을 GMT로 환산함
h			시각에 대한 GHA, d값을 천측력에서 구함
m s			분, 초에 대한 GHA, d값을 천측력에서 구함
GHA		d	시각, 분, 초에 대한 GHA, d값 합산함
aλ			가정경도. LHA값이 정수가 되도록 조정함 (동경이면 +, 서경이면 −)
LHA			천체의 지방시각임
t			자오선각, 관측자와 천체의 사이각을 가까운 쪽으로 측정(E or W)함
aL			

2. 태양방위각표의 이용

방위환을 통해 천체의 방위를 측정하고 측정 시각에 대하여 천측력과 계산을 통해 항해삼각형을 풀기 위한 3가지 값(LHA, d, aL)을 구한다. 그러나 태양방위각표(Red Azimuth Table)는 천체의 지방시각(LHA) 대신 시시를 이용하고 있으므로 자오선각(t)을 구한 후 시간으로 환산한 후 오전이면 (12−t)시각, 오후이면 t시각을 이용하여 계산방위를 구한다.

[예제] 2006년 2월 16일 09° 36'.6N, 110° 26'.3E에서 GMT 09시 18분 40초에 Gyro에서 측정한 방위가 254°일 때 태양방위각표에 의한 자이로 에러를 구하시오.

1) 먼저 천측력에서 2006년 2월 16일 09시 18분 40초에 대한 GHA 및 Dec를 구한다.

GHA 09시 311° 28'.8

Dec 09시 S 12° 19'.6

위의 값은 시간에 대한 Dec 및 GHA이므로 천측력 뒤쪽의 중분개정표에서 분 및 초에 대한 보정을 해준다.

GHA 18분 40초 4° 40′.0

Dec 18분 40초 − 0. ′3

GHA	Dec
09시 311° 28′.8	S 12° 19′.6
18분 40초 + 4° 40′.0	− 0′.3
316° 08'.8	S 12° 19′.3

2) GHA에 경도를 가감하여 LHA를 구한다.

GHA	315° 08′.8	
LONG	+ 110° 26′.3	
LHA	425° 35′.1	→ 65° 35′.1(t)

3) LHA를 자오선각으로 바꾼다.

LHA 65° 35′.1

t 65° 35′.1W (E - 오전, W- 오후)

4) 자오선각(t)을 시간으로 환산한다.

t 65° 35′.1 ⇒ 65° 4시 20분

35′.1 2.3분

즉, 오후 4시 22분 18초가 된다.

5) 위도 09°36.6′N, 적위 S 12° 19′.3, 시각 PM 4시 20분 18초에 대하여 보간하여 방위각을 구한다.

6) 즉, 적위 12°에서의 보간값 107° 11′.2 및 적위 13°에서의 보간값 107° 38′.6를 위도 9° 및 10°에 보간하면 Z(계산방위각) = 107° 27′.9가 된다.

〈표 2-28〉

	12°	12° 19′.3	13°
4h 30m	106° 17′	① 106° 36′.6	107° 18′
4h 20m 18s		③ 107° 11′.2	
4h 20m	106° 50′	② 107° 12′.2	107° 59′

① 60′ : 61′ = 19.3 : x, x = 19.6

② 60′ : 69′ = 19.3 : x, x = 22.2

③ 10m : −35.6′ = 0.3m : x, x = −1′.0

〈표 2-29〉

	12°	12° 19′.3	13°
4h 30m	106° 37′	④ 106° 56′.9	107° 39′
4h 20m 18s		⑥ 107° 38′.6	
4h 20m	107° 20′	⑤ 107° 39′.9	108° 22′

④ 60′ : 62′ = 19.3 : x, x = 19.9

⑤ 60′ : 62′ = 19.3 : x, x = 19.9

⑥ 10m : −43′ = 0.3m : x, x = −1′.3

⑦ ③과 ⑥의 값을 위도 9° 및 10°에 보간하면

60′ : 27.4′ = 36.6 : x, x = 16.7

7) 이 방위각을 방위로 환산하면 (N, 동명극) 107° 27′.9 (W, 오후)가 252°.5의 방위를 얻을 수 있다.

8) 따라서 Gyro Error는 252.5° − 254° = −1°.5, 즉 1.5(W)를 구할 수 있다.

〈표 2-30〉 2006. 2/15~2/17 (WED, THU, FRI)

UT d	UT h	SUN G.H.A. ° ′	SUN Dec. ° ′	MOON G.H.A. ° ′	MOON v ′	MOON Dec. ° ′	MOON d ′	MOON H.P. ′
15	0	176 27.7	S12 48.0	337 25.0	17.1	N 7 13.6	13.8	54.0
	1	191 27.7	S12 47.2	352 1.0	17.1	N 6 59.8	13.8	54.0
	2	206 27.7	S12 46.3	6 37.1	17.1	N 6 46.0	13.9	54.0
	3	221 27.8	S12 45.4	21 13.2	17.1	N 6 32.1	13.9	54.0
	4	236 27.8	S12 44.6	35 49.3	17.1	N 6 18.3	13.9	54.0
	5	251 27.8	S12 43.7	50 25.5	17.2	N 6 4.4	13.9	54.0
	6	266 27.9	S12 42.9	65 1.7	17.2	N 5 50.4	13.9	54.0
W	7	281 27.9	S12 42.0	79 37.8	17.2	N 5 36.5	14.0	54.0
E	8	296 27.9	S12 41.2	94 14.1	17.2	N 5 22.5	14.0	54.0
D	9	311 28.0	S12 40.3	108 50.3	17.2	N 5 8.5	14.0	54.0
N	10	326 28.0	S12 39.4	123 26.5	17.3	N 4 54.5	14.0	54.0
E	11	341 28.0	S12 38.6	138 2.8	17.3	N 4 40.5	14.0	54.1
S D	12	356 28.0	S12 37.7	152 39.0	17.3	N 4 26.4	14.1	54.1
A	13	11 28.1	S12 36.9	167 15.3	17.3	N 4 12.4	14.1	54.1
Y	14	26 28.1	S12 36.0	181 51.6	17.3	N 3 58.3	14.1	54.1
	15	41 28.1	S12 35.1	196 27.9	17.3	N 3 44.2	14.1	54.1
	16	56 28.2	S12 34.3	211 4.2	17.3	N 3 30.1	14.1	54.1
	17	71 28.2	S12 33.4	225 40.6	17.3	N 3 16.0	14.1	54.1
	18	86 28.2	S12 32.6	240 16.9	17.3	N 3 1.8	14.2	54.1
	19	101 28.3	S12 31.7	254 53.2	17.3	N 2 47.7	14.2	54.1
	20	116 28.3	S12 30.8	269 29.6	17.4	N 2 33.5	14.2	54.1
	21	131 28.3	S12 30.0	284 5.9	17.4	N 2 19.3	14.2	54.1
	22	146 28.4	S12 29.1	298 42.3	17.4	N 2 5.1	14.2	54.1
	23	161 28.4	S12 28.2	313 18.6	17.4	N 1 50.9	14.2	54.1
16	0	176 28.5	S12 27.4	327 55.0	17.4	N 1 36.7	14.2	54.1
	1	191 28.5	S12 26.5	342 31.4	17.4	N 1 22.5	14.2	54.1
	2	206 28.5	S12 25.7	357 7.7	17.4	N 1 8.3	14.2	54.2
	3	221 28.6	S12 24.8	11 44.1	17.4	N 0 54.0	14.2	54.2
	4	236 28.6	S12 23.9	26 20.4	17.4	N 0 39.8	14.2	54.2
	5	251 28.6	S12 23.1	40 56.8	17.3	N 0 25.6	14.3	54.2
	6	266 28.7	S12 22.2	55 33.1	17.3	N 0 11.3	14.3	54.2
T	7	281 28.7	S12 21.3	70 9.5	17.3	S 0 2.9	14.3	54.2
H	8	296 28.7	S12 20.5	84 45.8	17.3	S 0 17.2	14.3	54.2
U	9	311 28.8	S12 19.6	99 22.1	17.3	S 0 31.5	14.3	54.2
R	10	326 28.8	S12 18.7	113 58.5	17.3	S 0 45.7	14.3	54.2
S	11	341 28.9	S12 17.9	128 34.8	17.3	S 0 60.0	14.3	54.2
D A	12	356 28.9	S12 17.0	143 11.1	17.3	S 1 14.3	14.3	54.2
Y	13	11 28.9	S12 16.1	157 47.3	17.3	S 1 28.5	14.3	54.3
	14	26 29.0	S12 15.3	172 23.6	17.3	S 1 42.8	14.3	54.3
	15	41 29.0	S12 14.4	186 59.9	17.2	S 1 57.1	14.3	54.3
	16	56 29.1	S12 13.5	201 36.1	17.2	S 2 11.3	14.3	54.3
	17	71 29.1	S12 12.6	216 12.4	17.2	S 2 25.6	14.3	54.3
	18	86 29.1	S12 11.8	230 48.6	17.2	S 2 39.8	14.2	54.3
	19	101 29.2	S12 10.9	245 24.8	17.2	S 2 54.1	14.2	54.3
	20	116 29.2	S12 10.0	260 0.9	17.2	S 3 8.3	14.2	54.3
	21	131 29.3	S12 9.2	274 37.1	17.1	S 3 22.6	14.2	54.3
	22	146 29.3	S12 8.3	289 13.2	17.1	S 3 36.8	14.2	54.4
	23	161 29.4	S12 7.4	303 49.3	17.1	S 3 51.0	14.2	54.4
17	0	176 29.4	S12 6.6	318 25.4	17.1	S 4 5.2	14.2	54.4
	1	191 29.4	S12 5.7	333 1.5	17.0	S 4 19.4	14.2	54.4
	2	206 29.5	S12 4.8	347 37.5	17.0	S 4 33.6	14.2	54.4
	3	221 29.5	S12 3.9	2 13.5	17.0	S 4 47.8	14.2	54.4
	4	236 29.6	S12 3.1	16 49.5	16.9	S 5 2.0	14.2	54.4
	5	251 29.6	S12 2.2	31 25.4	16.9	S 5 16.2	14.2	54.4
	6	266 29.7	S12 1.3	46 1.3	16.9	S 5 30.3	14.1	54.5
	7	281 29.7	S12 0.4	60 37.2	16.8	S 5 44.4	14.1	54.5
F	8	296 29.7	S11 59.6	75 13.1	16.8	S 5 58.6	14.1	54.5
R	9	311 29.8	S11 58.7	89 48.9	16.8	S 6 12.7	14.1	54.5
I	10	326 29.8	S11 57.8	104 24.6	16.7	S 6 26.8	14.1	54.5
D	11	341 29.9	S11 56.9	119 0.4	16.7	S 6 40.8	14.1	54.5
A Y	12	356 29.9	S11 56.1	133 36.1	16.7	S 6 54.9	14.0	54.5
	13	11 30.0	S11 55.2	148 11.8	16.6	S 7 8.9	14.0	54.6
	14	26 30.0	S11 54.3	162 47.4	16.6	S 7 23.0	14.0	54.6
	15	41 30.1	S11 53.4	177 23.0	16.5	S 7 37.0	14.0	54.6
	16	56 30.1	S11 52.6	191 58.5	16.5	S 7 50.9	14.0	54.6
	17	71 30.2	S11 51.7	206 34.0	16.5	S 8 4.9	13.9	54.6
	18	86 30.2	S11 50.8	221 9.5	16.4	S 8 18.8	13.9	54.6
	19	101 30.3	S11 49.9	235 44.9	16.4	S 8 32.7	13.9	54.7
	20	116 30.3	S11 49.0	250 20.3	16.3	S 8 46.6	13.9	54.7
	21	131 30.4	S11 48.2	264 55.6	16.3	S 9 0.5	13.8	54.7
	22	146 30.4	S11 47.3	279 30.8	16.2	S 9 14.3	13.8	54.7
	23	161 30.5	S11 46.4	294 6.1	16.2	S 9 28.1	13.8	54.7
		S.D 16.2	d 0.9	S.D	14.7		14.8	14.9

Lat °	Twilight Naut h m	Twilight Civil h m	Sunrise h m	15 h m
N 72	6 11	7 31	8 49	19 29
N 70	6 10	7 21	8 30	19 33
68	6 8	7 14	8 15	19 36
66	6 7	7 7	8 2	19 39
64	6 6	7 1	7 52	19 41
62	6 5	6 57	7 43	19 43
60	6 4	6 52	7 36	19 44
N 58	6 2	6 48	7 29	19 46
56	6 1	6 45	7 23	19 47
54	6 0	6 42	7 18	19 48
52	5 59	6 39	7 13	19 49
50	5 58	6 36	7 9	19 50
45	5 55	6 30	7 0	19 52
N 40	5 53	6 24	6 52	19 54
35	5 50	6 19	6 45	19 55
30	5 47	6 15	6 39	19 56
20	5 40	6 6	6 29	19 59
10	5 33	5 58	6 19	20 1
0	5 25	5 50	6 11	20 2
S 10	5 15	5 40	6 2	20 4
20	5 3	5 29	5 52	20 6
30	4 47	5 16	5 41	20 9
35	4 36	5 8	5 35	20 10
40	4 24	4 58	5 27	20 11
45	4 9	4 47	5 18	20 13
S 50	3 49	4 33	5 8	20 15
52	3 39	4 26	5 3	20 16
54	3 27	4 18	4 58	20 17
56	3 14	4 9	4 52	20 18
58	2 58	4 0	4 45	20 19
60	2 39	3 48	4 37	20 20

Lat	Sunset h m	Twilight Civil h m	Twilight Naut h m	15 h m
N 72	15 41	16 59	18 19	8 48
N 70	16 0	17 9	18 21	8 41
68	16 15	17 16	18 22	8 35
66	16 27	17 23	18 23	8 31
64	16 37	17 28	18 24	8 27
62	16 46	17 33	18 25	8 23
60	16 53	17 37	18 26	8 20
N 58	17 0	17 41	18 27	8 17
56	17 6	17 44	18 28	8 15
54	17 11	17 47	18 29	8 12
52	17 16	17 50	18 30	8 10
50	17 20	17 53	18 31	8 8
45	17 29	17 59	18 34	8 4
N 40	17 37	18 4	18 36	8 1
35	17 43	18 9	18 39	7 58
30	17 49	18 14	18 42	7 55
20	18 0	18 22	18 48	7 50
10	18 9	18 30	18 55	7 46
0	18 17	18 39	19 3	7 42
S 10	18 26	18 48	19 13	7 38
20	18 36	18 58	19 25	7 34
30	18 47	19 12	19 41	7 29
35	18 53	19 19	19 51	7 26
40	19 0	19 29	20 3	7 23
45	19 9	19 40	20 18	7 19
S 50	19 19	19 54	20 38	7 15
52	19 24	20 1	20 48	7 13
54	19 29	20 8	20 59	7 10
56	19 35	20 17	21 12	7 8
58	19 42	20 27	21 27	7 5
60	19 49	20 38	21 46	7 2

Day	SUN Eqn. of Time 00h m s	SUN Eqn. of Time 12h m s	SUN Mer. Pass. h m	Mer. Pa Upper h m
15	14 9	14 8	12 14	1 33
16	14 6	14 4	12 14	2 12
17	14 2	14 0	12 14	2 51

〈표 2-31〉

LATITUDE 9°.

DECLINATION—CONTRARY NAME TO—LATITUDE.

Dec.	0°	1°	2°	3°	4°	5°	6°	7°	8°	9°	10°	11°	12°	13°	14°	15°	16°	17°	18°	19°	20°	21°	22°	23°	Dec.
Apparent Time. A. M.	March. 21	23	26	28	31	April. 3	5	8	11	13	16	19	22	25	28	May. 1	5	8	12	16	21	26	June. 1	10	Apparent Time. P. M.
	September. 23	21	18	16	13	10	8	5	2	August. 30	28	25	22	19	16	12	9	5	2	July. 28	24	19	12	3	
	September. 23	26	28	October. 1	4	6	9	11	14	17	19	22	25	28	31	November. 3	6	10	14	17	22	27	December. 3	11	
	March. 21	18	16	13	11	8	6	3	1	February. 26	23	20	18	15	12	9	5	2	January. 29	25	21	16	10	2	
h. m.	° ′	° ′	° ′	° ′	° ′	° ′	° ′	° ′	° ′	° ′	° ′	° ′	° ′	° ′	° ′	° ′	° ′	° ′	° ′	° ′	° ′	° ′	° ′	° ′	h. m.
VI 00	90 00	......	......	......	......	......	......	......	......	......	......	......	......	......	......	......	......	......	......	......	......	......	......	......	VI 00
10	90 23	91 23	92 22	93 21	94 21	95 20	96 19	97 18	98 18	99 17	100 16	101 15	102 14	103 14	104 13	......	......	......	......	......	......	......	......	......	50
20	90 47	91 47	92 46	93 45	94 45	95 44	96 44	97 43	98 42	99 41	100 41	101 40	102 39	103 38	104 37	105 36	106 36	107 35	108 34	109 33	110 32	111 31	112 30	113 29	40
30	91 11	92 11	93 10	94 10	95 09	96 09	97 09	98 08	99 08	100 07	101 07	102 06	103 05	104 04	105 03	106 02	107 02	108 01	109 00	109 59	110 58	111 57	112 56	113 55	30
40	91 35	92 35	93 35	94 35	95 35	96 35	97 35	98 34	99 34	100 34	101 33	102 33	103 32	104 32	105 31	106 30	107 30	108 29	109 28	110 27	111 26	112 25	113 24	114 23	20
50	91 59	93 00	94 00	95 01	96 01	97 01	98 02	99 02	100 02	101 02	102 02	103 02	104 01	105 01	106 01	107 00	108 00	108 59	109 58	110 58	111 57	112 56	113 55	114 54	10
VII 00	92 24	93 25	94 26	95 27	96 28	97 29	98 30	99 30	100 31	101 31	102 32	103 32	104 32	105 32	106 32	107 32	108 32	109 31	110 31	111 30	112 30	113 29	114 28	115 27	V 00
10	92 49	93 51	94 53	95 55	96 56	97 58	98 59	100 01	101 02	102 03	103 03	104 04	105 05	106 05	107 06	108 06	109 06	110 06	111 06	112 06	113 05	114 05	115 04	116 03	50
20	93 16	94 18	95 21	96 24	97 26	98 28	99 30	100 32	101 34	102 36	103 37	104 38	105 40	106 41	107 42	108 42	109 43	110 43	111 44	112 44	113 44	114 43	115 43	116 42	40
30	93 42	94 46	95 50	96 53	97 57	99 00	100 03	101 06	102 08	103 11	104 13	105 15	106 17	107 18	108 20	109 21	110 22	111 23	112 24	113 25	114 25	115 25	116 25	117 25	30
40	94 10	95 15	96 20	97 25	98 29	99 33	100 37	101 41	102 45	103 48	104 51	105 54	106 56	107 59	109 01	110 03	111 05	112 06	113 07	114 08	115 09	116 10	117 10	118 10	20
50	94 39	95 46	96 52	97 57	99 03	100 08	101 13	102 18	103 23	104 27	105 31	106 35	107 39	108 42	109 45	110 48	111 50	112 52	113 54	114 56	115 57	116 58	117 59	118 59	10
VIII 00	95 10	96 18	97 25	98 32	99 39	100 46	101 52	102 58	104 04	105 10	106 15	107 20	108 24	109 28	110 32	111 36	112 39	113 42	114 45	115 47	116 49	117 51	118 52	119 52	IV 00
10	95 42	96 51	98 00	99 09	100 17	101 26	102 33	103 41	104 48	105 55	107 01	108 07	109 13	110 18	111 23	112 28	113 32	114 36	115 39	116 42	117 44	118 46	119 48	120 50	50
20	96 15	97 26	98 37	99 48	100 58	102 08	103 18	104 27	105 35	106 44	107 51	108 59	110 06	111 12	112 18	113 24	114 29	115 34	116 38	117 41	118 44	119 47	120 49	121 51	40
30	96 51	98 04	99 17	100 30	101 42	102 54	104 05	105 16	106 26	107 36	108 45	109 54	111 02	112 10	113 17	114 24	115 30	116 36	117 41	118 45	119 49	120 53	121 55	122 58	30
40	97 29	98 45	100 00	101 15	102 29	103 43	104 56	106 09	107 21	108 33	109 44	110 54	112 04	113 13	114 21	115 29	116 37	117 43	118 49	119 54	120 59	122 03	123 07	124 10	20
50	98 09	99 28	100 46	102 03	103 20	104 36	105 52	107 07	108 21	109 34	110 47	111 59	113 11	114 21	115 31	116 40	117 49	118 56	120 03	121 09	122 15	123 20	124 24	125 27	10
IX 00	98 53	100 15	101 36	102 56	104 15	105 34	106 52	108 09	109 26	110 41	111 56	113 10	114 23	115 35	116 47	117 57	119 07	120 16	121 24	122 31	123 37	124 42	125 47	126 51	III 00
10	99 41	101 06	102 30	103 54	105 16	106 38	107 58	109 18	110 37	111 55	113 12	114 28	115 43	116 56	118 09	119 21	120 32	121 42	122 51	123 59	125 06	126 12	127 17	128 21	50
20	100 34	102 02	103 30	104 57	106 23	107 47	109 11	110 34	111 55	113 15	114 34	115 53	117 10	118 25	119 40	120 53	122 05	123 15	124 26	125 35	126 43	127 49	128 55	129 59	40
30	101 31	103 04	104 36	106 07	107 36	109 05	110 32	111 57	113 22	114 45	116 06	117 26	118 45	120 03	121 19	122 34	123 47	124 59	126 10	127 20	128 28	129 35	130 41	131 45	30
40	102 36	104 13	105 50	107 25	108 58	110 31	112 01	113 30	114 57	116 23	117 47	119 10	120 31	121 50	123 08	124 24	125 39	126 52	128 04	129 14	130 22	131 29	132 35	133 40	20
50	103 48	105 31	107 13	108 53	110 31	112 07	113 41	115 13	116 45	118 13	119 40	121 05	122 28	123 49	125 08	126 26	127 42	128 55	130 07	131 18	132 27	133 34	134 40	135 44	10
X 00	105 10	106 59	108 46	110 32	112 15	113 56	115 34	117 11	118 45	120 17	121 46	123 13	124 38	126 01	127 22	128 40	129 57	131 11	132 23	133 34	134 43	135 50	136 55	137 58	II 00
10	106 44	108 40	110 34	112 25	114 14	116 00	117 43	119 23	121 01	122 35	124 07	125 37	127 04	128 28	129 50	131 09	132 26	133 40	134 52	136 03	137 11	138 17	139 21	140 24	50
20	108 33	110 37	112 38	114 36	116 31	118 22	120 10	121 54	123 35	125 12	126 47	128 18	129 46	131 11	132 33	133 53	135 10	136 24	137 35	138 45	139 52	140 57	142 00	143 01	40
30	110 42	112 55	115 04	117 09	119 10	121 07	122 59	124 47	126 31	128 11	129 47	131 20	132 49	134 14	135 36	136 55	138 11	139 24	140 34	141 42	142 48	143 51	144 51	145 50	30
40	113 16	115 39	117 57	120 10	122 17	124 19	126 15	128 07	129 54	131 35	133 12	134 45	136 14	137 39	139 00	140 17	141 31	142 42	143 50	144 55	145 58	146 59	147 56	148 52	20
50	116 23	118 58	121 25	123 45	125 58	128 05	130 05	131 58	133 46	135 28	137 05	138 36	140 04	141 26	142 45	144 00	145 11	146 18	147 23	148 26	149 25	150 21	151 15	152 07	10
XI 00	120 17	123 02	125 39	128 05	130 23	132 32	134 34	136 28	138 15	139 55	141 30	142 58	144 22	145 41	146 55	148 06	149 12	150 16	151 16	152 13	153 07	153 59	154 49	155 36	I 00
10	125 13	128 09	130 52	133 22	135 42	137 51	139 51	141 42	143 25	145 00	146 29	147 53	149 10	150 23	151 31	152 35	153 36	154 33	155 27	156 17	157 06	157 51	158 35	159 16	50
20	131 35	134 37	137 21	139 50	142 06	144 10	146 03	147 46	149 20	150 47	152 07	153 21	154 30	155 34	156 33	157 28	158 20	159 09	159 55	160 38	161 19	161 58	162 34	163 09	40
30	139 55	142 52	145 27	147 45	149 47	151 37	153 15	154 43	156 03	157 16	158 23	159 23	160 19	161 11	161 59	162 44	163 25	164 04	164 40	165 14	165 46	166 16	166 45	167 12	30
40	150 47	153 16	155 24	157 14	158 49	160 12	161 27	162 32	163 31	164 24	165 12	165 56	166 36	167 12	167 46	168 17	168 46	169 13	169 38	170 02	170 24	170 45	171 04	171 23	20
XI 50	164 24	165 53	167 07	168 10	169 03	169 49	170 29	171 04	171 35	172 04	172 29	172 52	173 12	173 31	173 49	174 05	174 20	174 34	174 47	174 59	175 11	175 21	175 30	175 39	XII 10
	h. m.	*h. m.*	*h. m.*	*h. m.*	*h. m.*	*h. m.*	*h. m.*	*h. m.*	*h. m.*	*h. m.*	*h. m.*	*h. m.*	*h. m.*	*h. m.*	*h. m.*	*h. m.*	*h. m.*	*h. m.*	*h. m.*	*h. m.*	*h. m.*	*h. m.*	*h. m.*	*h. m.*	
Sun rises . .	6 00	6 01	6 01	6 02	6 03	6 03	6 04	6 04	6 05	6 06	6 06	6 07	6 08	6 08	6 09	6 10	6 10	6 11	6 12	6 13	6 13	6 14	6 15	6 15	Sun rises.
Sun sets . .	6 00	5 59	5 59	5 58	5 57	5 57	5 56	5 56	5 55	5 54	5 54	5 53	5 52	5 52	5 51	5 50	5 50	5 49	5 48	5 47	5 47	5 46	5 45	5 45	Sun sets.
Azimuth . .	90 00	91 01	92 01	93 02	94 03	95 04	96 05	97 05	98 06	99 07	100 08	101 08	102 09	103 10	104 11	105 11	106 12	107 13	108 14	109 15	110 16	111 16	112 17	113 18	Azimuth.

北緯度에 있어서, 天体가 뜨고 있거나 子午線의 東쪽에 있을 때에는 表記된 方位는 北에서 東으로 計算하고, 天体가 지고 있거나 子午線의 西쪽에 있을 때에는 表記된 方位는 北에서 西로 計算한다.

南緯度에 있어서, 天体가 뜨고 있거나 子午線의 東쪽에 있을 때에는 表記된 方位는 南에서 東으로 計算하고, 天体가 지고 있거나 子午線의 西쪽에 있을 때에는 表記된 方位는 南에서 西로 計算한다.

〈표 2-32〉

LATITUDE 10°.

DECLINATION—CONTRARY NAME TO—LATITUDE.

Dec.	0°	1°	2°	3°	4°	5°	6°	7°	8°	9°	10°	11°	12°	13°	14°	15°	16°	17°	18°	19°	20°	21°	22°	23°	Dec.
Apparent Time. A. M.	March. 21	23	26	28	31	April. 3	5	8	11	13	16	19	22	25	28	May. 1	5	8	12	16	21	26	June. 1	10	Apparent Time. P. M.
	September. 23	21	18	16	13	10	8	5	2	August. 30	28	25	22	19	16	12	9	5	2	July. 28	24	19	12	3	
	September. 23	26	28	October. 1	4	6	9	11	14	17	19	22	25	28	31	November. 3	6	10	14	17	22	27	December. 3	11	
	March. 21	18	16	13	11	8	6	3	1	February. 26	23	20	18	15	12	9	5	2	January. 29	25	21	16	10	2	
h. m.	° ′	° ′	° ′	° ′	° ′	° ′	° ′	° ′	° ′	° ′	° ′	° ′	° ′	° ′	° ′	° ′	° ′	° ′	° ′	° ′	° ′	° ′	° ′	° ′	h. m.
VI 00	90 00	......	......	......	......	......	......	......	......	......	......	......	......	......	......	......	......	......	......	......	......	......	......	......	VI 00
10	90 26	91 25	92 24	93 23	94 23	95 22	96 21	97 20	98 19	99 18	100 17	101 16	102 15	103 14	......	......	......	......	......	......	......	......	......	......	50
20	90 52	91 52	92 51	93 50	94 49	95 49	96 48	97 47	98 46	99 45	100 44	101 43	102 42	103 41	104 40	105 39	106 38	107 37	108 36	109 35	110 34	111 33	112 32	113 30	40
30	91 19	92 18	93 18	94 17	95 17	96 16	97 15	98 14	99 14	100 13	101 12	102 11	103 10	104 09	105 08	106 07	107 06	108 05	109 04	110 03	111 02	112 00	112 59	113 58	30
40	91 45	92 45	93 45	94 45	95 45	96 44	97 44	98 43	99 43	100 42	101 42	102 41	103 40	104 39	105 38	106 37	107 36	108 35	109 34	110 33	111 32	112 31	113 30	114 28	20
50	92 12	93 13	94 13	95 13	96 13	97 13	98 13	99 13	100 13	101 13	102 13	103 12	104 12	105 11	106 10	107 10	108 09	109 08	110 07	111 06	112 05	113 04	114 02	115 01	10
VII 00	92 40	93 41	94 41	95 42	96 43	97 44	98 44	99 45	100 45	101 45	102 45	103 45	104 45	105 45	106 44	107 44	108 43	109 43	110 42	111 41	112 40	113 39	114 38	115 37	V 00
10	93 08	94 10	95 11	96 13	97 14	98 15	99 16	100 17	101 18	102 19	103 19	104 20	105 20	106 20	107 20	108 20	109 20	110 20	111 19	112 19	113 18	114 17	115 16	116 15	50
20	93 37	94 40	95 42	96 44	97 46	98 48	99 50	100 52	101 53	102 55	103 56	104 57	105 58	106 58	107 59	108 59	109 59	110 59	111 59	112 59	113 58	114 58	115 57	116 56	40
30	94 07	95 10	96 14	97 17	98 20	99 23	100 25	101 28	102 30	103 32	104 34	105 36	106 37	107 39	108 40	109 40	110 41	111 42	112 42	113 42	114 42	115 42	116 41	117 40	30
40	94 38	95 42	96 47	97 51	98 55	99 59	101 03	102 06	103 09	104 12	105 15	106 17	107 20	108 22	109 23	110 25	111 26	112 27	113 28	114 29	115 29	116 29	117 28	118 28	20
50	95 10	96 16	97 22	98 27	99 32	100 37	101 42	102 46	103 51	104 55	105 58	107 01	108 04	109 07	110 10	111 12	112 14	113 16	114 17	115 18	116 19	117 19	118 19	119 19	10
VIII 00	95 44	96 51	97 58	99 05	100 12	101 18	102 24	103 30	104 35	105 40	106 45	107 49	108 53	109 57	111 00	112 03	113 05	114 08	115 10	116 11	117 13	118 14	119 14	120 14	IV 00
10	96 19	97 28	98 37	99 45	100 53	102 01	103 08	104 15	105 22	106 28	107 34	108 39	109 44	110 49	111 53	112 57	114 01	115 03	116 06	117 09	118 11	119 12	120 13	121 14	50
20	96 56	98 07	99 17	100 28	101 37	102 47	103 56	105 04	106 12	107 20	108 27	109 34	110 40	111 46	112 51	113 56	115 00	116 04	117 07	118 10	119 13	120 15	121 16	122 17	40
30	97 35	98 48	100 01	101 13	102 25	103 36	104 46	105 56	107 06	108 15	109 24	110 32	111 39	112 46	113 53	114 59	116 04	117 09	118 13	119 17	120 20	121 22	122 24	123 26	30
40	98 17	99 33	100 47	102 02	103 15	104 29	105 41	106 53	108 05	109 15	110 25	111 35	112 44	113 52	115 00	116 07	117 13	118 19	119 24	120 28	121 32	122 35	123 38	124 40	20
50	99 03	100 20	101 37	102 54	104 10	105 25	106 40	107 54	109 07	110 20	111 32	112 43	113 53	115 03	116 12	117 20	118 28	119 35	120 40	121 45	122 50	123 54	124 57	125 59	10
IX 00	99 51	101 12	102 32	103 51	105 10	106 28	107 45	109 01	110 16	111 31	112 45	113 58	115 10	116 21	117 31	118 40	119 49	120 56	122 03	123 09	124 14	125 19	126 22	127 25	III 00
10	100 44	102 08	103 31	104 53	106 15	107 35	108 55	110 14	111 31	112 48	114 04	115 18	116 32	117 45	118 56	120 07	121 17	122 25	123 33	124 40	125 45	126 50	127 54	128 57	50
20	101 42	103 09	104 36	106 02	107 26	108 50	110 12	111 33	112 54	114 13	115 30	116 47	118 02	119 17	120 30	121 42	122 52	124 02	125 10	126 18	127 24	128 29	129 33	130 36	40
30	102 45	104 17	105 47	107 17	108 45	110 12	111 37	113 01	114 24	115 46	117 06	118 24	119 41	120 57	122 12	123 25	124 37	125 47	126 57	128 04	129 11	130 16	131 21	132 24	30
40	103 56	105 32	107 07	108 40	110 12	111 43	113 12	114 39	116 04	117 28	118 51	120 11	121 30	122 48	124 04	125 18	126 31	127 42	128 52	130 00	131 07	132 13	133 17	134 20	20
50	105 15	106 56	108 36	110 14	111 50	113 25	114 57	116 27	117 56	119 22	120 47	122 10	123 31	124 50	126 07	127 22	128 36	129 48	130 58	132 06	133 13	134 19	135 22	136 25	10
X 00	106 45	108 32	110 17	112 00	113 41	115 19	116 55	118 29	120 01	121 30	122 57	124 21	125 44	127 05	128 23	129 39	130 53	132 05	133 15	134 23	135 30	136 35	137 38	138 40	II 00
10	108 27	110 21	112 12	114 00	115 46	117 29	119 09	120 46	122 21	123 53	125 22	126 48	128 12	129 33	130 52	132 09	133 23	134 35	135 45	136 53	137 59	139 03	140 05	141 05	50
20	110 25	112 27	114 24	116 19	118 10	119 57	121 41	123 22	124 59	126 33	128 04	129 32	130 57	132 19	133 38	134 54	136 08	137 20	138 29	139 36	140 40	141 43	142 44	143 42	40
30	112 44	114 54	116 59	118 59	120 56	122 48	124 36	126 19	127 59	129 38	131 07	132 36	134 01	135 23	136 41	137 57	139 10	140 20	141 28	142 33	143 35	144 36	145 34	146 31	30
40	115 30	117 48	120 01	122 07	124 09	126 06	127 57	129 43	131 24	133 01	134 34	136 03	137 27	138 48	140 05	141 19	142 29	143 37	144 42	145 45	146 45	147 42	148 38	149 31	20
50	118 51	121 18	123 38	125 51	127 57	129 56	131 50	133 37	135 19	136 55	138 27	139 54	141 16	142 35	143 49	145 00	146 08	147 12	148 14	149 13	150 09	151 03	151 55	152 44	10
XI 00	122 57	125 33	127 59	130 17	132 26	134 27	136 21	138 07	139 48	141 21	142 50	144 13	145 32	146 46	147 57	149 03	150 06	151 06	152 03	152 57	153 49	154 37	155 25	156 10	I 00
10	128 04	130 47	133 18	135 37	137 46	139 45	141 36	143 18	144 54	146 23	147 46	149 03	150 16	151 24	152 28	153 28	154 25	155 18	156 09	156 57	157 43	158 26	159 07	159 47	50
20	134 34	137 18	139 47	142 03	144 06	145 58	147 41	149 15	150 42	152 02	153 16	154 24	155 28	156 27	157 22	158 14	159 03	159 48	160 31	161 12	161 51	162 27	163 02	163 35	40
30	142 50	145 25	147 43	149 45	151 34	153 12	154 40	156 00	157 12	158 19	159 19	160 15	161 07	161 55	162 39	163 20	163 59	164 35	165 09	165 41	166 11	166 39	167 06	167 32	30
40	153 16	155 23	157 12	158 48	160 11	161 25	162 30	163 29	164 22	165 10	165 53	166 33	167 09	167 43	168 14	168 43	169 10	169 35	169 58	170 20	170 41	171 01	171 19	171 37	20
XI 50	165 53	167 07	168 09	169 02	169 48	170 28	171 03	171 34	172 02	172 27	172 50	173 11	173 30	173 47	174 03	174 18	174 32	174 45	174 57	175 08	175 19	175 29	175 38	175 47	XII 10
	h. m.	h. m.	h. m.	h. m.	h. m.	h. m.	h. m.	h. m.	h. m.	h. m.	h. m.	h. m.	h. m.	h. m.	h. m.	h. m.	h. m.	h. m.	h. m.	h. m.	h. m.	h. m.	h. m.	h. m.	
Sun rises . .	6 00	6 01	6 01	6 02	6 03	6 04	6 04	6 05	6 06	6 06	6 07	6 08	6 09	6 09	6 10	6 11	6 12	6 12	6 13	6 14	6 15	6 16	6 16	6 17	Sun rises.
Sun sets . .	6 00	5 59	5 59	5 58	5 57	5 56	5 56	5 55	5 54	5 54	5 53	5 52	5 51	5 51	5 50	5 49	5 48	5 48	5 47	5 46	5 45	5 44	5 44	5 43	Sun sets.
Azimuth . .	90 00	91 01	92 02	93 03	94 04	95 05	96 06	97 07	98 07	99 08	100 09	101 10	102 11	103 12	104 13	105 14	106 15	107 16	108 17	109 18	110 19	111 20	112 21	113 23	Azimuth.

北緯度에 있어서, 天体가 뜨고 있거나 子午線의 東쪽에 있을 때에는 表記된 方位는 北에서 東으로 計算하고, 天体가 지고 있거나 子午線의 西쪽에 있을 때에는 表記된 方位는 北에서 西로 計算한다.

南緯度에 있어서, 天体가 뜨고 있거나 子午線의 東쪽에 있을 때에는 表記된 方位는 南에서 東으로 計算하고, 天体가 지고 있거나 子午線의 西쪽에 있을 때에는 表記 [illegible]

3. 북극성 방위각법(Azimuth by polaris)

천측력의 북극성표는 관측자의 위도와 LHAΥ를 가지고 북극성의 계산방위를 쉽게 얻을 수 있도록 되어 있다. 계산방위와 측정방위를 상호 비교하여 자이로 오차를 구한다.

〈표 2-33〉 북극성표

POLARIS (POLE STAR) TABLES, 2006
FOR DETERMINING LATITUDE FROM SEXTANT ALTITUDE AND FOR AZIMUTH

LHA ARIES	240°–249°	250°–259°	260°–269°	270°–279°	280°–289°	290°–299°	300°–309°	310°–319°	320°–329°	330°–339°	340°–349°	350°–359°
	a_0	a_0	a_0	a_0	a_0	a_0	a_0	a_0	a_0	a_0	a_0	a_0
°	° ′	° ′	° ′	° ′	° ′	° ′	° ′	° ′	° ′	° ′	° ′	° ′
0	1 38.6	1 35.5	1 31.2	1 26.0	1 20.0	1 13.3	1 6.2	0 58.9	0 51.5	0 44.4	0 37.6	0 31.6
1	38.3	35.1	30.8	25.5	19.4	12.7	5.5	58.1	50.8	43.7	37.0	31.0
2	38.0	34.7	30.3	24.9	18.7	12.0	4.8	57.4	50.1	43.0	36.4	30.4
3	37.8	34.3	29.8	24.3	18.1	11.3	4.0	56.7	49.3	42.3	35.7	29.9
4	37.5	33.9	29.3	23.7	17.4	10.5	3.3	55.9	48.6	41.6	35.1	29.4
5	1 37.2	1 33.5	1 28.8	1 23.1	1 16.8	1 9.8	1 2.6	0 55.2	0 47.9	0 40.9	0 34.5	0 28.8
6	36.8	33.1	28.2	22.5	16.1	9.1	1.8	54.4	47.2	40.3	33.9	28.3
7	36.5	32.6	27.7	21.9	15.4	8.4	1.1	53.7	46.5	39.6	33.3	27.8
8	36.2	32.2	27.2	21.3	14.7	7.7	0.4	53.0	45.8	38.9	32.7	27.3
9	35.8	31.7	26.6	20.7	14.0	7.0	0 59.6	52.2	45.1	38.3	32.1	26.8
10	1 35.5	1 31.2	1 26.0	1 20.0	1 13.3	1 6.2	0 58.9	0 51.5	0 44.4	0 37.6	0 31.6	0 26.3
Lat.	a_1	a_1	a_1	a_1	a_1	a_1	a_1	a_1	a_1	a_1	a_1	a_1
°	′	′	′	′	′	′	′	′	′	′	′	′
0	0.6	0.5	0.5	0.4	0.4	0.3	0.3	0.3	0.3	0.3	0.4	0.4
10	.6	.5	.5	.4	.4	.4	.3	.3	.3	.4	.4	.5
20	.6	.5	.5	.5	.4	.4	.4	.4	.4	.4	.4	.5
30	.6	.6	.5	.5	.5	.5	.4	.4	.4	.5	.5	.5
40	0.6	0.6	0.6	0.5	0.5	0.5	0.5	0.5	0.5	0.5	0.5	0.5
45	.6	.6	.6	.6	.6	.6	.6	.5	.6	.6	.6	.6
50	.6	.6	.6	.6	.6	.6	.6	.6	.6	.6	.6	.6
55	.6	.6	.6	.6	.6	.7	.7	.7	.7	.7	.6	.6
60	.6	.6	.7	.7	.7	.7	.7	.7	.7	.7	.7	.7
62	0.6	0.6	0.7	0.7	0.7	0.8	0.8	0.8	0.8	0.8	0.7	0.7
64	.6	.7	.7	.7	.8	.8	.8	.8	.8	.8	.8	.7
66	.6	.7	.7	.8	.8	.8	.9	.9	.9	.8	.8	.8
68	0.6	0.7	0.7	0.8	0.9	0.9	0.9	0.9	0.9	0.9	0.8	0.8
Month	a_2	a_2	a_2	a_2	a_2	a_2	a_2	a_2	a_2	a_2	a_2	a_2
	′	′	′	′	′	′	′	′	′	′	′	′
Jan.	0.4	0.4	0.4	0.5	0.5	0.5	0.6	0.6	0.6	0.7	0.7	0.7
Feb.	.3	.3	.3	.3	.3	.4	.4	.4	.5	.5	.6	.6
Mar.	.4	.3	.3	.3	.3	.3	.3	.3	.3	.4	.4	.5
Apr.	0.5	0.4	0.3	0.3	0.3	0.2	0.2	0.2	0.2	0.3	0.3	0.3
May	.6	.5	.5	.4	.4	.3	.3	.2	.2	.2	.2	.2
June	.7	.7	.6	.6	.5	.5	.4	.4	.3	.3	.3	.3
July	0.8	0.8	0.8	0.7	0.7	0.6	0.6	0.5	0.5	0.4	0.4	0.3
Aug.	.9	.9	.9	.8	.8	.8	.7	.7	.6	.6	.5	.5
Sep.	.8	.9	.9	.9	.9	.9	.9	.9	.8	.8	.7	.7
Oct.	0.7	0.8	0.8	0.9	0.9	1.0	1.0	1.0	1.0	0.9	0.9	0.9
Nov.	.6	.6	.7	.8	.9	0.9	.0	.0	.0	1.1	1.1	1.0
Dec.	0.4	0.5	0.6	0.7	0.7	0.8	0.9	1.0	1.0	1.1	1.1	1.1
Lat.						AZIMUTH						
°	°	°	°	°	°	°	°	°	°	°	°	°
0	0.3	0.4	0.5	0.6	0.6	0.7	0.7	0.7	0.7	0.6	0.6	0.5
20	0.3	0.4	0.5	0.6	0.7	0.7	0.7	0.7	0.7	0.7	0.6	0.5
40	0.4	0.5	0.6	0.7	0.8	0.9	0.9	0.9	0.9	0.8	0.8	0.7
50	0.5	0.6	0.8	0.9	1.0	1.1	1.1	1.1	1.1	1.0	0.9	0.8
55	0.5	0.7	0.9	1.0	1.1	1.2	1.2	1.2	1.2	1.1	1.0	0.9
60	0.6	0.8	1.0	1.1	1.3	1.4	1.4	1.4	1.4	1.3	1.2	1.0
65	0.7	0.9	1.2	1.4	1.5	1.6	1.7	1.7	1.6	1.5	1.4	1.2

위도 = 시고도(굴절을 보정한 값) $- 1^\circ + a_0 + a_1 + a_2$

위의 첫번째 표에서 춘분점의 L.H.A.를 가지고 a_0의 값을 찾고, 다음 두번째 표에서는 위도와 춘분점의 L.H.A. 로서, a_1의 값을, 세번째 표에서는 관측월과 춘분점의 L.H.A.로서 a_2의 값을 각각 취한다.
a_0, a_1, a_2의 값은 언제나 정(+)이다. 마지막 표에는 북극성의 방위각를 수록하였다.

[예제] 2006년 11월 13일 18시 16분 57초에 D.R. Lat. 35° 10′, Long. 129° 10′에서 북극성 방위를 360°로 관측하였다(항성 관측자료 참조).

〈표 2-34〉

	계산방법		내 용
GMT	GHA*r*	Dec.(적위)	고도 관측시간을 GMT로 환산함
09h	187° 20′.5	N23° 29′.8	시각에 대한 GHA*r*, d값을 천측력에서 구함
16m 57s	4° 14′.9	-	분, 초에 대한 GHA*r*, d값을 천측력에서 구함
GHA*r*	191° 35′.4	d N23° 29′.8	시각, 분, 초에 대한 GHA, d값 합산함
aλ	129° 18′.2		가정경도. LHA값이 정수가 되도록 조정함 (동경이면 +, 서경이면 −)
LHA*r*	320° 53′.7		관측자와 춘분점의 사이각을 서쪽으로 측정함

① LHA*r* 320°−330°의 값과 위도 35° 10′에 해당하는 값을 찾는다.

② 위도 20° 및 40°에 관한 값에서 위도 35° 10′에 해당하는 값을 보간하면 북극성의 계산방위는 0.85°가 된다. 따라서 Gyro Error는 0.85°(E)가 된다.

4. 고도 방위각법(Altitude azimuth method)

천체의 고도, 적위, 가정위도로 항해삼각형을 풀어 천체의 진방위를 산출하는 방법으로 산출표가 없어 직접 계산해야 하고 시수평이 보이지 않으면 이용할 수 없다.

2.13 위도측정법

2.13.1 북극성 위도법

또한 북극성표에서는 북극성의 시고도(기차를 수정한 시고도)를 통해 위도를 구할 수 있다. 이론적으로 북극에서 북극성은 수평면에 대해 090°로 볼 것이고 적도에서는 수평선(000°)에서 보일 것이다. 그러나 여러 가지 값의 개정이 필요하므로 북극성표에서는 a0, a1, a2 값을 제공하였다.

$$위도 = 시고도(기차\ 수정\ 후) - 1^\circ + a0 + a1 + a2$$

북극성의 고도를 측정하여 위도를 결정하는 방법으로 계산이 간단하고 정확하나, 북극성과 같은 2등성은 고도가 적어도 10° 이상이 되어야 관측되므로 북위 10° 이상의 지점에서만 이용가능하다.

〈천측력 북극성표에 의한 계산〉

북극성표는 북극성의 고도를 극의 고도로 개정하기 위한 개정치를 구하는 표로써 a0, a1 및 a2 개정표로 구별되어 있다.

① a0 개정표 : 관측시의 LSHr에 대한 표 값을 구하면 된다.

② a1 개정표 : 위도 L과 LSHr를 변수로 계산한 값을 구하면 된다.

③ a2 개정표 : 적위 1년에 약 0.'3, 항성시각은 약 25'.0쯤 변하고 있으므로 이에 대한 수정을 하기 이한 개정표이다.

이와 같이 각 개정표들에는 도합 1°가 더해져 있으므로 추가개정으로 1°를 빼주어야 한다.

[예제] 2006년 11월 22일 ZT 18시 40분경 DR 35° 40'.0N, 140° 15'.5E인 지점에서 Chronometer가 9시 46분 52초를 가르킬 때 북극성의 고도를 35° 30'.5 으로 측정하였다. 이때의 위도를 구하여라(단, 이 Chronometer 시는 GMT보다 1분 15초 빠르며 IC(+) 2'.0이고 Dip 19.5m임).

Polaris

NOV. 22ND

C 09h 46m 52s

CE (−)00h 01m 15s → Chronometer가 GMT보다 빠르므로 (−)해 준다.

GMT 09h 45m 37s → NOV. 22ND

09h 196° 12'.7 → 천측력표에서 11월 22일의 ARIES를 구한다.

45m 37s 11° 26'.1 → 천측력의 증분 및 개정표에서 해당되는 시간의 ARIES를 구한다.

GHAY 207° 38'.8

λ (+)140° 15'.5 → E일 경우 (+), W일 경우 (−)해 준다.

LHAY 347° 54'.3

〈표 2-35〉

	+	★	−	내 용
I.C.	2'.0			육분의 기차
Dip			7'.8	천측력에서 안고차 개정값을 확인
★- P			1'.4	Alt. 33'.43과 35'.38사이이므로
a0	32'.8			북극성표에서 LHAY 값
a1	0'.5			북극성표에서 Lat. 값
a2	1'.1			북극성표에서 Month 값
add'I			60'.0	개정표값에는 1°씩 더해져 있으므로
sum	36'.4		69'.2	+, − 값을 따로 계산
corr.			32'.8	수정값 표시
Hs	35° 30'.5			북극성 관측고도
L	34° 57'.7			실제 위도

〈표 2-36〉 2006. 11/21~11/23 (TUE, WED, THU)

UT		ARIES	VENUS -3.9		MARS 1.6		JUPITER -1.7		SATURN 0.5		STARS		
		G.H.A.	G.H.A.	Dec.	G.H.A.	Dec.	G.H.A.	Dec.	G.H.A.	Dec.	Name	S.H.A.	Dec.
d	h	° ′	° ′	° ′	° ′	° ′	° ′	° ′	° ′	° ′		° ′	° ′
21	0	59 51.4	177 12.0	S20 57.9	192 54.8	S17 27.3	182 36.7	S19 17.5	272 20.7	N14 14.4	Acamar	315 21.6	S40 16.5
	1	74 53.9	192 11.2	S20 58.6	207 55.5	S17 27.8	197 38.6	S19 17.6	287 23.1	N14 14.4	Achernar	335 29.7	S57 12.2
	2	89 56.3	207 10.3	S20 59.2	222 56.3	S17 28.2	212 40.5	S19 17.8	302 25.5	N14 14.3	Acrux	173 15.9	S63 8.0
	3	104 58.8	222 9.5	S20 59.9	237 57.0	S17 28.7	227 42.4	S19 17.9	317 27.9	N14 14.3	Adhara	255 16.3	S28 58.6
	4	120 1.3	237 8.7	S21 0.6	252 57.7	S17 29.2	242 44.3	S19 18.0	332 30.3	N14 14.3	Aldebaran	290 54.9	N16 31.6
	5	135 3.7	252 7.9	S21 1.2	267 58.4	S17 29.7	257 46.2	S19 18.1	347 32.7	N14 14.3			
	6	150 6.2	267 7.1	S21 1.9	282 59.1	S17 30.2	272 48.1	S19 18.3	2 35.1	N14 14.3	Alioth	166 25.2	N55 55.1
	7	165 8.7	282 6.2	S21 2.6	297 59.9	S17 30.7	287 50.0	S19 18.4	17 37.5	N14 14.3	Alkaid	153 3.1	N49 16.6
T	8	180 11.1	297 5.4	S21 3.2	313 0.6	S17 31.2	302 51.8	S19 18.5	32 39.9	N14 14.2	Al Na'ir	27 49.9	S46 55.9
U	9	195 13.6	312 4.6	S21 3.9	328 1.3	S17 31.7	317 53.7	S19 18.6	47 42.3	N14 14.2	Alnilam	275 51.2	S 1 11.7
E	10	210 16.0	327 3.8	S21 4.5	343 2.0	S17 32.2	332 55.6	S19 18.7	62 44.7	N14 14.2	Alphard	218 1.1	S 8 41.1
S	11	225 18.5	342 2.9	S21 5.2	358 2.7	S17 32.7	347 57.5	S19 18.9	77 47.1	N14 14.2			
D A	12	240 21.0	357 2.1	S21 5.9	13 3.5	S17 33.2	2 59.4	S19 19.0	92 49.5	N14 14.2	Alphecca	126 15.6	N26 41.4
Y	13	255 23.4	12 1.3	S21 6.5	28 4.2	S17 33.7	18 1.3	S19 19.1	107 51.9	N14 14.2	Alpheratz	357 48.7	N29 7.9
	14	270 25.9	27 0.4	S21 7.2	43 4.9	S17 34.2	33 3.2	S19 19.2	122 54.3	N14 14.2	Altair	62 13.4	N 8 53.2
	15	285 28.4	41 59.6	S21 7.8	58 5.6	S17 34.7	48 5.1	S19 19.3	137 56.6	N14 14.1	Ankaa	353 20.2	S42 16.2
	16	300 30.8	56 58.8	S21 8.5	73 6.3	S17 35.1	63 7.0	S19 19.5	152 59.0	N14 14.1	Antares	112 32.9	S26 26.9
	17	315 33.3	71 58.0	S21 9.2	88 7.0	S17 35.6	78 8.8	S19 19.6	168 1.4	N14 14.1			
	18	330 35.8	86 57.1	S21 9.8	103 7.7	S17 36.1	93 10.7	S19 19.7	183 3.8	N14 14.1	Arcturus	146 0.6	N19 8.7
	19	345 38.2	101 56.3	S21 10.5	118 8.5	S17 36.6	108 12.6	S19 19.8	198 6.2	N14 14.1	Atria	107 40.0	S69 2.5
	20	0 40.7	116 55.5	S21 11.1	133 9.2	S17 37.1	123 14.5	S19 20.0	213 8.6	N14 14.1	Avior	234 20.0	S59 31.5
	21	15 43.2	131 54.6	S21 11.8	148 9.9	S17 37.6	138 16.4	S19 20.1	228 11.0	N14 14.1	Bellatrix	278 37.2	N 6 21.5
	22	30 45.6	146 53.8	S21 12.4	163 10.6	S17 38.1	153 18.3	S19 20.2	243 13.4	N14 14.0	Betelgeuse	271 6.5	N 7 24.7
	23	45 48.1	161 53.0	S21 13.1	178 11.3	S17 38.6	168 20.2	S19 20.3	258 15.8	N14 14.0			
22	0	60 50.5	176 52.1	S21 13.7	193 12.0	S17 39.1	183 22.1	S19 20.4	273 18.2	N14 14.0	Canopus	263 58.0	S52 41.6
	1	75 53.0	191 51.3	S21 14.4	208 12.8	S17 39.6	198 24.0	S19 20.6	288 20.6	N14 14.0	Capella	280 41.6	N46 0.4
	2	90 55.5	206 50.5	S21 15.0	223 13.5	S17 40.0	213 25.8	S19 20.7	303 23.0	N14 14.0	Deneb	49 35.3	N45 18.5
	3	105 57.9	221 49.6	S21 15.6	238 14.2	S17 40.5	228 27.7	S19 20.8	318 25.4	N14 14.0	Denebola	182 38.9	N14 32.0
	4	121 0.4	236 48.8	S21 16.3	253 14.9	S17 41.0	243 29.6	S19 20.9	333 27.8	N14 14.0	Diphda	349 0.7	S17 56.9
	5	136 2.9	251 48.0	S21 16.9	268 15.6	S17 41.5	258 31.5	S19 21.0	348 30.2	N14 13.9			
	6	151 5.3	266 47.1	S21 17.6	283 16.3	S17 42.0	273 33.4	S19 21.2	3 32.6	N14 13.9	Dubhe	193 57.7	N61 42.6
W	7	166 7.8	281 46.3	S21 18.2	298 17.0	S17 42.5	288 35.3	S19 21.3	18 35.0	N14 13.9	Elnath	278 18.7	N28 36.9
E	8	181 10.3	296 45.5	S21 18.8	313 17.7	S17 43.0	303 37.2	S19 21.4	33 37.4	N14 13.9	Eltanin	90 49.0	N51 29.3
D	9	196 12.7	311 44.6	S21 19.5	328 18.5	S17 43.5	318 39.1	S19 21.5	48 39.8	N14 13.9	Enif	33 52.2	N 9 54.5
N	10	211 15.2	326 43.8	S21 20.1	343 19.2	S17 43.9	333 41.0	S19 21.7	63 42.2	N14 13.9	Fomalhaut	15 29.4	S29 35.3
E	11	226 17.7	341 43.0	S21 20.8	358 19.9	S17 44.4	348 42.9	S19 21.8	78 44.6	N14 13.9			
S D	12	241 20.1	356 42.1	S21 21.4	13 20.6	S17 44.9	3 44.7	S19 21.9	93 47.0	N14 13.9	Gacrux	172 7.3	S57 8.9
A	13	256 22.6	11 41.3	S21 22.0	28 21.3	S17 45.4	18 46.6	S19 22.0	108 49.4	N14 13.8	Gienah	175 57.8	S17 34.7
Y	14	271 25.0	26 40.4	S21 22.7	43 22.0	S17 45.9	33 48.5	S19 22.1	123 51.8	N14 13.8	Hadar	148 56.1	S60 24.3
	15	286 27.5	41 39.6	S21 23.3	58 22.7	S17 46.4	48 50.4	S19 22.3	138 54.2	N14 13.8	Hamal	328 6.3	N23 29.9
	16	301 30.0	56 38.8	S21 23.9	73 23.4	S17 46.9	63 52.3	S19 22.4	153 56.6	N14 13.8	Kaus Aust.	83 50.9	S34 23.0
	17	316 32.4	71 37.9	S21 24.5	88 24.2	S17 47.3	78 54.2	S19 22.5	168 59.0	N14 13.8			
	18	331 34.9	86 37.1	S21 25.2	103 24.9	S17 47.8	93 56.1	S19 22.6	184 1.4	N14 13.8	Kochab	137 20.4	N74 7.5
	19	346 37.4	101 36.2	S21 25.8	118 25.6	S17 48.3	108 58.0	S19 22.8	199 3.8	N14 13.8	Markab	13 43.4	N15 14.7
	20	1 39.8	116 35.4	S21 26.4	133 26.3	S17 48.8	123 59.8	S19 22.9	214 6.2	N14 13.7	Menkar	314 20.1	N 4 7.2
	21	16 42.3	131 34.5	S21 27.1	148 27.0	S17 49.3	139 1.7	S19 23.0	229 8.6	N14 13.7	Menkent	148 14.1	S36 24.1
	22	31 44.8	146 33.7	S21 27.7	163 27.7	S17 49.8	154 3.6	S19 23.1	244 11.0	N14 13.7	Miaplacidus	221 41.0	S69 44.3
	23	46 47.2	161 32.9	S21 28.3	178 28.4	S17 50.3	169 5.5	S19 23.2	259 13.4	N14 13.7			
23	0	61 49.7	176 32.0	S21 28.9	193 29.1	S17 50.7	184 7.4	S19 23.4	274 15.8	N14 13.7	Mirfak	308 47.3	N49 53.3
	1	76 52.2	191 31.2	S21 29.5	208 29.8	S17 51.2	199 9.3	S19 23.5	289 18.2	N14 13.7	Nunki	76 4.9	S26 17.4
	2	91 54.6	206 30.3	S21 30.2	223 30.5	S17 51.7	214 11.2	S19 23.6	304 20.6	N14 13.7	Peacock	53 27.3	S56 43.1
	3	106 57.1	221 29.5	S21 30.8	238 31.2	S17 52.2	229 13.1	S19 23.7	319 23.0	N14 13.7	Pollux	243 33.7	N28 0.6
	4	121 59.5	236 28.6	S21 31.4	253 32.0	S17 52.7	244 15.0	S19 23.8	334 25.4	N14 13.6	Procyon	245 4.8	N 5 12.6
	5	137 2.0	251 27.8	S21 32.0	268 32.7	S17 53.2	259 16.8	S19 24.0	349 27.8	N14 13.6			
	6	152 4.5	266 26.9	S21 32.6	283 33.4	S17 53.6	274 18.7	S19 24.1	4 30.2	N14 13.6	Rasalhague	96 11.5	N12 33.3
T	7	167 6.9	281 26.1	S21 33.2	298 34.1	S17 54.1	289 20.6	S19 24.2	19 32.6	N14 13.6	Regulus	207 48.9	N11 56.1
H	8	182 9.4	296 25.2	S21 33.9	313 34.8	S17 54.6	304 22.5	S19 24.3	34 35.0	N14 13.6	Rigel	281 16.6	S 8 11.4
U	9	197 11.9	311 24.4	S21 34.5	328 35.5	S17 55.1	319 24.4	S19 24.4	49 37.4	N14 13.6	Rigil Kent.	139 59.8	S60 51.8
R	10	212 14.3	326 23.6	S21 35.1	343 36.2	S17 55.6	334 26.3	S19 24.6	64 39.8	N14 13.6	Sabik	102 18.7	S15 44.1
S	11	227 16.8	341 22.7	S21 35.7	358 36.9	S17 56.0	349 28.2	S19 24.7	79 42.2	N14 13.6			
D A	12	242 19.3	356 21.9	S21 36.3	13 37.6	S17 56.5	4 30.1	S19 24.8	94 44.6	N14 13.6	Schedar	349 46.3	N56 34.8
Y	13	257 21.7	11 21.0	S21 36.9	28 38.3	S17 57.0	19 32.0	S19 24.9	109 47.0	N14 13.5	Shaula	96 29.2	S37 6.7
	14	272 24.2	26 20.2	S21 37.5	43 39.0	S17 57.5	34 33.8	S19 25.0	124 49.5	N14 13.5	Sirius	258 38.0	S16 43.3
	15	287 26.7	41 19.3	S21 38.1	58 39.7	S17 58.0	49 35.7	S19 25.2	139 51.9	N14 13.5	Spica	158 36.9	S11 11.8
	16	302 29.1	56 18.4	S21 38.7	73 40.4	S17 58.4	64 37.6	S19 25.3	154 54.3	N14 13.5	Suhail	222 56.2	S43 27.3
	17	317 31.6	71 17.6	S21 39.3	88 41.1	S17 58.9	79 39.5	S19 25.4	169 56.7	N14 13.5			
	18	332 34.0	86 16.7	S21 39.9	103 41.8	S17 59.4	94 41.4	S19 25.5	184 59.1	N14 13.5	Vega	80 42.8	N38 47.5
	19	347 36.5	101 15.9	S21 40.5	118 42.5	S17 59.9	109 43.3	S19 25.7	200 1.5	N14 13.5	Zuben'ubi	137 11.4	S16 4.2
	20	2 39.0	116 15.0	S21 41.1	133 43.3	S18 0.4	124 45.2	S19 25.8	215 3.9	N14 13.5			
	21	17 41.4	131 14.2	S21 41.7	148 44.0	S18 0.8	139 47.1	S19 25.9	230 6.3	N14 13.4			
	22	32 43.9	146 13.3	S21 42.3	163 44.7	S18 1.3	154 48.9	S19 26.0	245 8.7	N14 13.4			
	23	47 46.4	161 12.5	S21 42.9	178 45.4	S18 1.8	169 50.8	S19 26.1	260 11.1	N14 13.4			
Mer. Pass.		h m 19 53.4	v -0.8	d 0.6	v 0.7	d 0.5	v 1.9	d 0.1	v 2.4	d 0.0			

	S.H.A. ° ′	Mer.Pass h m
Venus	116 1.6	12 13
Mars	132 21.5	11 7
Jupiter	122 31.5	11 45
Saturn	212 27.7	5 46

2.13.2 자오선 고도 위도법

천체의 자오선 정중시의 고도를 측정하여 위도를 구하는 방법을 자오선 고도 위도법이라고 한다. 천체가 극상정중이나 극하정중시에 천체의 방위는 진북 혹은 진남에 있으므로 천의 극, 천정, 천체가 자오선상에 일직선으로 표시되므로 천체의 정거와 적위를 알면 양자의 합 또는 차에 의해 위도를 구할 수 있다.

1. 자오선 고도 위도법의 계산

1) 극상정중

L=z+d(L, d 동명으로 L 〉 d인 경우)

L=d−z(L, d 동명으로 L 〈 d인 경우)

L=z−d(L, d 이명인 경우)

2) 극하정중

L=h+p

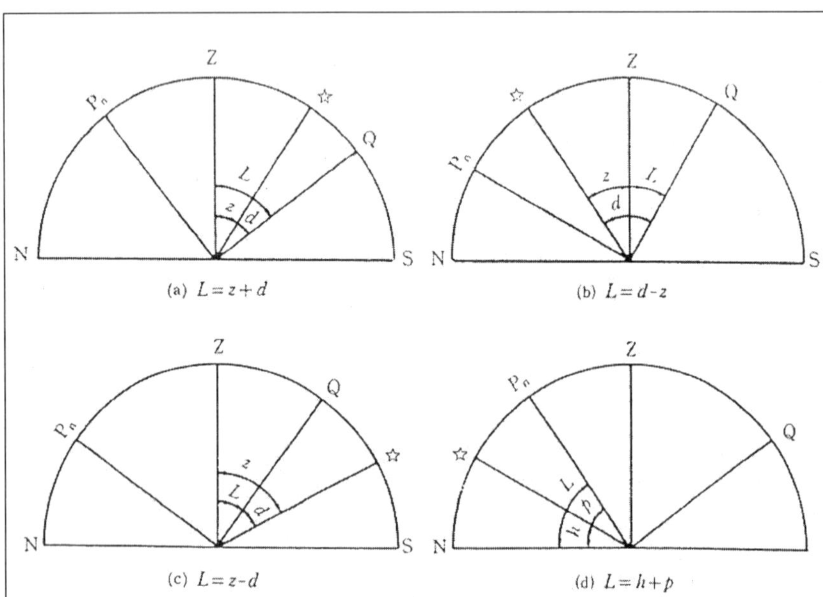

[그림 2-95] 자오선 고도와 위도와의 관계

2. 자오선 고도 위도법의 계산절차

① 천체의 자오선 정중시 및 이에 대한 적위를 구한다.

② 육분의 고도를 개정하여 관측고도(Ho)를 구한다.

③ 관측고도를 90°에서 감하여 정거(z)를 구한다.

정거의 부호 붙이는 방법은 천체를 기준으로 하여 관측자가 천체의 북쪽에 있으면 N을 붙이고, 천체의 남쪽에 있으면 S를 붙인다.

④ 정거(z)와 적위(d)가 동명일 때는 그 합을 구하여 동명의 부호를 붙이고 위도로 하며, 이명일 때는 그 차를 구하여 큰 편의 부호를 붙이어 위도로 한다.

⑤ 극하정중의 경우는 관측고도(Ho)에 극거(90°-d)를 더하여 적위와 동명의 위도로 한다.

2.13.3 근오고도 위도법

자오선 정중시 천체가 구름에 가리거나 다른 사정에 의해 관측하지 못한 경우 자오선각이 크지 않는 범위내에서 자오선 근방에 있는 천체의 고도를 관측하여 위도를 구하는 방법을 근오고도 위도법이라고 한다. 정확한 위도계산을 위해서는 위치삼각형을 푸는 대수 계산을 해야 하므로 실용성이 낮다.

실습과제

〈2.1〉 항해의 기초

1. 거리표를 이용하여 부산에서 홍콩까지의 거리를 확인하고, 홍콩 ETA를 구해보자.
2. Gyro Compass의 Error를 방파제 통과시 또는 중시선을 이용하여 구해보고, Repeater의 연동상태를 점검해서 수정해 보자.
3. 진풍향 및 진풍속을 이용 가능한 여러 방법을 이용해서 구해보자.
4. 임의 항만의 조석을 표준항의 조시 및 조고를 이용하여 구해보자.

〈2.2〉 해도 및 수로서지

1. 실습 중인 선박에서 보유하고 있는 해도의 종류를 알아보자.
2. 한국해도 202에서 해도도식에 해당되는 내용을 20개 이상 찾아보자.
3. 수로서지의 종류에는 어떤 것들이 있으며, 그 중 실습 중인 선박에 비치된 수로서지는 어떤 것들이 있는지 찾아보고 그 내용을 정리해 보자.
4. 항행통보를 받아서 해도를 직접 소개정 해보자.

〈2.3〉 항로표지

1. 항로표지의 종류에 대해서 알아보고, 한국해도 202에서 항로표지의 도식을 살펴보자.
2. IALA 해상부표방식이 A방식인 나라와 B방식인 나라의 해도를 찾아 비교해 보자.

〈2.4〉 연안 항법

1. 교차방위법에 의한 선위결정시 물표선정 및 방위측정시 주의사항에 대해 알아보자.
2. 부산항을 출발하여 인천항에 입항하는 항해계획을 수립해 보고, 필요한 해도를 찾아 Course Line을 그어 보자.

〈2.5〉 대양 항법

1. 부산항을 출발하여 싱가포르항에 입항하는 항해계획을 수립해 보고, 필요한 해도를 찾아 Course Line을 그어 보자.
2. 부산항을 출발하여 쓰가루 해협을 경유하여 LA항에 입항하는 항해계획을 수립해 보고, 필요한 해도를 찾아 Course Line을 그어 보자.

제3장

해상안전

3.1 SOLAS 협약
3.2 FSS code
3.3 LSA code

3.1 SOLAS 협약

3.1.1 SOLAS 협약의 도입배경 및 의의

1. SOLAS 협약의 도입배경

SOLAS(International Convention for the Safety Of Life At Sea, 1974)는 해상에서의 인명안전에 관한 협약으로 과거 불침선이라고 자부하면서 건조된 'Titanic'호의 1912년 북대서양 항해 중 빙산과 충돌하여 약 1,500여명의 여객과 선원이 사망하는 사고의 발생으로 말미암아 도입되게 된 협약이다. 이러한 사고 이후 1914년 런던에서 국제회의가 개최되었고, 역사상 최초의 SOLAS 협약이 채택되었다.

당시 채택된 협약에는 항해안전문제, 여객선의 수밀구획, 방화벽, 구명설비 및 소화설비에 대한 내용과 더불어 사고 당시 조난신호에 대한 다른 선박과의 의사소통 문제가 발생한 점을 고려하여 무선설비 비치 의무도 포함되었다.

선박기술개발에 따라 1929년, 1948년, 1960년 SOLAS 협약이 채택되었고, 지속적인 개정 및 시행절차상의 문제점을 해결하고, 미발효 상태에 있던 협약의 개정분을 수용함으로써 1974년 SOLAS 협약이 채택, 이후 지속적인 국제회의를 통하여 1980년 동 협약을 발효하게 되었다.

2. SOLAS 협약의 의의

SOLAS 협약은 해상에서의 인명안전 증진을 위해 채택 및 발효된 국제협약 선박의 안전을 위한 선박의 구조, 설비 및 운항에 관한 최저기준을 설정한 협약이다.

협약의 규정에 따른 기준은 협약의 실효성 확보와 더불어 최소한의 해상에서의 인명안전을 확보하기 위한 최저기준으로 이러한 기준마저 준수하지 못하는 선박은 상당한 위험이 내재되어 있는 선박이라 할 수 있다.

SOLAS 협약은 개정절차에서 있어서도 과거 협약을 채택, 발효하거나 개정하고자 할 때에 가장 난관이었던 명시수락(explicit acceptance) 절차를 개정하여 묵시수락(tacit acceptance) 절차를 수용했다는 점에서도 상당한 의미가 있다.

3.1.2 SOLAS 협약의 적용과 구성

1. SOLAS 협약의 적용

SOLAS는 실효성을 높이고, 보다 많은 선박에 적용시키기 위하여 현존선 및 신조선, 여객선과 화물선 등으로 구분하여 여러 규정을 적용시키고 있다.

즉, 각 장별로 별도의 명문규정을 두어 적용범위를 정하고 있으며, 별도의 명문규정이 없는 한 모든 여객선 및 총톤수 500톤 이상의 선박으로 국제항해에 종사하는 모든 선박에 적용이 된다. 이러한 SOLAS 협약의 관련 국내법으로는 선박안전법, 전파법, 항만법 및 해사안전법 등에 수용되어 국내 연안선이나 내항선에도 적용이 되고 있다.

2. SOLAS 협약의 구성

SOLAS 협약은 지속적으로 국제회의를 통하여 개정작업이 진행되고 있으며, 이러한 개정작업은 앞으로도 계속될 것으로 예상된다. 선박의 구조변화, 해상에서의 안전에 대한 인식변화, 해상환경변화 등이 협약개정에 영향을 미치고 있다할 수 있는바 이러한 변화는 앞으로도 계속될 것으로 이에 따라 협약개정도 지속적으로 이루어질 것으로 사료된다.

언급한 바와 같이 SOLAS 협약은 향후에도 지속적으로 이루어질 것으로 예상되지만, 지금(2014년 개정협약)까지 개정된 협약은 14개의 장과 1개의 부록으로 구성되어 다음과 같다.

Chapter I	General provision
Chapter II-1	Construction - Structure, subdivision and stability, machinery and electric installations
Chapter II-2	Construction - Fire protection, fire detection and fire extinction
Chapter III	Life-saving appliances and arrangements
Chapter IV	Radiocommunications
Chapter V	Safety of navigation
Chapter VI	Carriage of cargoes and oil fuels
Chapter VII	Carriage of dangerous goods
Chapter VIII	Nuclear ships
Chapter IX	Management for the safe operation of ships
Chapter X	Safety measures for high-speed craft

Chapter XI -1	Special measures to enhance maritime safety
Chapter XI -2	Special measures to enhance maritime security
Chapter XII	Additional safety measures for bulk carriers
Chapter XIII	Verification of compliance
Chapter XIV	Safety measures for ships operating in polar waters
Appendix	Certificate

각 장의 주요 내용을 간단히 살펴보면 다음과 같다.

① Chapter I : 일반 규정으로 협약에 사용되는 용어의 정의와 각종 선박의 검사 및 선박이 협약요건에 적합한지를 증명하는 증서에 관한 내용을 담고 있으며, 항만국 통제와 해난에 관한 일반사항을 규정하고 있다.

② Chapter II-1 : 선박건조와 관련하여 선박의 구조, 구획 및 복원성, 기관설비, 전기설비, 정기적으로 무인이 되는 기관구역에 대한 부가사항 및 기관 및 전기설비에 대한 대체설계 및 배치에 관한 사항을 규정하고 있다.

③ Chapter II-2 : 선박건조와 관련하여 방화, 화재탐지 및 소화와 관련된 사항을 담고 있는데, 화재 및 폭발의 예방, 화재의 진압, 탈출, 방화장치 및 소화장치의 운전요건, 화재안전에 대한 대체설계 및 배치 및 헬리콥터, 위험물 및 차량, 특수분류 및 로로(ro-ro)구역을 가지고 있는 선박에 대한 특별 요건에 관한 사항을 규정하고 있다.

④ Chapter III : 구명설비 및 장치에 관한 내용을 담고 있는데, 선박의 구명설비 요건, 여객선에 있어서 관련 추가요건, 화물선에 있어서의 관련 추가요건, 기타 사항으로 훈련지침서, 비상배치표 및 비상지침 및 구명설비에 대한 대체설계 및 배치에 관한 사항을 규정하고 있다.

⑤ Chapter IV : 무선통신에 관한 내용을 담고 있는데, 선박의 무선설비의 기능 요건, 각 당사국 정부의 약속 및 선박에 구축해야 할 무선설비 요건에 관한 사항을 규정하고 있다.

⑥ Chapter V : 항해안전에 관한 내용을 담고 있는데, 항해안전을 도모하기 위하여 당사국 정부가 취해야 할 사항과 선박이 안전항해를 하는데 운항측면에서 지켜야 되고, 해야 할 사항들을 규정하고 있다.

⑦ Chapter VI : 연료유 및 화물의 운송에 관한 내용을 담고 있는데, 곡물류를 운송하는 화물선의 요건, 고체 산적화물의 적하, 양하, 화물정보, 물질안전 보건자료 등

화물을 운송에 필요한 요건에 관한 사항을 규정하고 있다.

⑧ Chapter VII : 위험물의 운송에 관한 내용을 담고 있는데, 위험물을 운송하는데 있어 포장된 형태 또는 고체 산적화물형태로 운송, 위험액체 화학물을 산적 운송하는 선박의 구조 및 설비, 액화가스를 산적 운송하는 선박의 구조 및 설비 및 포장된 형태의 사용 후 핵연료, 플루토늄 및 고준위 방사성 폐기물의 선박운송을 위한 특별 요건에 관한 사항을 규정하고 있다.

⑨ Chapter VIII : 원자력선에 관한 내용을 담고 있는데, 원자력선의 안전에 관한 기본 요건, 방사능 위험성 및 원자력선에 대한 특별한 감독 및 해양사고시의 조치사항을 규정하고 있다.

⑩ Chapter IX : 선박의 안전운항을 위한 관리에 관한 내용을 담고 있는데, 1993년 11월 IMO 총회결의 A.741(18)로 채택된 ISM code(International Safety Management code)를 수용하여 그 내용을 규정하고 있다.

⑪ Chapter X : 고속선의 안전조치에 관한 내용을 담고 있는데, MSC(Maritime Safety Committee ; IMO 산하기구 중 하나) 제63차 회의에서 채택한 HSC code(International Code of Safety for High-speed Craft)를 수용하여 그 내용을 규정하고 있다.

⑫ Chapter XI-1 : 해상안전 강화를 위한 특별조치에 관한 내용을 담고 있는데, 인정단체에 대한 위임사항, 선박검사의 강화, 해상안전(海上安全) 강화를 위한 선박식별번호, 운항요건에 관한 항만국통제, 선박이력기록부 및 해양사고에 조사에 관한 추가 요건에 관한 사항을 규정하고 있다.

⑬ Chapter XI-2 : 해상보안 강화를 위한 특별조치에 관한 내용을 담고 있는데, 해상보안과 관련된 당사국 정부의 의무사항, 관련 선박, 회사의 요건 및 책임사항, 선박보안경보시스템, 선박의 안전과 보안을 위한 선장의 재량권, 항만시설 요건 및 관련 정보의 통보에 관한 사항을 규정하고 있다.

⑭ Chapter XII : 산적화물선에 대한 추가 조치에 관한 내용을 담고 있는데, 산적화물선에 적용되는 손상복원성 요건, 구조강도, 검사 및 정비, 화물창의 설계배치, 적하지침기기, 수위감지기, 배출 시스템 및 화물창을 비운 상태에서의 항해 제한사항을 규정하고 있다.

⑮ Chapter XIII : 준수에 대한 검증에 관한 내용을 담고 있는데, 각 체약국 정부가 현행 협약의 준수 및 이행여부를 검증하기 위하여 감사기준에 따라 기구로부터 정기적인

감사를 받아야 하는 사항을 규정하고 있다.

⑯ Chapter XIV : 극지에서 항행하는 선박들의 안전을 확보하기 위한 여러 요건과 수단에 관한 사항을 규정하고 있다.

⑰ Appendix : 각 증서 형식들을 담고 있다.

3.1.3 SOLAS 협약의 주요 규정

1. 범 위

SOLAS 협약의 주요 내용 중에 있어 해상에서의 선박 및 인명안전에 직접적으로 가장 영향을 많이 미치고 있는 부분으로 동 장에서는 SOLAS Chapter I~III(일반 규정, 화재 및 인명안전에 관련된 설비)에 관해서만 살펴보기로 한다.

물론 다른 장의 내용 또한 선박 및 인명안전과 직결되는 내용이지만 시간적인 제약이 감안하고, 일부 장들은 특정 선박에만 적용되는 내용을 담고 있으며, 몇몇 장은 다른 과정에서 다루어진다는 점을 고려하여 범위를 제한한다.

2. 일반규정(Chapter I)의 주요 내용

1) Regulation 1 : Application

(a) Unless expressly provided otherwise, the present Regulations apply only to ships engaged on international voyages.

SOLAS 협약은 국제협약으로써 국제항해에 종사하는 선박에 대해서만 적용이 된다. 앞서 설명한 바와 같이 내수 및 연안에서만 항해하는 내항선 및 연안선은 이러한 SOLAS 협약의 내용을 유사하게 담고 있는 국내법인 선박안전법, 해사안전법, 전파법 등의 적용을 받게 된다.

(b) The classes of ships to which each Chapter applies are more precisely defined, and the extent of the application is shown, in each Chapter.

각 장의 규정이 적용되는 선박의 종류는 각 장에서 더 명확히 정하여지며 또한 이들 규정의 적용범위는 각 장에서 제시한다.

SOLAS 협약의 내용을 지속적으로 개정이 이루어지고 있는 협약으로 모든 규정을 동일하게 모든 선박에 일괄적으로 적용한다는 것은 불가능하며 실효성이 떨어질 수 있다. 이러한 점을 고려하여 SOLAS 협약의 각 규정이 적용되어야 할 선박의 종류 및 적용범위가 명시되어 있다.

2) Regulation 2 : Definitions

(b) Administration means the Government of the State whose flag the ship is entitled to fly.

SOLAS 협약 본문에서 언급되고 있는 주관청은 선박이 그 국가의 국기를 게양할 자격이 있는 국가의 정부(즉, 기국 정부)를 의미한다.

[그림 3-1] 선박의 선미에 게양되어 있는 기국 정부의 기

(e) A passenger is every person other than :

(i) the master and the members of the crew or other persons employed or engaged in any capacity on board a ship on the business of that ship and

(ii) a child under one year of age.

SOLAS 협약 대다수의 규정이 화물선과 여객선으로 분리하여 명시되어 있는데 이러한 화물선과 여객선을 구분하는 기준은 여객을 몇 명이상 운송하느냐에 따라 결정이 된다. 그렇기 때문에 여객은 협약에 있어 중요한 의미를 갖고 있다.

협약에서 정의하고 있는 여객의 의미는 선장과 선원 또는 자격여하를 불문하고 선박에 승선하여 선박의 업무에 고용되거나 종사하는 기타의 사람 및 1세 미만의 유아를 제외하고 해당 선박에 승선하고 있는 다른 모든 사람을 의미한다.

일예로 선박의 수리나 검사를 위해 선박에 승선하고 있는 사람은 여객으로 간주되지 않는다.

(f) A passenger ship is a ship which carries more than twelve passengers.

앞서 언급한 바와 같이 SOLAS 협약은 화물선과 여객선으로 대다수의 규정이 구분되어 있는데, 여객을 1명이라도 승선시킨다고 해서 여객선이 되는 것은 아니며, 반대로 화물선은 한명의 여객이라도 승선시킬 수 없는 것은 아니다.

협약에서 정의하고 있는 여객선의 정의는 12명 초과(13명 이상)하는 여객을 운송하는 선박을 의미한다고 규정하고 있다. 따라서 12명 이하의 여객을 운송하는 선박은 여객선으로 간주되지 않으며, 화물선의 경우 12명 이하의 여객을 해당 선박의 최대 승선정원을 초과하지 않는 범위 내에서 승선시킬 수 있다.

일예로 화물선에서 종종 승무원의 가족을 일정기간 동승시키거나 촬영 등을 위해 위부인을 동승시키는 경우가 있는데, 이러한 경우 동승가족이나 촬영 관계자들은 여객에 해당됨에도 불구하고 화물선에 동승이 가능하다.

[그림 3-2] 여객선

(g) A cargo ship is any ship which is not a passenger ship.

화물선이라 함은 여객선이 아닌 선박이라고 협약에서는 정의하고 있는데, 이는 앞서 언급한 바와 같이 협약의 규정들이 화물선과 여객선으로 구분되어 있는 결과에서 비롯되어 여객선이 아닌 다른 모든 선박들은 화물선으로 간주하여 화물선의 규정을 적용받도록 하기 위함이다.

(h) A tanker is a cargo ship constructed or adapted for the carriage in bulk of liquid cargoes of an inflammable nature.

화물선 중에 인화성 액체화물을 산적운송하는 선박을 탱커라고 정의하고 있는데, 인화성의 액체화물은 다른 화물들에 비해 위험성이 높으며, 산적상태로 운송이 이루어지기 때문에 선박의 감항성에도 영향을 미치는 바 이에 대한 별도의 추가 안전조치들이 필요하다하겠다. 이러한 점을 고려하여 협약에서는 화물선 중에서도 탱커를 별도로 명시하고 이에 대한 추가적인 안전조치 규정을 두고 있다.

아울러 정의에 명시되고 있는 '인화성이 있는'이란 단어로 사용되고 있는 'Inflammable'은 'Flammable' 단어와 같은 의미로 사용된다.

[그림 3-3] 탱커

(k) New ship means a ship the keel of which is laid or which is at a similar stage of construction on or after 25 May 1980.

(l) Existing ship means a ship which is not a new ship.

신조선(신선)이라 함은 1980년 5월 25일 이후에 용골을 거치하거나 동등한 건조 단계에 있는 선박을 의미하며, 그 이전에 건조된 선박은 현존선(existing ship)이라고 명시하고 있다.

이러한 특정 날짜를 기준으로 신조선과 현존선을 구분한 이유는 SOLAS 협약을 발효하기 이전에 건조된 선박들이 상당수 운항을 하고 있었고, 이러한 선박들에게 발효된 SOLAS 협약의 모든 규정들을 일률적으로 적용하기란 현실적으로 불가능하였기 때문에,

이러한 점을 감안하여 협약이 발효되기 전에 이미 건조되어 운항하고 있는 선박들에게 적용할 수 있는 규정 및 신조선에게만 적용하는 규정을 별도로 명시하여 규정함으로써 국제항해에 종사하는 모든 선박들이 SOLAS 협약의 적용 대상이 될 수 있게 하였다.

3) Regulation 3 : Exceptions

(a) The present Regulations, unless expressly provided otherwise, do not apply to :

(i) Ships of war and troopships.

(ii) Cargo ships of less than 500 tons gross tonnage.

(iii) Ships not propelled by mechanical means.

(iv) Wooden ships of primitive build.

(v) Pleasure yachts not engaged in trade.

(vi) Fishing vessels.

SOLAS 협약은 앞서 언급한 바와 같이 국제항해에 종사하는 모든 선박에 적용이 된다고 할 수 있다. 다만, 현실적으로 적용이 불가능한 다음과 같은 선박에 있어서는 적용을 제외하고 있다.

(i) 군함 및 군대수송선

(ii) 총톤수 500톤 미만의 화물선

(iii) 기계로 추진되지 아니하는 선박

(iv) 원시적 구조의 목선

(v) 무역에 종사하지 아니하는 유람요트

(vi) 어선

제외 대상 선박들을 보면 협약을 왜 적용시킬 수 없는지 이해할 수 있을 것으로 판단된다. 다만, 어선의 경우는 원양어선과 같이 대형어선도 있으며 국제적으로 운항하는 선박이 상당수 있음에도 불구하고, 제외 대상에 포함시킨 이유는 어선들을 위한 협약, 일명 트레몰리노스(Torremolinos) 협약이 별도로 채택되어 있기 때문이다. 하지만 국제항해에 종사하는 화물선이나 여객선과 달리 어선은 국가마다 그 형태나 구조가 상당히 상이하여 일정한 기준을 만들어 이를 협약으로 발효시킨다는 것이 쉬운 일은 아니며, 각 국가간의 이해관계가 복잡하게 형성되어 아직까지 비준을 하지 않은 국가가 많아 정식적인 발효는 되어 있지 않은 상황이다.

4) Regulation 6 : Inspection and Survey

(a) The inspection and survey of ships, so far as regards the enforcement of the provisions of the present regulations and the granting of exemptions therefrom, shall be carried out by officers of the Administration. The Administration may, however, entrust the inspections and surveys either to surveyors nominated for the purpose or to organizations recognized by it.

협약의 준수 및 시행 여부를 판단하기 위한 선박의 검사, 협약의 시행 및 그 적용의 면제에 관한 한 주관청의 직원이 행한다. 다만, 주관청은 검사를 위하여 지명하는 검사원 또는 주관청이 인정하는 단체에 위탁할 수 있다고 규정되어 있다.

이에 따라 우리나라 정부는 선박안전법상의 선박검사관 및 선박검사를 수행할 수 있는 단체에 위탁하여 선박검사를 시행하고 있는데 정부 인정 단체로 한국선급(KR)이 있다.

5) Regulation 7 : Survey of passenger ships

(a) A passenger ship shall be subjected to the surveys specified below :

(i) an initial survey before the ship is put in service;

(ii) a renewal survey once every 12 months

(iii) additional surveys, as occasion arises.

여객선의 검사는 선박취항 전 최초검사, 12개월마다 정기검사 및 임시의 추가검사로 이루어진다.

[그림 3-4] 여객선 선저 검사

선박취항 전 최초검사에는 선저의 외부 및 보일러 내외부를 포함한 선체, 기관 및 설비의 완전한 검사를 포함하여야 하고, 12개월마다 시행하는 정기검사는 선저외부를 포함한 선체, 보일러 및 기타 압력용기, 기관 및 설비의 검사를 포함하여야 한다.

임시의 추가검사는 전반적이든 부분적이든 간에 조사결과에 따른 수리 후 또는 어떠한 중요한 수리나 신환이 있을 때마다 행해져야 한다.

6) Regulation 8 : Survey of life-saving appliances and other equipment of cargo ships

(a) The life-saving appliances and other equipment of cargo ships of 500 tons gross tonnage and upwards

(i) an initial survey before the ship is put in service;

(ii) a renewal survey at intervals specified by the Administration but not exceeding 5 years

(iii) a periodical survey within three months before or after the second anniversary date or within three months before or after the third anniversary date of the Cargo Ship Safety Equipment Certificate

(iv) an annual survey within 3 months before or after each anniversary date of the Cargo Ship Safety Equipment Certificate;

(v) an additional survey

화물선의 검사는 여객선과 달리 대분류, 즉 구명설비, 무선설비 및 구조, 기관 및 설비 검사로 구분되어 규정되어 있으며, 각각 검사에 따른 검사항목도 별도로 명시되어 있다.

우선 화물선 구명설비에 관한 검사는 취항되기 전 최초검사, 주관청이 5년을 초과하지 아니하는 간격으로 규정한 정기검사, 화물선안전설비증서의 제2년차일의 3개월 전후 이내의 정기적 검사 또는 제3년차일의 3개월 전후 이내의 정기적 검사, 화물선안전설비증서의 매 연차일 전후 3개월 이내의 연차검사 및 추가검사로 이루어져 있다.

이러한 화물선 구명설비에 관한 검사 중에 추가검사를 제외한 모든 검사는 화재안전장치 및 설비, 구명설비(무선 장치는 제외), 선박용 항해설비, 도선사용 승하선장치 및 기타 설비에 대한 완전한 검사를 포함하고, 추가검사에 있어서는 여객선과 같이 전반적이든 부분적이든 간에 조사결과에 따른 수리 후 또는 어떠한 중요한 수리나 신환이 있을 때마다 행해져야 한다.

7) Regulation 9 : Survey of radio installations of cargo ships

(a) The radio installations shall be subject to the surveys specified below :

(i) an initial survey before the ship is put in service;

(ii) a renewal survey at intervals specified by the Administration but not exceeding five years

(iii) a periodical survey within three months before or after each anniversary date of the Cargo Ship Safety Radio Certificate;

(iv) an additional survey

화물선 무선설비에 관한 검사는 취항되기 전 최초검사, 주관청이 5년을 초과하지 아니하는 간격으로 규정한 정기검사, 화물선안전설비증서의 매 연차일 전후 3개월 이내의 연차검사 및 추가검사로 이루어져 있다.

이러한 화물선 무선설비에 관한 검사 중에 추가검사를 제외한 모든 검사는 구명설비에 사용되는 것을 포함한 화물선의 무선설비의 완전한 검사를 포함하고, 추가검사의 경우는 여객선과 같이 전반적이든 부분적이든 간에 조사결과에 따른 수리 후 또는 어떠한 중요한 수리나 신환이 있을 때마다 행해져야 한다.

8) Regulation 10 : Surveys of structure, machinery and equipment cargo ships

(a) The structure, machinery and equipment (other than items in respect of which a Cargo Ship Safety Equipment Certificate and a Cargo Ship Safety Radio Certificate are issued) of a cargo ship

(i) an initial survey including an inspection of the outside of the ship's bottom before the ship is put in service

(ii) a renewal survey at intervals specified by the Administration but not exceeding 5 years

(iii) an intermediate survey within three months before or after the second anniversary date or within three months before or after the third anniversary date of the Cargo Ship Safety Construction Certificate

(iv) an annual survey within 3 months before or after each anniversary date of the Cargo Ship Safety Construction Certificate;

(v) an additional survey

화물선의 구조, 기관 및 설비에 관한 검사는 선박이 취항되기 전 선저외판의 검사를 포함한 최초검사, 5년을 초과하지 아니하는 범위내에서 주관청이 정한 간격의 정기검사, 화물선안전구조증서의 제2년차일 전후 3개월 이내 또는 제3년차 일 전후 3개월 이내의 중간검사, 화물선안전구조증서의 매 연차일 전후 3개월 이내의 연차검사 및 추가검사로 이루어져 있다.

최초검사는 및 정기검사는 구조, 기관 및 설비의 완전한 검사를 포함하여야 하고, 중간검사는 선체, 보일러 및 기타 압력용기, 기관 및 설비, 조타설비 및 관련 제어계통과 전기설비의 검사를 포함한다. 연차검사 항목은 정기검사 항목과 같으며, 추가검사는 앞서 언급되었던 추가검사와 동일하게 시행된다.

9) Regulation 19 : Control

(a) Every ship when in a port of another Contracting Government is subject to control by officers duly authorized by such Government.

(b) Such certificates, if valid, shall be accepted unless there are clear grounds for believing that the condition of the ship or of its equipment does not correspond substantially with the particulars of any of the certificates or that the ship and its equipment are not in compliance with the provisions of regulation.

항만국통제(Port State Control, PSC)에 관한 사항으로 다른 당사국 정부의 항구에 있는 모든 선박은 그 정부로부터 정당하게 권리를 부여받은 직원(PSCO)의 통제를 받아야 하는데 선박이 갖고 있는 증서의 유효성 여부만을 확인해야 한다.

(c) Where a certificate has expired or ceased to be valid, the officer carrying out the control shall take steps to ensure that the ship shall not sail until it can proceed to sea or leave the port for the purpose of proceeding to the appropriate repair yard without danger to the ship or persons on board.

만약 선박 또는 그 설비의 상태가 실질적으로 증서의 기재사항과 일치하지 아니하거나 또는 선박 또는 그 설비가 규정에 적합하지 아니하다고 믿을만한 명확한 근거가 있는 경우에는 선박 또는 승선자에게 위험을 미치지 아니하고 적당한 수리조선소에 갈 목적으로 항해 또는 출항할 수 있을 때까지 그 선박의 출항금지 조치를 취하여야 한다.

(d) In the event of this control giving rise to an intervention of any kind, the officer carrying out the control shall forthwith inform, in writing, the Consul or, in his absence, the nearest diplomatic representative of the State whose flag the ship is entitled to fly of all the circumstances in which intervention was deemed necessary. In addition, nominated surveyors or recognized organizations responsible for the issue of the certificates shall also be notified. The facts concerning the intervention shall be reported to the Organization.

이러한 통제 과정에서 어떠한 종류의 조치를 취하여야 할 필요가 있는 경우에는 통제를 행하는 직원은 선박이 그 국기를 게양할 자격이 있는 국가의 영사 또는 그의 부재시에는 가장 가까운 외교대표에게 즉시 서면으로 조치가 필요하다고 인정되는 상황을 통보하여야 한다. 추가로 증서를 발급한 지명된 검사원 또는 인정된 단체에 대하여도 통보하여야 하고, 조치에 관한 사실을 기구에 보고하여야 한다.

(f) When exercising control under this regulation all possible efforts shall be made to avoid a ship being unduly detained or delayed. If a ship is thereby unduly detained or delayed it shall be entitled to compensation for any loss or damage suffered.

규칙에 의하여 통제를 행하는 경우에는 선박이 부당하게 억류되거나 지체되지 아니하도록 가능한 한 모든 노력을 하여야 한다. 선박이 부당하게 억류 또는 지체되는 경우에는 피해를 입은 손실 및 손상에 대한 보상을 받을 수 있는 권리가 있다.

3. 건조, 구조, 구획 및 복원성, 기관 및 전기설비(Chapter II-1)의 주요 내용

1) Regulation 3-2 : Protective coatings of dedicated seawater ballast tanks in all types of ships and double-side skin spaces of bulk carriers

All dedicated seawater ballast tanks arranged in ships and double-side skin spaces arranged in bulk carriers of 150 m in length and upwards shall be coated during construction in accordance with the Performance standard for protective coatings for dedicated seawater ballast tanks in all types of ships and double-side skin spaces of bulk carriers.

선박에 배치된 모든 해수전용 밸러스트 탱크와 길이 150m 이상의 산적화물선에 배치된 이중선측 구역은 선박의 전용 해수 밸러스트 탱크 및 산적화물선의 이중선측 구역의 보호도장에 대한 성능기준에 따라 건조시에 도장되어야 한다.

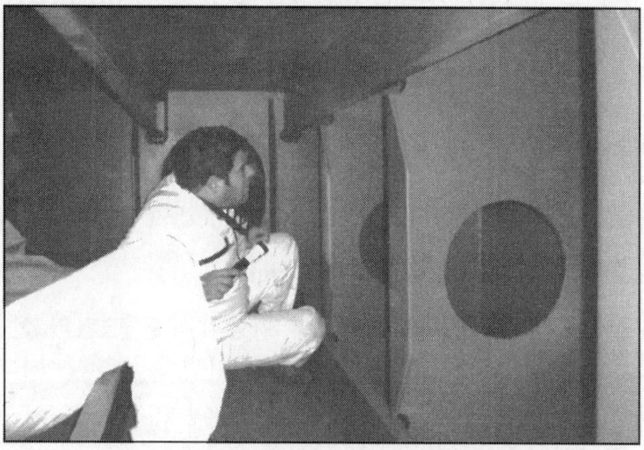

[그림 3-5] 밸러스트 탱크 내부

2) Regulation 3-3 : Safe access to tanker bows

Every tanker shall be provided with the means to enable the crew to gain safe access to the bow even in severe weather conditions.

[그림 3-6] 탱커선에 설치되어 있는 선수부까지의 안전통로

모든 탱커선에는 악천후에도 선원들이 안전하게 선수까지 접근할 수 있게 설비를 갖추도록 규정하고 있다. 탱커선의 액체상태의 위험화물을 운송하는 선박이기 때문에 항상 선박의 선수부를 포함한 모든 구역의 점검이 수시로 이루어져 이상 여부를 조기에 발견할 수 있도록 해야 한다. 만약 화물구역에 파공 등이 생겼다면 이를 조속히 발견하고 대응하지 못한다면 그 피해규모가 엄청날 것이다.

3) Regulation 3-4 : Emergency towing arrangements and procedures

Emergency towing arrangements shall be fitted at both ends on board every tanker of not less than 20,000 tonnes deadweight.

비상예인장치를 재화중량톤수 20,000톤 이상의 모든 탱커의 선수 및 선미 양쪽에 설치하여야 한다.

탱커선에 설치되어 있는 비상예인장치 중 선수부에 있는 비상예인장치는 다른 선박이 본선을 예인할 때에 사용되고, 선미부에 있는 비상예인장치는 본선이 다른 선박을 예인하고자 할 경우에 일반적으로 사용된다.

[그림 3-7] 선수 비상예인장치

[그림 3-8] 선미 비상예인장치

4) Regulation 9 : Double bottoms in passenger ships and cargo ships other than tankers

A double bottom shall be fitted extending from the collision bulkhead to the afterpeak bulkhead.

이중저는 충돌격벽으로부터 선미격벽까지 설치하여야 한다.

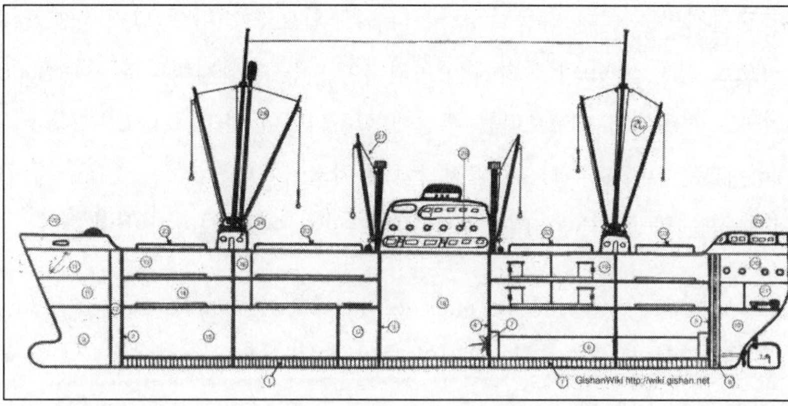

[그림 3-9] 이중저가 설치되어 있는 선박

5) Regulation 12 : Peak and machinery space bulkheads, shaft tunnels, etc.

A collision bulkhead shall be fitted which shall be watertight up to the bulkhead deck. This bulkhead shall be located at a distance from the forward perpendicular of not less than 0.05L or 10m, whichever is the less, and, except as may be permitted by the Administration, not more than 0.08L or 0.05L+3m, whichever is the greater.

선박에는 격벽 갑판까지 수밀의 충돌격벽을 설치하여야 한다. 충돌격벽은 전부수선으로부터의 거리가 0.05L 또는 10m 중 작은 쪽의 거리보다 적지 아니한 곳에 설치하여야 한다. 주관청이 인정하는 경우를 제외하고 0.08L 또는 0.05L +3m 중 큰 쪽의 거리 이내가 되도록 설치한다.

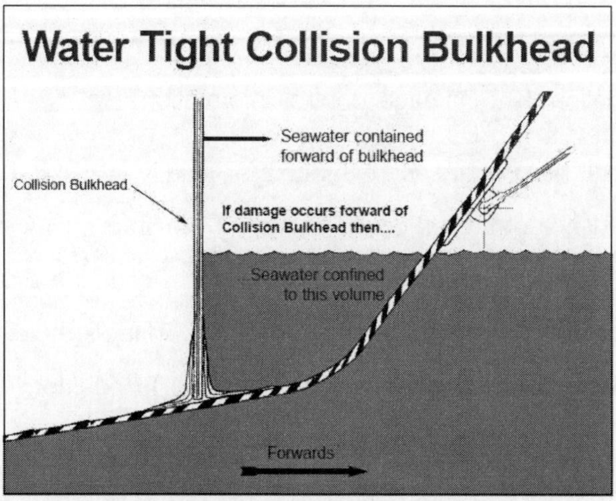

[그림 3-10] 충돌격벽

이러한 충돌격벽은 선박 충돌발생 시 다음 구획으로 범람이 되지 않도록 하기 위해 설치된 것으로 길이 300m의 선박에 설치되어야 하는 충돌격벽의 위치는 300m의 0.05L은 15m이므로 선수에서 10m보다는 후면에 설치되어야 하고, 0.08L은 24m이고 0.05L+3m은 18m이므로 선수로부터 24m 전면에 설치되어야 한다. 즉, 300m 길이의 선박에 설치되어야 하는 충돌격벽은 선수에서 10~24m 사이에 설치되어야 한다.

In all cases stern tubes shall be enclosed in watertight spaces of moderate volume.

어떤 경우에도 선미관은 적절한 용적의 수밀구역에 설치하여야 한다. 이는 프로펠러에 연결되어 있는 축(shaft)이 선체를 관통하고 있고, 이러한 관통부를 통해 외부로부터의 해수가 유입되지 않게 봉입(seal)설비가 되어 있는데 자칫 봉입설비가 잘못되어 해수가 유입되는 경우에 대비하기 위하여 수밀구역에 선미관을 설치해야 하다.

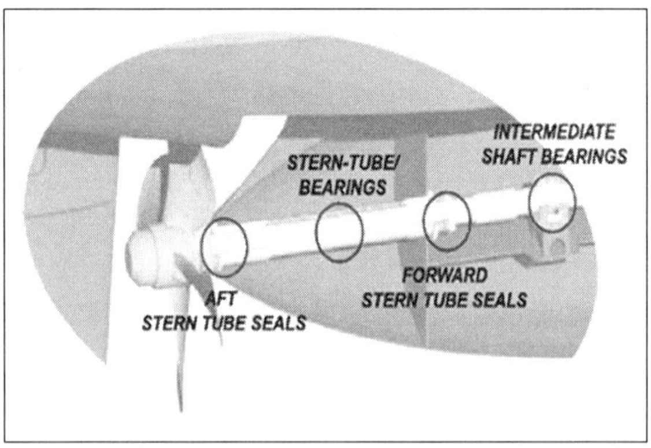

[그림 3-11] 선미관

6) Regulation 19 : Damage control information

There shall be permanently exhibited, or readily available on the navigation bridge, for the guidance of the officer in charge of the ship, plans showing clearly for each deck and hold the boundaries of the watertight compartments, the openings therein with the means of closure and position of any controls thereof, and the arrangements for the correction of any list due to flooding. In addition, booklets containing the aforementioned information shall be made available to the officers of the ship.

선박의 담당사관의 지침용으로 각 갑판 및 선창에 대하여 수밀구획실 경계 및 그 개구

(폐쇄장치와 그 제어장치의 위치포함)와 침수로 인한 선박의 횡경사 수정계획을 명시하는 도면을 항해선교에 영구적으로 게시하거나 또는 항해선교에서 쉽게 이용할 수 있어야 한다. 더욱이 이 자료를 포함한 소책자를 선박사관들이 이용할 수 있도록 하여야 한다.

손상제어에 관한 소책자에는 선박에서 발생할 수 있는 여러 경우의 침수상황에 대하여 선박상태 변화를 미리 계산해 두어 선박에서 침수상황이 발생했을 때에 신속하게 대응할 수 있도록 하고 있다.

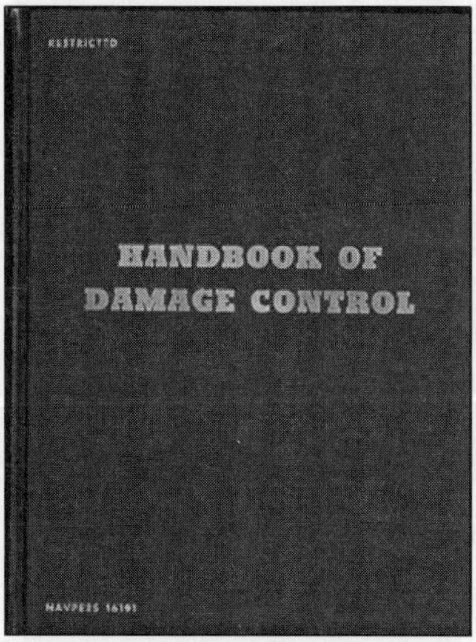

[그림 3-12] 손상제어 책자

7) Regulation 28 : Means of going astern

Sufficient power for going astern shall be provided to secure proper control of the ship in all normal circumstances.

선박은 모든 상태에서 선박의 조종을 확보하기 위하여 충분한 후진력을 가지고 있어야 한다.

The stopping times, ship headings and distances recorded on trials, together with the results of trials to determine the ability of ships having multiple propellers to navigate and manoeuvre with one or more propellers inoperative, shall be available on board for the use of the master or designated personnel.

해상시운전시의 선박의 정지시간, 선수방향 및 정지거리, 다축선에서 1개 또는 2개 이상의 프로펠러를 사용하지 않은 상태의 항해 및 조선성능을 결정하는 해상시운전 결과와 함께 선장 또는 기타 지정된 승무원이 이용할 수 있도록 선내에 비치하여야 한다.

통상 선박에는 'Result of Sea Trial'이라는 제목의 해상시운전을 통해 얻은 결과를 담고 있는 책자가 비치되어 있으며, 이중 선박조정 성능에 관한 'Turing circle'은 선교에 별도로 게시되어 있다.

[그림 3-13] 시운전 자료

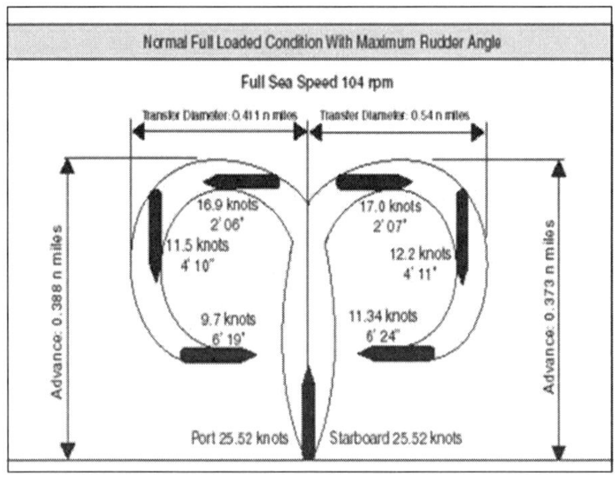

[그림 3-14] Turing circle

8) Regulation 29 : Steering gear

Unless expressly provided otherwise, every ship shall be provided with a main steering gear and an auxiliary steering gear to the satisfaction of the Administration. The main steering gear and the auxiliary steering gear shall be so arranged that the failure of one of them will not render the other one inoperative.

선박에는 주관청이 인정하는 주조타장치 및 보조조타장치를 설치하여야 한다. 주조타장치 및 보조조타장치는 그 장치 중 한 쪽이 고장으로 인하여 남은 다른 부분의 기능이 정지하는 일이 없도록 배치하여야 한다. 조타장치는 선박에 있어서 가장 중요한 설비 중에 하나로 주조타장치 및 보조조타장치로 구분하여 완전히 독립적으로 설치되어 어느 것 하나에 문제가 발생하여 작동 불가하더라도 다른쪽 조타장치가 정상적으로 작동하여 언제든지 조타가 가능하도록 되어 있어야 한다.

Relief valves shall be fitted to any part of the hydraulic system which can be isolated and in which pressure can be generated from the power source or from external forces. The setting of the relief valves shall not exceed the design pressure. The valves shall be of adequate size and so arranged as to avoid an undue rise in pressure above the design pressure.

유압계통 중 격리될 가능성이 있는 부분으로서 동력원 또는 외력에 의하여 압력이 발생할 가능성이 있는 부분에 안전밸브를 부착하여야 한다. 안전밸브의 설정압력은 설계압력을 초과하여서는 아니된다. 안전밸브는 적당한 크기의 것이어야 하며 설계압력을 초과하는 과도의 압력상승을 피할 수 있도록 배치하여야 한다. 이는 자칫 과도한 압력 상승으로 인해 유압계통 라인이나 장치가 파손되어 조타를 할 수 없는 상황이 발생될 수 있기 때문에 이를 막기 위한 장치이다.

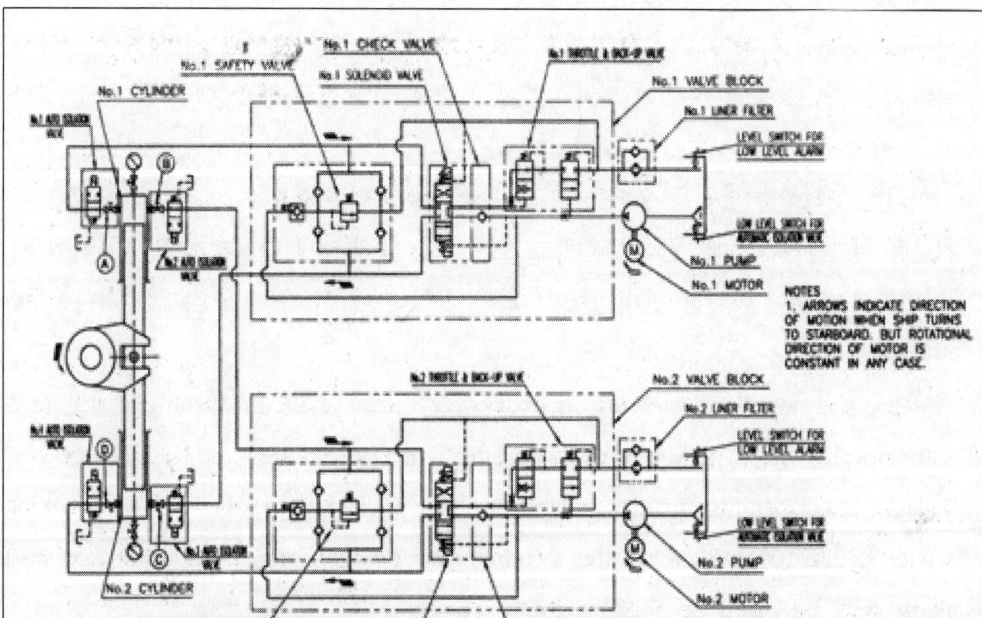

[그림 3-15] 조타장치 유압계통도

The main steering gear and rudder stock shall be capable of putting the rudder over from 35° on one side to 35° on the other side with the ship at its deepest seagoing draught and running ahead at maximum ahead service speed and, under the same conditions, from 35° on either side to 30° on the other side in not more than 28 seconds.

The auxiliary steering gear shall be capable of putting the rudder over from 15° on one side to 15° on the other side in not more than 60 seconds with the ship at its deepest seagoing draught and running ahead at one half of the maximum ahead service speed or 7 knots, whichever is the greater.

주조타장치 및 타두재는 최대 항해흘수에서 최대 항해속력으로 전진 중에, 타를 한쪽 현 35°로부터 반대 현 35°까지 조작할 수 있는 것이어야 하며, 또한 이와 동일한 조건하에 어느 현으로부터도 한쪽 현 35°에서 반대 현 30°까지 28초 이내에 조작할 수 있는 것이어야 하고, 보조조타장치는 최대 항해흘수에서 최대 항해전진속력의 2분의 1에 상당하는 속력 또는 7노트 중 빠른 쪽의 속력으로 전진 중에 타를 한쪽 현 15°에서 반대 현 15°까지 60초 이내에 조작할 수 있는 것이어야 한다.

이러한 조타성능 점검을 수행할 때에 한쪽 현에서 다른쪽 현으로 조타를 수행한 후에 시간을 체크하게 되는데, 반대쪽 현으로 조타를 할 때에 해당 각도로 타를 사용하여 시간을 체크하는 것이 아니라 반대 현으로 타는 완전히 다 쓰고, 해당 각도를 지나갈 때에 시간을 체크한다.

Main and auxiliary steering gear power units shall be arranged to restart automatically when power is restored after a power failure; and capable of being brought into operation from a position on the navigating bridge. In the event of a power failure to any one of the steering gear power units, an audible and visual alarm shall be given on the navigating bridge.

주조타장치 및 보조조타장치의 동력장치는 동력공급이 정지된 후에 다시 회복되면 자동적으로 기동되어야 하고, 선교에서 조작할 수 있는 것이어야 한다. 조타장치나 동력장치중의 어느 1개가 고장난 경우에는 선교에 가시, 가청의 경보를 발할 수 있는 것이어야 한다.

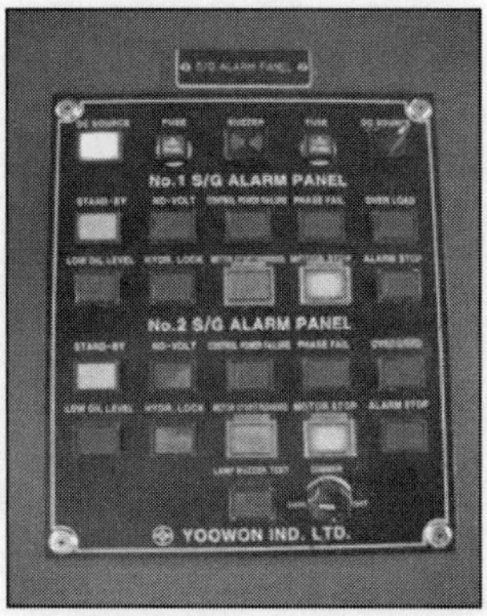

[그림 3-16] 조타장치 알람 패널

The main steering gear is so arranged that after a single failure in its piping system or in one of the power units the defect can be isolated so that steering capability can be maintained or speedily regained.

주조타장치는 배관계통 또는 1개의 동력장치에 고장이 발생한 경우에는 조타 능력을 유지 또는 신속히 회복시키기 위하여 결함부분을 격리시키기 위한 장치가 되어 있어야 하는데 일병 'Isolation valve'가 설치되어 있다.

Steering gear control shall be provided for the main steering gear, both on the navigating bridge and in the steering gear compartment.

조타장치의 제어에 있어 주조타장치에 대하여는 선교 및 조타기실에서 할 수 있어야 하는데, 실질적적으로 자동조타장치(autopilot)에 문제가 발생하여 조타기실에서 비상조타를 수행하여야 하기 때문에 조타기실에도 조타장치를 제어할 수 있어야 한다.

A means of communication shall be provided between the navigating bridge and the steering gear compartment.

선교와 조타기실간에는 통신장치를 설치하여야 하는데, 비상조타시에 선교의 지시를 받아 조타를 수행하기 위해서는 통신장치가 필요하기 때문이다. 일반적으로 무선통신장치인 워키토기를 사용하기도 하지만 유선통신장치로 선내전화(automatic telephone) 및 자가전원전화(sound powered telephone)이 설치되어 있다.

[그림 3-17] 조타기실 통신장치

Hydraulic power-operated steering gear shall be provided with a low level alarm for each hydraulic fluid reservoir to give the earliest practicable indication of hydraulic fluid leakage. Audible and visual alarms shall be given on the navigating bridge and in the machinery space where they can be readily observed.

각 작동유탱크에서 작동유가 누설하는 것을 가능한 한 신속히 알 수 있도록 저유면 상태가 경보되어야 하며, 가시가청의 경보는 선교 및 기관구역의 확인되기 쉬운 장소에 발하는 것이어야 한다. 일반적으로 작동유탱크에는 LOW 알람과 LOW-LOW 알람 장치가 되어 있으며, LOW-LOW 알람이 발생하면 기기보호를 위하여 자동으로 작동이 멈추게 된다.

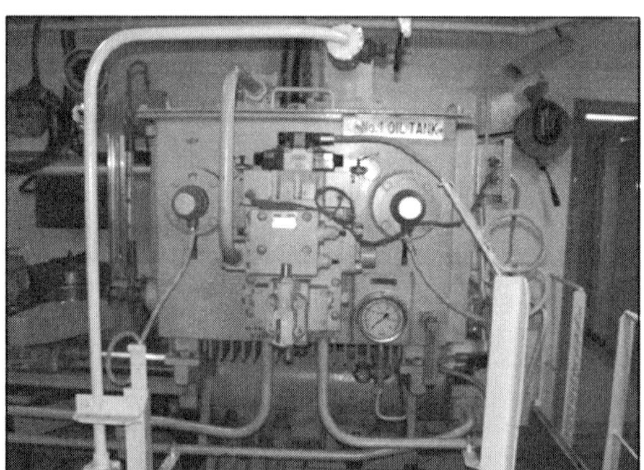

[그림 3-18] Low 및 Low-Low 알람 장치

The steering gear compartment shall be readily accessible and, as far as practicable, separated from machinery spaces.

조타기실은 항상 접근할 수 있어야 하며 가능한 한 기관구역으로부터 분리 설치하여야 하는데 이는 기관실 침수와 같이 사고가 발생하더라도 언제라도 조타는 가능하도록 하기 위함이다.

In every tanker, chemical tanker or gas carrier of 10,000 tons gross tonnage and upwards and in every other ship of 70,000 tons gross tonnage and upwards, the main steering gear shall comprise two or more identical power units.

총톤수 10,000톤 이상의 모든 유탱커, 위험화학품운반선 또는 액화가스운반선 및 총톤수 70,000톤 이상의 기타 선박에서 주조타장치는 2개 이상의 동등한 동력장치를 가지는 것이어야 한다. 모든 선박에 있어 조타장치는 가장 중요한 설비이다. 다만, 상대적으로 위험성이 높고, 사고발생 시 그 피해가 클 것으로 예상되는 선박에 대하여 특별히 안전을 확보하기 위함이다.

9) Regulation 31 : Machinery controls

The remote control of propulsion machinery from the navigating bridge is provided with an emergency stopping device.

선교에서 추진기관을 원격제어할 수 있어야 하며, 주추진기관에는 선교의 제어장치와는 별도로 비상정지장치를 선교에 설치하여야 한다. 일반적으로 선교에서 추진기관을 제어하기 위하여 'Engine Telegraph'가 설치되어 있으며, 비상정지장치로 일명 'Emergency stop' 스위치가 설비되어 있다.

The remote control of the propulsion machinery shall be possible only from one location at a time; at such locations interconnected control positions are permitted. At each location there shall be an indicator showing which location is in control of the propulsion machinery. The transfer of control between the navigating bridge and machinery spaces shall be possible only in the main machinery space or the main machinery control room. This system shall include means to prevent the propelling thrust from altering significantly when transferring control from one location to another.

선교로부터의 추진기관에 관한 명령은 주기관제어실 또는 조종장소의 적당한 곳에 표시되어야 하며, 추진기관의 원격제어는 한번에 1개 장소에서만 가능하여야 한다. 이는 각각의 장소에서 원격제어를 하게 되면 신호체계의 혼란 및 잘못된 기관사용이 이루어질 수 있기 때문에 이를 방지하기 위한 방안으로 기관제어는 항상 1개의 장소에서만 가능하도록 되어 있으며, 다른 장소에서는 추진기관 사용상태만 확인 가능하다. 제어장치는 상호 연결하는 것은 허용되며, 각 제어장소에는 현재 어느 장소에서 추진기관의 제어를 하고 있는가를 나타내는 표시기를 설치하여야 한다.

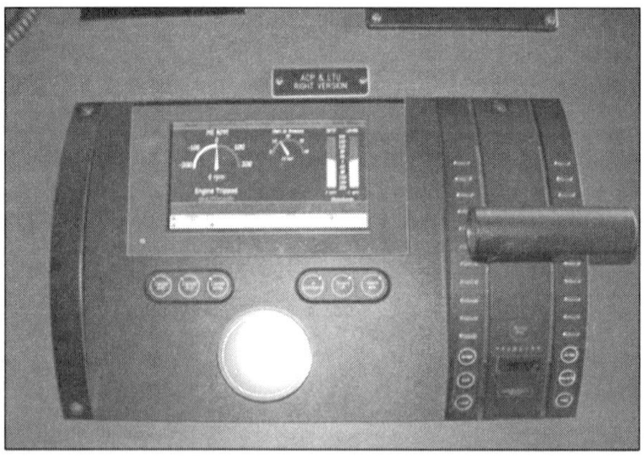

[그림 3-19] Engine telegraph

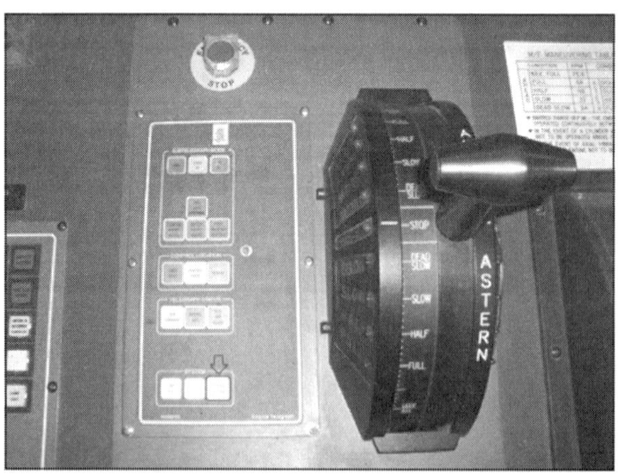

[그림 3-20] Engine telegraph 제어장치

The remote control of the propulsion machinery shall be possible to control the propulsion machinery locally, even in the case of failure in any part of the remote control system.

원격제어장치의 어떤 부분이 고장난 경우에도 기관측에서 추진기관을 제어할 수 있는 것이어야 한다.

The indicators shall be fitted on the navigating bridge for propeller speed and direction of rotation in the case of fixed pitch propellers.

선교에는 고정 피치프로펠러의 경우 프로펠러의 회전수 및 회전방향 표시기를 설치하여야 한다.

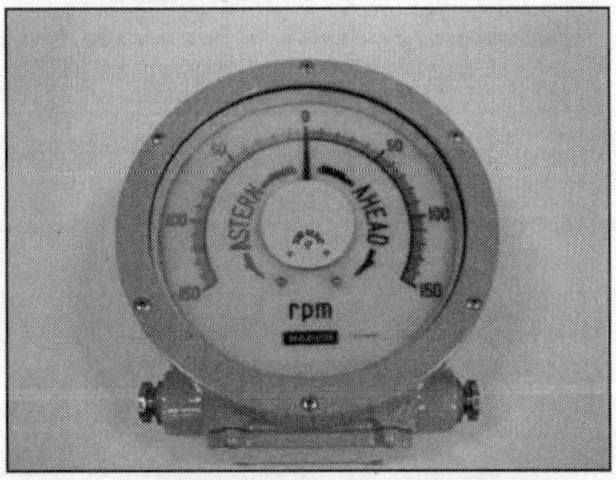

[그림 3-21] RPM indicator

10) Regulation 35 : Ventilation systems in machinery spaces

Machinery spaces of category A shall be adequately ventilated so as to ensure that when machinery or boilers therein are operating at full power in all weather conditions including heavy weather, an adequate supply of air is maintained to the spaces for the safety and comfort of personnel and the operation of the machinery. Any other machinery space shall be adequately ventilated appropriate for the purpose of that machinery space.

기관구역은 악천후를 포함한 모든 기상상태하에서 이 구역 내의 기관 또는 보일러를 전 부하로 운전하고 있는 경우에 인명의 안전, 쾌적 및 기관운전을 위한 충분한 공기공급을 유지하기 위하여 충분히 환기하여야 한다. 기타의 모든 기관구역은 그 구역의 목적에 적합하도록 충분히 환기하여야 한다. 이러한 기관구역의 통풍장치는 충분한 공기의 순환을 위하여 공기공급(air supply) 및 공기배출(air exhaust) 통풍장치로 이루어져 있다.

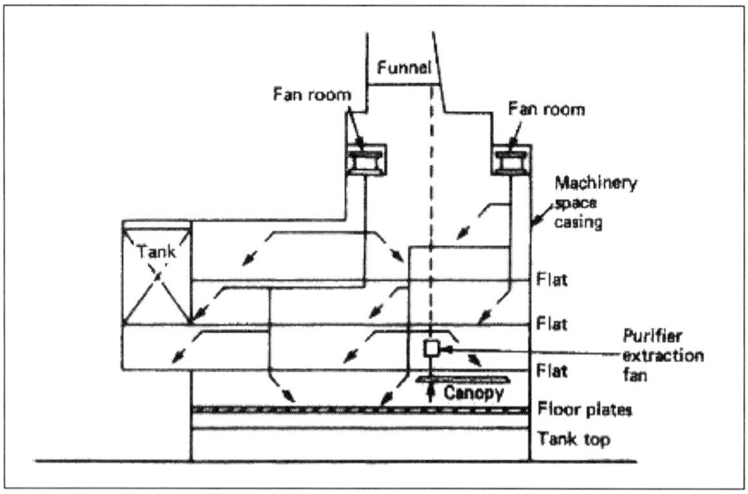

[그림 3-22] Fan room

11) Regulation 35-1 : Bilge pumping arrangements

An efficient bilge pumping system shall be provided, capable of pumping from and draining any watertight compartment other than a space permanently appropriated for the carriage of fresh water, water ballast, oil fuel or liquid cargo and for which other efficient means of pumping are provided, under all practical conditions. Efficient means shall be provided for draining water from insulated holds.

여객선 및 화물선은 모든 실제적인 조건에서 청수, 물 밸러스트, 연료유 또는 액체화물의 운송에 영구적으로 할당된 구역으로서 이에 대하여 다른 유효한 배출수단이 설치된 이외의 모든 수밀구획으로부터 배출 및 배수가 가능한 유효한 빌지 배출 시스템을 갖추어야 한다. 단열된 화물창으로부터 배수하기 위한 유효한 수단을 갖추어야 한다.

Sanitary, ballast and general service pumps may be accepted as independent power bilge pumps if fitted with the necessary connections to the bilge pumping system.

위생, 밸러스트 및 일반용도의 펌프도 빌지 배출 시스템에 필요한 연결을 장착한 경우 독립된 동력 빌지 펌프로 인정될 수 있다. 일반적으로 이러한 펌프와 함께 'Ejector'가 설치되어 빌지의 원활한 배출이 이루어지도록 하고 있다.

[그림 3-23] GS 펌프

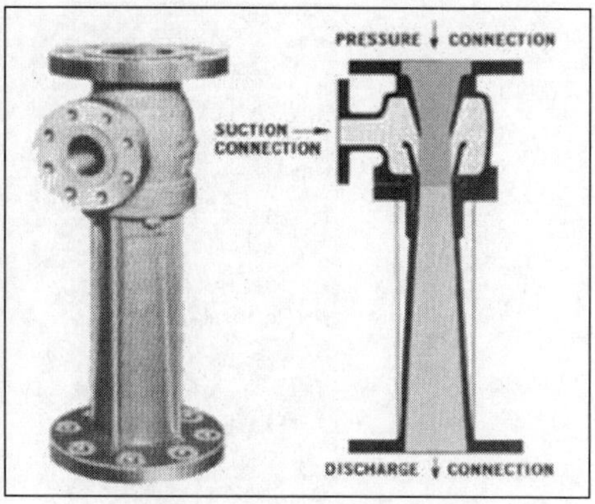

[그림 3-24] Ejector

12) Regulation 37 : Communication between navigation bridge and machinery space

At least two independent means shall be provided for communicating orders from the navigating bridge to the position in the machinery space or in the control room from which the engines are normally controlled : one of these shall be an engine room telegraph which provides visual indication of the orders and responses both in the machinery space and on the navigating bridge. Appropriate means of communication shall be provided to any other positions from which the engines may be controlled.

프로펠라 추력의 속도 및 방향이 통상적으로 제어되는 기관구역 또는 제어실내의 장소로 선교에서 명령을 전달할 수 있는 적어도 2개의 독립된 통신수단을 구비하여야 한다. 이들 중 1개는 기관구역 및 선교양쪽에서 명령 및 응답을 볼 수 있게 표시하는 엔진텔레그라프이어야 한다. 선교 및 기관구역에서 프로펠라 추력의 속도 또는 방향을 제어할 수 있는 기타 모든 장소와 적절한 통신수단을 구비하여야 한다. 일반적으로 기관구역과 선교의 통신수단으로 선내전화(automatic telephone)와 자가전원전화(sound powered telephone)가 설치되어 있다.

[그림 3-25] 통신장치

13) Regulation 38 : Engineer's alarm

An engineer's alarm shall be provided to be operated from the engine control room or at the manoeuvring platform as appropriate, and shall be clearly audible in the engineers' accommodation.

기관제어실이나 적당한 기관조종장소 내에서 조작되고 또한 기관사 거주구역에서 명확히 청취할 수 있는 호출장치를 설치하여야 한다. 일반적으로 기관사 침실, 선교, 식당 및 휴게실에 호출장치가 설치되어 있다.

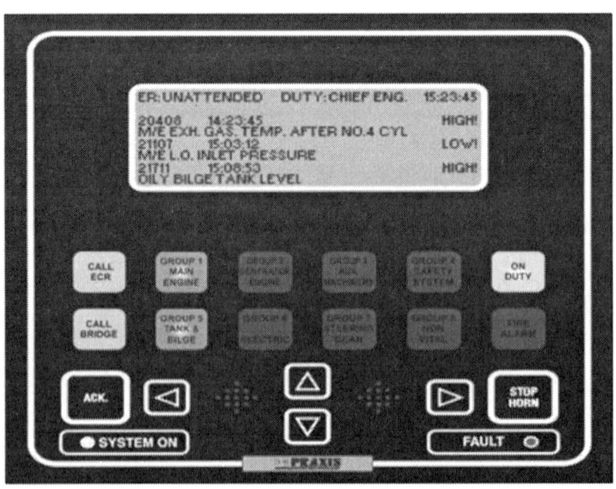

[그림 3-26] 기관사 호출장치

14) Regulation 41 : Main source of electrical power and lighting systems

A main source of electrical power of sufficient capacity to supply all those services shall be provided. This main source of electrical power shall consist of at least two generating sets.

모든 설비에 급전하는데 충분한 용량을 가지는 주전원을 설치하여야 하며, 이 주전원은 적어도 2개의 발전장치로 구성되어야 한다. 일반적으로 전기소모가 많은 냉동 컨테이너를 많이 선적하는 대형 컨테이너 운반선을 제외하고는 통상 3대의 발전기를 설비하고 있다.

The capacity of these generating sets shall be such that in the event of any one generating set being stopped it will still be possible to supply those services necessary to provide normal operational conditions of propulsion and safety. Minimum comfortable conditions of habitability shall also be ensured which include at least adequate services for cooking, heating, domestic refrigeration, mechanical ventilation, sanitary and fresh water.

발전장치의 용량은 어느 1개의 발전장치가 정지한 경우에도 통상의 추진 및 안전성을 유지하기 위하여 필요한 전기설비에 급전할 수 있는 것이어야 한다. 또한 적어도 취사, 난방, 일상용 냉장, 기계식 통풍, 위생수 및 청수를 사용하는 것을 포함한 최저한의 쾌적한 거주생활을 유지하도록 급전하여야 한다.

[그림 3-27] Diesel Generator

A main electric lighting system which shall provide illumination throughout those parts of the ship normally accessible to and used by passengers or crew shall be supplied from the main source of electrical power.

선내에서 통상 접근할 수 있고 여객 또는 선원에 의하여 사용되는 장소의 조명에 이용되는 주조명장치는 주전원으로부터 급전되어야 한다.

The main source of electrical power is necessary for propulsion and steering of the ship, the system shall be so arranged that the electrical supply to equipment necessary for propulsion and steering and to ensure safety of the ship will be maintained or immediately restored in the case of loss of any one of the generators in service.

주전원이 선박의 추진 및 조종에 필요한 경우, 그 장치는 추진과 조종에 필요한 기기 및 선박의 안전을 확보하기 위하여 공급되는 전원은 작동 중인 어느 하나의 발전기가 고장이 발생한 때에도 전원이 계속 공급되거나 즉시 복구될 수 있도록 배치되어야 한다.

이를 위하여 일반적으로 설치되어 있는 선박의 발전기는 1대가 작동되고 있을 때에 나머지 발전기들은 'Stand-by' 상태로 되어 있어 작동되고 있는 발전기에 어떤 문제가 발생하여 정지하게 되면 다른 발전기가 자동으로 작동되어 전원을 공급하도록 되어 있다.

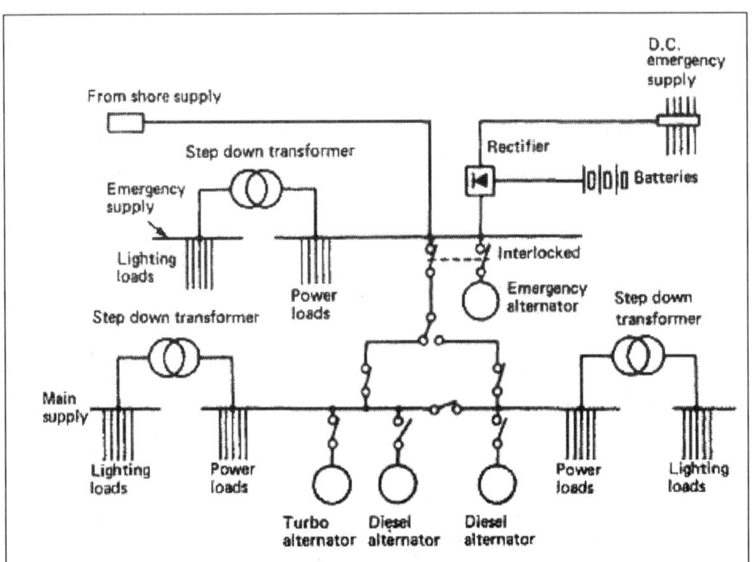

[그림 3-28] 발전기 계통도

15) Regulation 43 : Emergency source of electrical power in cargo ships

A self-contained emergency source of electrical power shall be provided.
자기기전식의 비상전원을 비치하여야 한다.

The emergency source of electrical power, associated transforming equipment, if any, transitional source of emergency power, emergency switchboard and emergency lighting switchboard shall be located above the uppermost continuous deck and shall be readily accessible from the open deck.

비상전원은 최상층 전통갑판상에 설치하여야 하며 또한 개방갑판으로부터 쉽게 접근할 수 있어야 한다. 이러한 연유로 일반적으로 선박에는 전통갑판상 개방갑판에 '비상발전기실'을 별도로 마련해 두고 있다.

The electrical power available shall be sufficient to supply all those services that are essential for safety in an emergency, due regard being paid to such services as may have to be operated simultaneously.

비상전원으로부터의 전력은 비상시 안전확보를 위하여 필수적인 모든 설비에, 동시에 작동하는 것을 고려하여 모든 설비에 충분한 용량이어야 한다.

[그림 3-29] 비상발전기

16) Regulation 46 : General(Additional requirements for periodically unattended machinery spaces)

Every ship shall be provided with documentary evidence, to the satisfaction of the

Administration, of its fitness to operate with periodically unattended machinery spaces.

모든 선박에는 기관구역에 정기적으로 선원이 배치되지 아니하고 항해하는데 적합한 것을 주관청이 인정하는 증명서류를 비치하여야 한다.

4. 건조, 방화, 화재탐지 및 소화(Chapter II-2)의 주요 내용

1) Regulation 3 : Definitions

Accommodation spaces are those spaces used for public spaces, corridors, lavatories, cabins, offices, hospitals, cinemas, game and hobby rooms, barber shops, pantries containing no cooking appliances and similar spaces.

거주구역이라 함은 공용구역, 통로, 화장실, 선실, 사무실, 병실, 영화관, 오락실, 이발소, 조리기가 없는 배식실 및 기타 이와 유사한 장소로 이용하는 장소를 말한다. 이러한 거주구역의 최상부에는 선교가 위치하고 있는데 선교는 거주구역에 포함되지 않는다.

[그림 3-30] 선박의 거주구역

Central control station is a control station in which the following control and indicator functions are centralized:

(a) fixed fire detection and fire alarm systems;

(b) automatic sprinkler, fire detection and fire alarm systems;

(c) fire door indicator panels;

(d) fire door closure;

(e) watertight door indicator panels;

(f) watertight door closures;

(g) ventilation fans;

(h) general/fire alarms;

(i) communication systems including telephones; and

(j) microphones to public address systems.

중앙제어장소라 함은 다음의 제어 및 지시장치 기능들이 집중되어 있는 제어장소를 말한다.

(a) 고정식 화재탐지 및 경보장치

(b) 자동 스프링클러, 화재탐지 및 화재경보장치

(c) 방화문 표시반

(d) 방화문 폐쇄장치

(e) 수밀문 표시반

(f) 수밀문 폐쇄장치

(g) 통풍기

(h) 일반 / 화재경보장치

(i) 전화를 포함한 통신장치 및

(j) 선내방송장치와 연결된 마이크

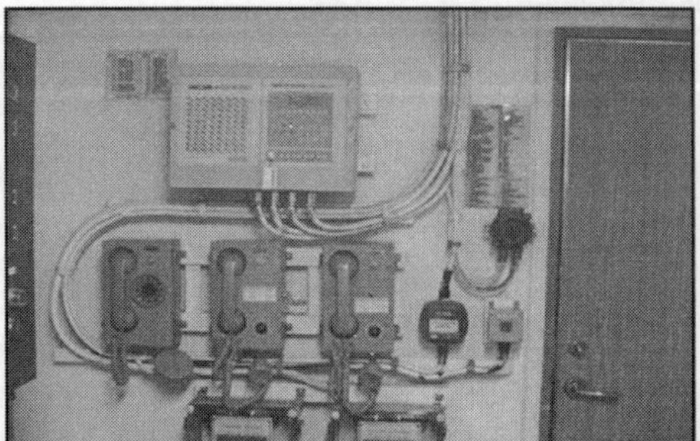

[그림 3-31] 중앙제어장소내부(1)

[그림 3-32] 중앙제어장소내부(2)

2) Regulation 4 : Probability of ignition

The ventilation of machinery spaces shall be sufficient under normal conditions to prevent accumulation of oil vapour.

기관구역의 통풍은 통상상태에서 기름증기의 축적을 방지하기에 충분한 것이어야 한다. 기관실의 원활한 통풍을 위하여 앞서 언급한 바와 같이 통풍장치로 공급을 흡입하여 공급하거나 기관실 내부의 공기를 배출시킬 수 있도록 설비되어 있다.

Fuel oil, lubrication oil and other flammable oils shall not be carried in forepeak tanks.

연료유, 윤활유 및 기타 인화성 기름을 선수탱크에 적재하여서는 아니된다. 선박의 선체 중에서 충격이나 다른 이유로 인하여 가장 손상을 입은 부분이 선수부이기 때문에 이러한 부분에 연료유나 인화성 기름을 적재하는 것은 그 만큼 사고로 인한 해양오염을 가중시킬 가능성이 크기 때문에 선수탱크에 적재하는 것은 금지되어야 한다.

Electric radiators, if used, shall be fixed in position and so constructed as to reduce fire risks to a minimum. No such radiators shall be fitted with an element so exposed that clothing, curtains, or other similar materials can be scorched or set on fire by heat from the element.

전기난방기를 사용하는 경우에는 한 위치에 고정하여야 하며, 화재위험을 최소로 줄일 수 있는 구조이어야 한다. 그러한 난방기에 부속된 부품에서 발생된 열에 의하여 의류, 커튼 또는 유사한 재료가 그을리거나 불이 붙도록 부품이 노출되어서는 아니된다. 선박에서는 가능한 전기난방기를 사용하고 있지 않다.

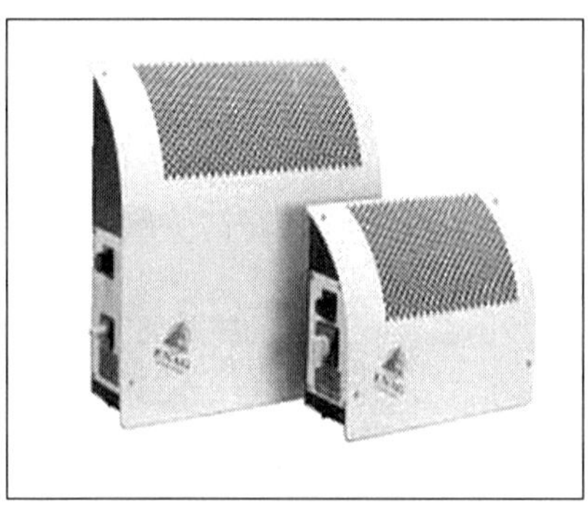

[그림 3-33] 전기난방기

다만, 필요에 따라 건조실에 건조를 돕기 위하여 난방기가 설치되어 있다. 이러한 난방기도 규정에 맞게 설비되어 있다. 현재 운항하고 있는 대다수의 선박에서 난방을 위하여 전기난방기는 사용되지 않고 있으며, 선내에 난방이 필요할 경우는 통풍구를 통하여 더운 공기를 공급함으로써 난방을 시행하고 있다.

Waste receptacles shall be constructed of non-combustible materials with no openings in the sides or bottom.

쓰레기통은 불연성 재료로 만들어야 하며 옆이나 바닥에 개구가 없어야 한다. 선내에 비치되어 있는 쓰레기통은 여러 곳에 비치가 되어 있어 정기적인 점검이 불가하며, 선내에서 발생한 쓰레기의 상당부분은 기름이 묻어 있는 경우도 많기 때문에 자칫 온도상승으로 인한 자연발화의 개연성도 있기 때문에 이에 대한 대비가 필요하고, 그 일환으로 동 규정을 두고 있다고 할 수 있다.

[그림 3-34] 쓰레기통

3) Regulation 4-1 : Cargo areas of tankers

Tankers operating with a cargo tank cleaning procedure using crude oil washing shall be fitted with an inert gas system complying with the Fire Safety Systems Code and with fixed tank washing machines.

원유세정에 의한 화물탱크 세정방식을 사용하는 탱커에는 화재안전시스템 코드의 요건에 적합한 이너트가스 시스템 및 고정식 탱크 세정기를 설치하여야 한다.

원유세정시 해당 탱크에 충분히 이너트가스가 공급되어 있지 않다면, 자칫 원유세정시 정전기나 스파크가 발생하여 폭발에 가능이 있기 때문에 반드시 원유세정이 수행되어야 할 탱크는 충분히 이너트가스가 공급되어 있어야 한다.

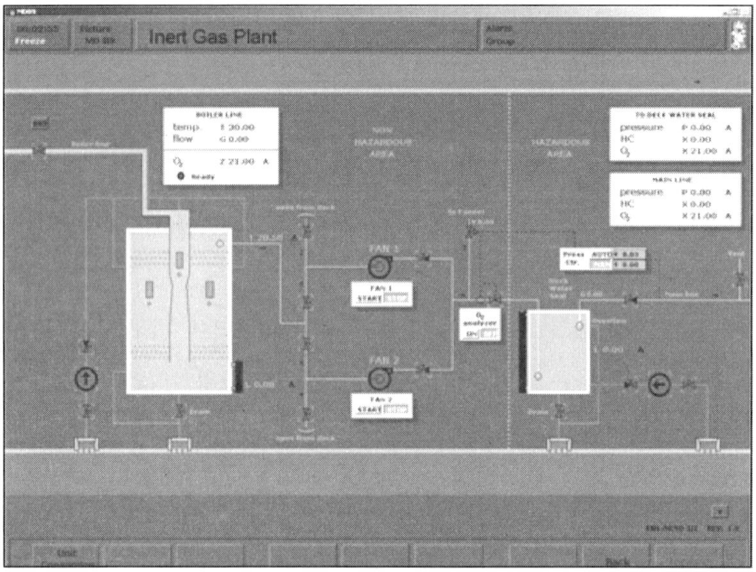

[그림 3-35] 이너트가스 장치계통도

The inert gas system shall be capable of inerting, purging and gas-freeing empty tanks and maintaining the atmosphere in cargo tanks with the required oxygen content.

이너트가스 장치는 빈 탱크의 이너팅, 퍼징 및 가스프리 할 수 있어야 하고, 화물 탱크 내부를 요구되는 산소농도로 유지할 수 있어야 한다.

Tankers fitted with a fixed inert gas system shall be provided with a closed ullage system.

고정식 이너트가스 장치를 설치한 탱커는 밀폐식 액면 계측장치를 설치하여야 한다.

필요에 따라 화물탱크에 사람이 진입하여 작업을 해야 할 경우도 있기 때문에 이에 대한 준비를 하려면 이너트가스 장치가 규정에 명시된 기능 수행이 가능하여야 하며, 화물작업시 해당 탱크에 이너트가스를 주입했을 경우 탱크내의 화물적재상태를 확인하기가 쉽지 않기 때문에 이에 대한 대안으로 밀폐식 액면 계측장치를 설치함으로써 화물적재상태를 쉽게 파악할 수 있다.

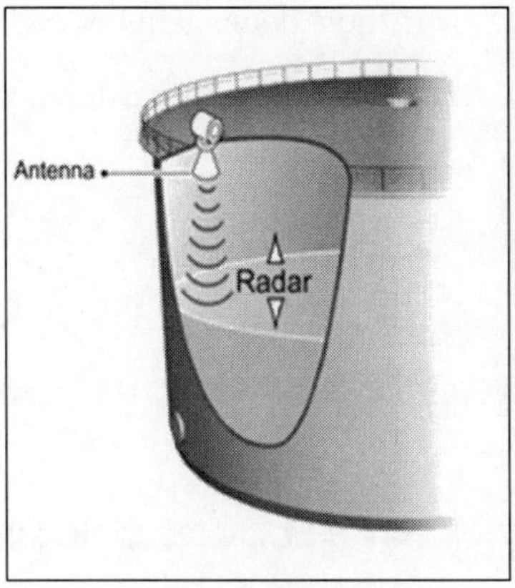

[그림 3-36] 액면 계측장치

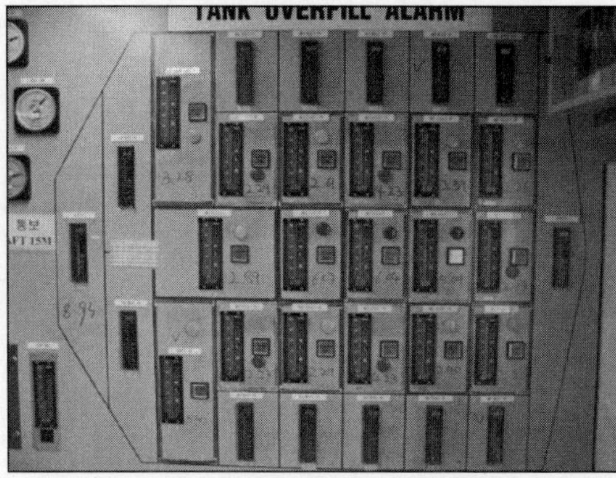

[그림 3-37] 액면 계측장치 표시기

Tankers shall be equipped with at least one portable instrument for measuring oxygen and one for measuring flammable vapour concentrations, together with a sufficient set of spares. Suitable means shall be provided for the calibration of such instruments.

탱커에는 산소농도를 측정하기 위한 최소한 1개의 휴대식 기기와 가연성 폭발성 증기의 농도를 측정하기 위한 최소한 한 개의 휴대식 기기를 비치하여야 하며 이와 함께 충분한 수의 예비품을 비치하여야 한다. 그러한 기기들의 교정을 위한 적절한 방법이 제공되어야 한다.

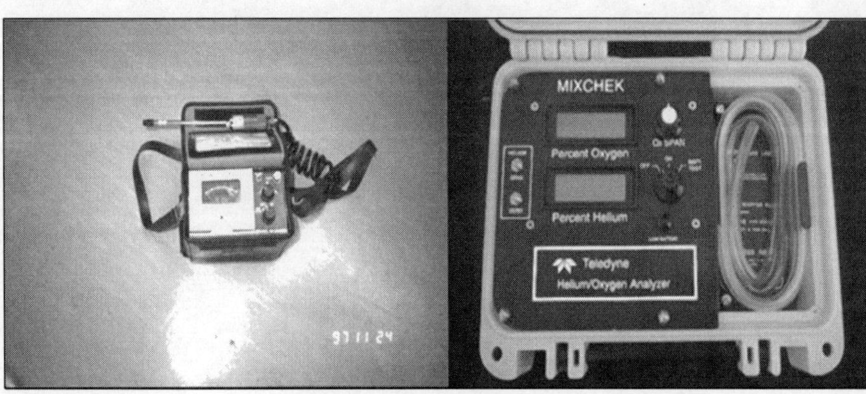

[그림 3-38] 휴대식 가스 측정기

Drip pans for collecting cargo residues in cargo lines and hoses shall be provided in the area of pipe and hose connections under the manifold area. Cargo hoses and tank washing hoses shall have electrical continuity over their entire lengths including couplings and flanges (except shore connections) and should be earthed for removal of electrostatic charges.

화물관 및 호스에서의 잔류물을 모으기 위해 매니폴드 하방의 파이프 및 호스 연결부에는 넘침팬을 설치하여야 한다. 화물호스 및 탱크세정 호스는 커플링과 플랜지(육상연결구 제외)를 포함한 전체 길이에 대하여 전기적 연결성이 유지되어야 하며, 정전기의 대전을 제거하기 위해 접지되어야 한다.

호스 연결부는 기름이 유출될 수 있는 가능성이 가장 높은 곳으로 이에 대한 대비책으로 넘침팬을 설치하여 만약의 경우에 유출된 기름을 가두어 해상으로의 유출을 막고자함이다. 또한 화물 관련 호스에 화물이 이송되면서 마찰이나 기타 연유로 정전기가 축적될 수 있고, 이러한 축적된 정전기가 방출되면서 스파크를 유발할 가능성이 있다. 만에 하나 폭발성 가스가 존재하는 곳에서 스파크가 발생한다면 화재나 폭발이 발생될 수 있기 때문에 적절히 정전기의 대전을 제거하기 위하여 접지되어야 한다.

[그림 3-39] Drip pan

4) Regulation 5 : Fire growth potential

The main inlets and outlets of all ventilation systems shall be capable of being closed from outside the spaces being ventilated. The means of closing shall be easily accessible as well as prominently and permanently marked and shall indicate whether the shutoff is open or closed.

모든 통풍장치의 주흡기구와 주배기구는 통풍되는 장소의 외부로부터 폐쇄할 수 있어야 한다. 폐쇄장치는 쉽게 접근할 수 있으며 항구적으로 잘 보이도록 표시되어야 하며, 열림 또는 닫힘이 표시되어야 한다.

화재발생시 화재 확산을 막거나, 화재진압을 위해 통풍장치의 폐쇄를 시행하고자 할 경우 통풍장치의 폐쇄장치가 접근하기 어려운 장소에 위치하거나 통풍되는 장소의 내부에 위치하고 있다면 화재발생시 접근이 불가하다면 통풍장치의 폐쇄가 이루어지지 않아 화재확산을 막거나 화재를 진압하기 힘들 것이다.

5) Regulation 7 : Detection and alarm

The purpose of this regulation is to detect a fire in the space of origin and to provide alarm for safe escape and fire-fighting activity. For this purpose, the following functional requirements shall be met :

(a) fixed fire detection and fire alarm system installations shall be suitable for the nature of the space, fire growth potential and potential generation of smoke and gases;

(b) manually operated call points shall be placed effectively to ensure a readily accessible means of notification; and

(c) fire patrols shall provide an effective means of detecting and locating fires and alerting the navigation bridge and fire teams.

이 규칙의 목적은 발화장소에서 화재를 탐지하고 안전탈출과 소화활동을 위한 경보를 울리는 것이다. 이러한 목적을 위하여 다음의 기능 요건들을 만족하여야 한다.

(a) 고정식 화재탐지 및 화재경보장치의 설치는 구역의 상태, 화재성장 가능성 및 연기 및 가스 발생가능성에 대하여 적합하여야 한다.

(b) 수동조작 콜포인트는 쉽게 접근할 수 있는 통보의 방법을 확보하기 위하여 배치를 효과적으로 하여야 한다.

(c) 화재순찰은 탐지, 화재위치파악 및 항해선교 및 소화대원에게 경보를 울리는 효과적인 수단을 제공하여야 한다.

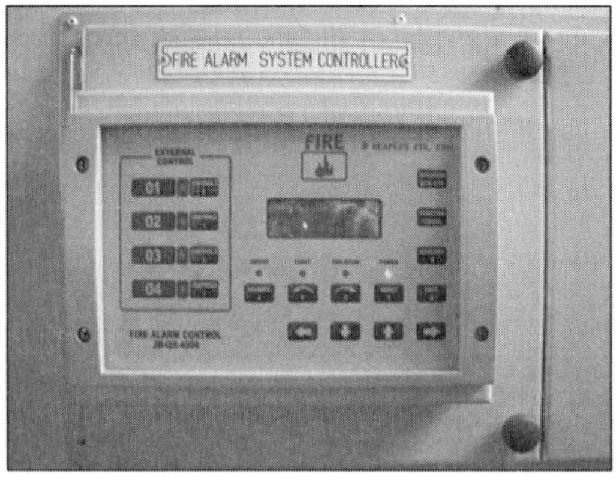

[그림 3-40] 화재경보장치

Smoke detectors shall be installed in all stairways, corridors and escape routes within accommodation spaces.

연기탐지기는 거주구역 내의 모든 계단, 통로 및 탈출로에 설치되어야 한다.

[그림 3-41] 연기탐지기

Manually operated call points complying with the Fire Safety Systems Code shall be installed throughout the accommodation spaces, service spaces and control stations. One manually operated call point shall be located at each exit. Manually operated call points shall be readily accessible in the corridors of each deck such that no part of the corridor is more than 20 m from a manually operated call point.

화재안전장치 코드에 만족하는 수동조작 콜 포인트를 거주구역, 업무구역 및 제어장소에 걸쳐 설치하여야 한다. 각 출입구에는 하나의 수동조작 콜 포인트가 설치되어야 한다. 수동조작 콜 포인트는 각 갑판의 통로에서 쉽게 접근할 수 있어야 하며 통로의 어느 부분도 수동조작 콜 포인트에서 20m 이상 떨어져서는 아니된다.

이러한 장치의 작동법은 제조사별로 다소 다른데, 전면의 유리를 깨거나 누르거나 또는 둘 다를 행하더라도 작동되는 타입이 있기 때문에 필히 작동법을 확인하여야 한다.

[그림 3-42] 수동조작 콜 포인트

6) Regulation 10 : Fire Fighting

The number and position of hydrants shall be such that at least two jets of water not emanating from the same hydrant, one of which shall be from a single length of hose, may reach any part of the ship normally accessible to the passengers or crew while the ship is being navigated and any part of any cargo space when empty, any ro-ro space or any vehicle space in which latter case the two jets shall reach any part of the space, each from a single length of hose. Furthermore, such hydrants shall be positioned near the accesses to the protected spaces.

소화전의 수 및 위치는 같은 소화전으로부터 방출되지 아니하는 최소한 두 줄기의 사수가(그중의 한 줄기는 단일 소화호스로부터 방출되는) 선박이 항행하는 동안 여객 또는 선원이 통상 접근할 수 있는 선박의 어느 부분, 비어 있을 때의 화물구역의 어느 부분, 및 어느 로로구역 또는 차량구역에도 도달할 수 있어야 한다. 상기 로로구역 및 차량구역에 대하여는 각각 단일 호스로부터 방출되는 두 줄기의 사수가 구역의 어느 부분에도 도달할 수 있어야 한 다. 또한 이와 같은 소화전은 보호구역으로의 입구 근처에 배치되어야 한다.

이러한 소화전 수와 관련된 규정으로 인해 통상 선박에서 화재 관련 수소화 작업훈련을 수행할 때에 적어도 2개 이상의 소화호스 및 소화전을 설치하여 시행하고 있다.

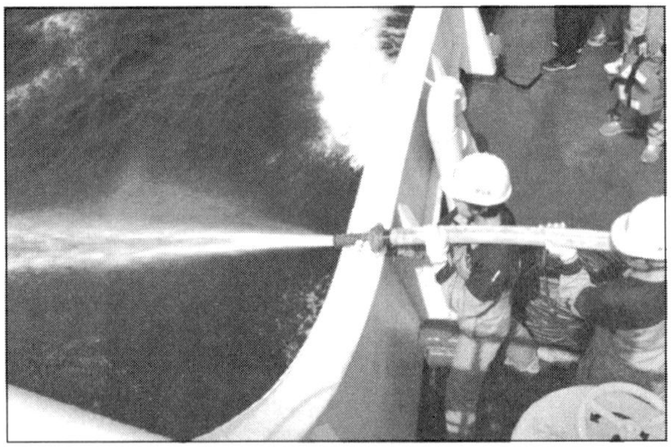

[그림 3-43] 수소화 작업모습

Ships of 500 gross tonnage and upwards shall be provided with at least one international shore connection complying with the Fire Safety Systems Code.

Facilities shall be available enabling such a connection to be used on either side of the ship.

총톤수 500톤 이상의 선박에는 화재안전장치 코드에 적합한 최소한 1개의 국제육상시설연결구를 비치하여야 한다. 그러한 국제육상시설연결구를 선박의 양측에서 사용하는 것이 가능하도록 설비가 되어야 한다.

협약규정상 선박에는 최소한 1개의 국제육상시설연결구를 비치하고 있어야 하며, 선박의 양측에서 사용 가능하도록 설비되어 있어야 한다고 명시되어 있어 이를 위하여 양현에 각각 연결구를 설치하고 있는 선박도 있으나, 거주구 내부 중앙이나 중앙통제장소에 1개를 비치하여 필요한 현으로 가지고 가서 설치 후 사용하기도 한다.

[그림 3-44] 국제육상시설 연결구

Number of fire pumps

Ships shall be provided with independently driven fire pumps as follows:

(a) in passenger ships of:

4,000 gross tonnage and upwards at least three

less than 4,000 gross tonnage at least two

(b) in cargo ships of:

1,000 gross tonnage and upwards at least two

less than 1,000 gross tonnage at least two power driven pumps, one of which shall be independently driven.

소화펌프의 수에 있어 선박은 다음과 같이 독립적으로 구동되는 소화펌프가 설치되어야 한다.

(a) 여객선

- 총톤수 4,000톤 이상 최소한 3대
- 총톤수 4,000톤 미만 최소한 2대

(b) 화물선

- 총톤수 1,000톤 이상 최소한 2대
- 총톤수 1,000톤 미만 최소한 2대의 동력구동펌프, 그 중에 1대는 독립동력으로 구동이 되어야 한다.

[그림 3-45] 소화펌프

일반적으로 선박에 설치되어 있는 소화펌프는 소화펌프용으로 별도로 설비하기 보다는 G/S(General Service) 펌프를 설치하여 소화작업용으로도 사용할 수 있도록 설비하고 있다.

The space containing the emergency fire pump shall not be contiguous to the boundaries of machinery spaces of category A or those spaces containing main fire pumps. Where this is not practicable, the common bulkhead between the two spaces shall be insulated to a standard of structural fire protection equivalent to that required for a control station.

비상소화펌프가 있는 장소는 A류 기관구역 또는 주소화펌프가 있는 장소에 인접하지 않아야 한다. 이것이 실행 불가능할 경우, 두 구역 사이의 공통격벽은 제어장소에 대하여 요구되는 것과 동등한 방화구조기준으로 방열되어야 한다.

비상소화펌프는 기관 수역이 침수되거나 선내 전원이 공급되지 않더라도 작동이 될 수 있게 설비규정을 두고 있다. 이러한 연유로 일반적으로 화물선에서 비상소화펌프는 기관실과 완전히 분리된 장소로 조타기실 내부 가장 하단부에 설비하고 있다.

[그림 3-46] 비상소화펌프

Accommodation spaces, service spaces and control stations shall be provided with portable fire extinguishers of appropriate types and in sufficient number to the satisfaction of the Administration. Ships of 1,000 gross tonnage and upwards shall carry at least five portable fire extinguishers.

거주구역, 업무구역 및 제어장소에는 주관청이 만족하는 적합한 형식과 충분한 수의 휴대식 소화기가 비치되어야 한다. 총톤수 1,000톤 이상의 선박에는 최소한 5개의 휴대식 소화기를 비치하여야 한다.

One of the portable fire extinguishers intended for use in any space shall be stowed near the entrance to that space.

어떠한 장소에서도 그 곳에서 사용하는 휴대식 소화기들 중 1개는 그 장소의 출입구 부근에 비치되어야 한다.

화재발생 시 신속하게 화재구역에 접근하여 소화기를 사용할 수 있도록 하기 위함이다.

Carbon dioxide fire extinguishers shall not be placed in accommodation spaces. In control stations and other spaces containing electrical or electronic equipment or appliances necessary for the safety of the ship, fire extinguishers should be provided whose extinguishing media are neither electrically conductive nor harmful to the equipment and appliances.

탄산가스소화기는 거주구역에 비치될 수 없다. 선박의 안전에 필요한 전기 또는 전자 장비 또는 설비가 있는 제어장소 및 다른 구역에서는 소화매체가 전기적으로 전도성이 없고 장비 및 설비에 해롭지 않은 소화기가 비치되어야 한다.

탄산가스소화기의 소화원리는 질식이기 때문에 거주구에 화재발생시 동 소화기를 사용하면 화재진압을 하고 있거나 인근에 있는 사람들에게 질식을 초해할 수 있기 때문에 거주구 비치는 금지하고 있다.

Fire extinguishers shall be situated ready for use at easily visible places, which can be reached quickly and easily at any time in the event of a fire, and in such a way that their serviceability is not impaired by the weather, vibration or other external factors. Portable fire extinguishers shall be provided with devices which indicate whether they have been used.

소화기는 쉽게 보이는 장소로 화재가 발생할 경우 어느 때나 빠르고 쉽게 접근하여 사용할 수 있으며 그리고 날씨, 진동 또는 외적인 요소에 의해 편의성이 손상되지 않도록 비치가 되어야 한다. 휴대식 소화기는 소화기가 사용되었는지를 나타내는 장치가 제공되어야 한다.

선박은 외부 요소로 인하여 많은 움직임이 있기 때문에 소화기를 적절하게 비치하지 않으면 손상되기 쉽다하겠다. 하지만 이러한 연유로 소화기를 너무 고박을 단단하게 하면 자칫 화재발생 시 원활하게 소화기를 사용할 수 없는 경우도 발생할 수 있기 때문에 선박의 움직임으로 인한 소화기의 손상이 되지 않을 정도로만 고박을 하여야 한다.

아울러 소화기의 사용상태를 표시해 두어야 하는데 특별한 장치를 통하여 표시하기 보다는 일반적으로 소화기의 안전한 관리 및 사용상태 표시용으로 안전핀을 소화기 레버에 설치하고 있다.

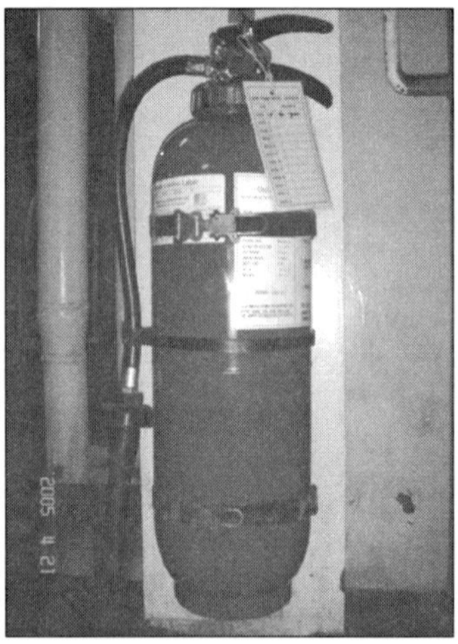

[그림 3-47] 휴대용 폼소화기

[그림 3-48] 휴대용 분말소화기

Spare charges shall be provided for 100% of the first 10 extinguishers and 50% of the remaining fire extinguishers capable of being recharged on board. Not more than 60 total spare charges are required. Instructions for recharging shall be carried on board.

예비소화제는 처음 10개의 소화기에 대하여는 100퍼센트와 선내에서 재충전이 가능한 나머지 소화기의 50퍼센트가 비치되어야 한다. 예비소화제는 총 60개를 넘을 필요는 없다. 재충전을 위한 지침서가 선내에 있어야 한다.

예를 들어 선내 재충전이 가능한 소화기가 100개 비치되어 있다면 처음 10개의 소화기의 예비소화제 10개 그리고 나머지 90개의 소화기의 50퍼센트에 해당하는 45개의 예비소화제를 비치해야 하기 때문에 합계 55개의 예비소화제를 비치하면 된다.

For fire extinguishers which cannot be recharged onboard, additional portable fire extinguishers of the same quantity, type, capacity and number.

선내에서 재충전할 수 없는 소화기에 대하여는 같은 양, 형식, 용량 및 수의 추가의 휴대식 소화기가 예비소화제를 대신하여 비치되어야 한다.

선박에 비치되어 있는 재충전이 불가한 소화기는 대표적으로 탄산가스 소화기이다.

[그림 3-49] 예비소화제 비치모습

Extinguishing systems for Machinery spaces containing oil-fired boilers or oil fuel units

(a) Fixed fire-extinguishing systems

(b) At least one portable foam applicator unit in each boiler room or at an entrance outside of the boiler room

(c) At least two portable foam extinguishers in each firing space in each boiler room and in each space in which a part of the oil fuel installation is situated.

(d) Not less than one approved foam-type extinguisher of at least 135 ltrs capacity in each boiler room.

기름보일러 또는 연료유장치가 있는 기관구역의 소화장치

(a) 고정식 소화장치

(b) 각 보일러실 또는 보일러실의 출입구 바깥에는 최소한 1개의 휴대식 포말방사기

(c) 각 보일러실의 각 점화장소 및 연료유설비의 일부가 있는 각 장소에는 최소한 2개의 휴대식 포말소화기

(d) 보일러실에는 최소한 135리터 용량의 승인된 포말소화기

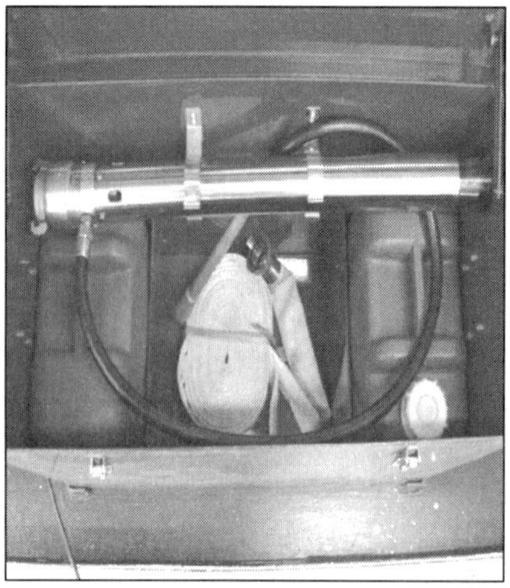

[그림 3-50] 휴대식 포말방사기

[그림 3-51] 135리터 포말소화기

Extinguishing systems for Machinery spaces containing internal combustion machinery.

(a) Fixed fire-extinguishing systems

(b) At least one portable foam applicator unit

(c) Approved foam-type at least 45 ltrs fire extinguishers sufficient in number in each space

(d) A sufficient number of portable foam extinguishers which shall be so located that no point in the space is more than 10 m walking distance from an extinguisher and that there are at least two such extinguishers in each such space.

내연기관이 있는 기관구역의 소화장치

(a) 고정식 소화장치

(b) 최소한 1개의 휴대식 포말방사기

(c) 충분한 수의 각각 45리터 이상의 용량의 승인된 포말소화기

(d) 소화기로부터 도보 10미터를 넘는 장소가 없도록 배치된 충분한 수의 휴대식 포말소화기 또는 그 동등물이 비치되어야 하며, 그러한 장소마다 최소한 2개의 그러한 휴대식 소화기가 있어야 한다.

Fixed local application fire-fighting systems shall apply to passenger ships of 500 gross tonnage and above and cargo ships of 2000 gross tonnage and above.

총톤수 500톤 이상의 여객선과 총톤수 2,000톤 이상의 화물선에 적용한다.

Fixed local application fire-fighting systems are to protect areas such as the following without the necessity of engine shutdown, personnel evacuation, or sealing of the spaces.

(a) the fire hazard portions of internal combustion machinery or, for ships constructed before 1 July 2014, the fire hazard portions of internal combustion machinery used for the ship's main propulsion and power generation.

(b) boiler fronts.

(c) the fire hazard portions of incinerators.

(d) purifiers for heated fuel oil.

고정식 국부소화장치는 엔진의 정지, 인원의 대피 또는 구역을 밀폐할 필요가 없이 다음과 같은 장소들을 보호하여야 한다.

(a) 내연기관의 화재위험부분 또는 2014년 7월 1일 전에 건조된 선박의 경우에는 선박의 주추진 및 동력발생을 위하여 사용되는 내연기관의 화재위험부분

(b) 보일러 전부

(c) 소각기의 화재위험부분

(d) 가열된 연료유를 위한 청정기

선박에서 고정식 국부 소화장치로 가장 많이 채택하여 사용하고 있는 것으로 분무(Mist)방식가 있다.

[그림 3-52] 고정식 국부 소화장치(Mist)

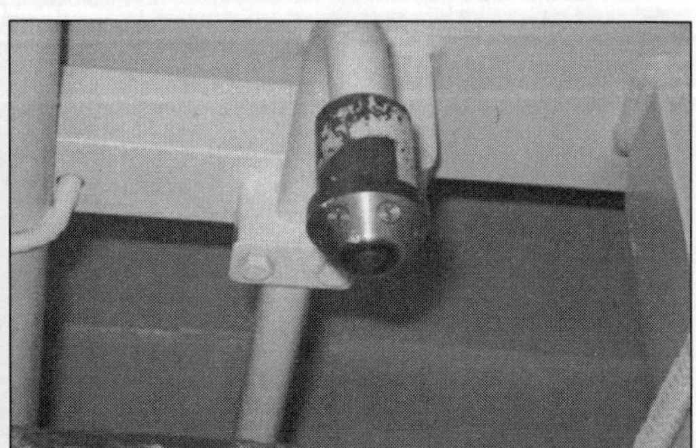

[그림 3-53] Mist 생성 Nozzle

분무는 물만 있으면 언제든지 그리고 여러번 사용할 수 있다는 장점이 있기 때문에 많이 채택되어 사용되고 있다할 수 있다. 분무장치는 특별한 설비 없이 화재탐지장치와 연결되어 화재가 탐지되었을 경우 분무 발생을 위한 펌프가 기동되고, 해당 구역으로 물을 공급하면 분무 생성 Nozzle을 통하여 물이 분사되면서 분무가 형성되어 있다.

Paint lockers shall be protected by ;

(a) a carbon dioxide system

(b) a dry powder system

(c) a water spraying or sprinkler system,

(d) a system providing equivalent protection,

In any case, the system shall be operable from outside the protected space."

페인트 창고는 다음의 장치에 의하여 보호가 되어야 한다.

(a) 탄산가스장치

(b) 분말소화장치

(c) 물분무 또는 스프링클러장치

(d) 주관청에 의하여 결정된 동등한 보호를 제공하는 장치

어떠한 경우에도 장치는 보호되는 장소 외부에서 작동될 수 있어야 한다.

페인트 창고는 유류물질인 페인트와 더불어 연소성이 강한 페인트 희석제 등이 많이 보관되어 있어 화재의 위험성이 높은 장소 중에 하나이다. 이러한 연유로 페인트 창고에 소화장비를 규정하고 있으며, 규정에서 언급하고 있는 소화장치 중에 하나 이상만 설비하고 있으면 된다.

[그림 3-54] 페인트 창고 소화장치(이산화탄화)

아울러 페인트 창고는 일반적으로 외부에 위치하고 있으며, 특별한 경우를 제외하고는 페인트 창고로 통하는 문은 1개만 설치되어 있다. 그리고 페인트 창고에 화재가 발생할 경우는 화재확산이 급격하게 될 수 있다는 점을 고려하여 규정에서 언급되고 있는 소화장치들은 모두 페인트 창고 외부에서 작동될 수 있도록 설비되어 있다.

Deep-fat cooking equipment installed in enclosed spaces or on open decks shall be fitted with the following :

(a) an automatic or manual extinguishing system

(b) a primary and backup thermostat with an alarm

(c) arrangements for automatically shutting off the electrical power upon activation of the extinguishing system

(d) an alarm for indicating operation of the extinguishing system in the galley

(e) controls for manual operation of the extinguishing system

폐위구역 또는 노출갑판에 설치된 튀김 기름을 사용하는 요리설비는 다음의 사항들과 같이 설치되어야 한다.

(a) 자동 또는 수동 소화장치

(b) 주 및 예비 온도조절장치

(c) 소화장치의 작동에 따라 전원을 자동적으로 차단하는 설비

(d) 조리실에 있는 소화장치의 작동을 나타내는 경보

(e) 소화장치의 수동작동을 위한 제어장치

거주구내에 위치하고 있는 주방에는 열을 사용하여 음식을 조리하기 때문에 다른 장소에 비해 상대적으로 화재의 위험성이 크다하겠다. 이러한 주방설비 중에 튀김 기름을 사용하는 기구는 상당히 화재의 위험성이 크기 때문에 이에 대비하여 별도의 소화장치의 설치를 규정하고 있다.

Cargo spaces on cargo ships of 2,000 gross tonnage and upwards shall be protected by a fixed carbon dioxide or inert gas fire-extinguishing system.

총톤수 2,000톤 이상의 화물선의 화물구역은 고정식 탄산가스 또는 불활성가스 소화장치를 제공하는 소화장치에 의하여 보호되어야 한다.

For tankers of 20,000 tonnes deadweight and upwards, a fixed deck foam system shall be provided in accordance with the requirements of the Fire Safety Systems Code

[그림 3-55] 튀김 기름 주방설비 소화장치

재화중량 20,000톤 이상의 탱커에는 화재안전장치 코드 요건에 적합한 고정식 갑판포말장치가 설치되어야 한다.

탱커의 경우는 인화성의 액체화물을 운송하기 때문에 누출된 기름 화재에 대비하기 위하여 외부 갑판상에 일정한 간격으로 포말분사장치(일명, foam monitor)를 설비하고 있다.

[그림 3-56] 일명 Foam monitor를 설치하고 있다.

Ships shall carry at least two fire-fighter's outfits

선박에는 최소한 2조의 소방원 장구가 있어야 한다.

In addition, in passenger ships there shall be provided for every 80m.

추가로, 여객선에는 길이 80m마다 2조의 소방원 장구가 비치하여야 한다.

In tankers, two fire-fighter's outfits shall be provided.

추가로, 탱커에는 2조의 소방원 장구가 비치되어야 한다.

Two spare charges shall be provided for each required breathing apparatus.

각 요구되는 호흡구마다 2개의 예비 공기병이 비치되어야 한다.

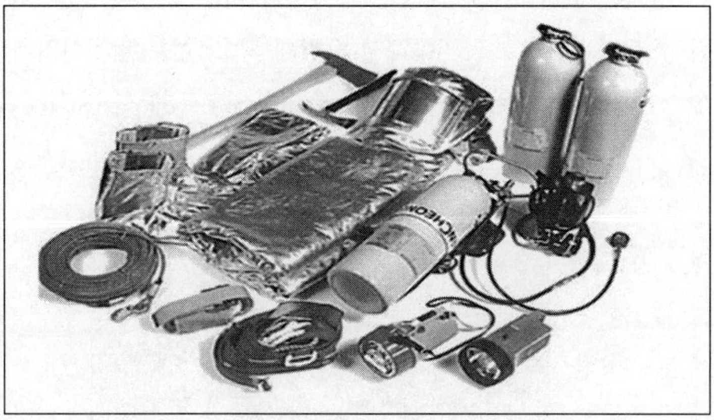

[그림 3-57] 소방원 장구

For ships constructed on or after 1 July 2014, a minimum of two two-way portable radiotelephone apparatus for each fire party for fire-fighter's communication shall be carried on board. Those two-way portable radiotelephone apparatus shall be of an explosion-proof type or intrinsically safe. Ships constructed before 1 July 2014 shall comply with the requirements of this paragraph not later than the first survey after 1 July 2018.

2014년 7월 1일 이후 건조된 선박들에는 각 소화반별로 최소한 2대씩의 쌍방향 휴대식 무전기가 제공되어야 한다. 그러한 쌍방향 휴대식 무전기는 방폭형 또는 본질안전형이어야 한다. 2014년 7월 1일 전에 건조된 선박들은 2018년 7월 1일 후에 도래하는 첫번째 검사시까지 같은 요건에 적합하여야 한다.

기존에 소방원 장구 착용자간의 통신은 Life line을 통하여 간단한 교신을 하였으나 이러한 통신은 실효성이 없으며, 아주 간단한 통신 밖에 이루어지지 않기 때문에 화재 현장의 상황파악이 어렵고, 소방원 장구 착용자와 원활한 교신을 할 수 없었다. 이러한 점을 보완하기 규정이라 볼 수 있다.

7) Regulation 12 : Notification of crew and passengers

A general emergency alarm system shall be used for notifying crew and passengers of a fire.

총비상경보장치는 선원 및 여객에게 화재를 알리기 위하여 사용되어야 한다.

총비상경보장치는 일반적으로 대다수의 선박에 자동 및 수동으로 이루어져 있으며, 자동은 한번 누르면 비상신호를 지속해서 울리게 되고, 이를 정지시키기 위해서는 다시 한번 누르면 된다. 수동은 버튼을 누리고 있는 동안만 경보를 울리게 된다.

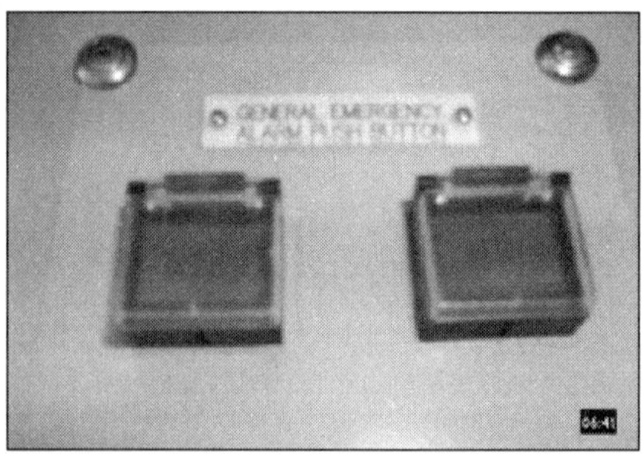

[그림 3-58] 총비상경보장치

8) Regulation 13 : Means of escape

Stairways and ladders shall be so arranged as to provide ready means of escape to the lifeboat and liferaft embarkation deck from passenger and crew accommodation spaces and from spaces in which the crew is normally employed, other than machinery spaces.

여객 및 선원 거주구역과 선원이 통상업무에 종사하는 장소(기관구역 제외)에서 구명정 및 구명뗏목의 승정갑판까지의 상설 탈출설비로서 계단 및 사다리를 설치하여야 한다.

Means of escape in cargo ships

(a) At all levels of accommodation there shall be provided at least two widely separated means of escape from each restricted space or group of spaces.

(b) Below the lowest open deck the main means of escape shall be a stairway and the second escape may be a trunk or a stairway.

(c) Above the lowest open deck the means of escape shall be stairways or doors to an open deck or a combination."

화물선의 탈출설비

(a) 거주구역의 모든 층에서는 각각 폐위된 1개 또는 일단의 장소로부터 상호 멀리 떨어진 2개 이상의 탈출설비를 설치하여야 한다.

(b) 최하층의 개방갑판보다 하층의 주탈출설비는 계단으로 하여야 하며 제2의 탈출설비는 트렁크 또는 계단으로 할 수 있다.

(c) 최하층의 개방갑판보다 상층의 탈출설비는 개방갑판으로 통하는 계단 또는 문 또는 두 가지의 조합으로 되어야 한다.

All ships shall carry at least two emergency escape breathing devices within accommodation spaces.

모든 선박은 거주구역 내에 최소 2개의 비상탈출용 호흡구를 비치하여야 한다.

[그림 3-59] 비상탈출용 호흡구

Means of escape from each machinery space in cargo ships shall comply with the following provisions.

(a) Two means of escape shall be provided from each machinery space

(b) Two sets of steel ladders as widely separated as possible leading to doors in the upper part of the space similarly separated and from which access is provided to the open deck. One of these ladders shall be located within a protected enclosure that satisfies regulation from the lower part of the space it serves to a safe position outside the space. Self-closing fire doors of the same fire integrity standards shall be fitted in the enclosure. The ladder shall be fixed in such a way that heat is not transferred into the enclosure through non-insulated fixing points. The enclosure shall have minimum internal dimensions of at least 800 mm x 800 mm, and shall have emergency lighting provisions

화물선의 각 기관구역의 탈출설비는 다음 요건에 적합하여야 한다.

(a) 기관구역으로부터 2개의 탈출설비를 설치하여야 한다.

(b) 2개의 강재 사다리는 가능한 한 멀리 떨어진 것으로 이와 비슷하게 떨어진 해당 장소 상부에 있는 문으로 유도하고 그 문으로부터 적절한 구명정 및 구명뗏목의 승정갑판으로 통하는 통로가 설치된 것이어야 한다. 이들 사다리 중의 1개는 규칙을 만족하는 보호된 폐위구역 내에 위치되어야 하며, 계단이 있는 해당 장소의 하부로부터 외부의 안전한 위치까지 설치하여야 한다. 보호된 폐위구역은 최소 800mm×800mm의 내부치수와 비상조명이 있어야 한다.

특히, 폐외구역 내의 탈출통로에 설치된 비상조명은 주기적으로 점검을 하여 비상시에 탈출로 확보에 원활하게 될 수 있도록 하여야 한다.

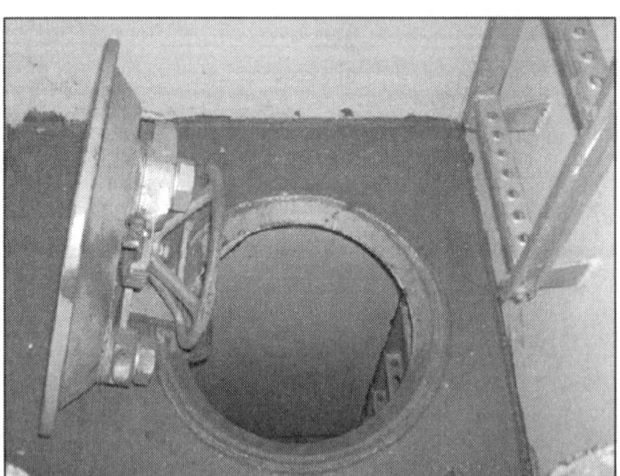

[그림 3-60] Emergency escape truck

[그림 3-61] 기관실 내 탈출통로

On all ships, within the machinery spaces, emergency escape breathing devices shall be situated ready for use at easily visible places, which can be reached quickly and easily at any time in the event of fire. The location of emergency escape breathing devices shall take into account the layout of the machinery space and the number of persons normally working in the spaces.

모든 선박의 기관구역에는 비상탈출용 호흡구가 항상 사용 가능하도록 화재시 언제나 신속히 쉽게 접근할 수 있는 눈에 쉽게 띄는 장소에 있어야 한다. 비상탈출용 호흡구는 해당 기관구역의 통상업무에 종사하는 인원과 기관구역의 배치를 고려하여 배치하여야 한다.

9) Regulation 15 : Instructions, onboard training and drills

Crew members shall receive instruction on fire safety onboard the ship.

Crew members shall receive instructions on their assigned duties.

Parties responsible for fire-extinguishing shall be organized. These parties shall have the capability to complete their duties at all times while the ship is in service.

선원은 선상의 화재안전에 대한 교육을 받아야 한다.

선원은 그들의 부여된 임무에 대한 교육을 받아야 한다.

소화에 대한 책임이 있는 조직이 구성되어야 한다. 이 조직은 선박이 운항 중인 동안 언제라도 그들의 임무를 완수할 수 있는 능력을 가져야 한다.

이러한 선원의 교육은 선박에서 주기적으로 시행되고 있으며, 로그북에 관련 기사가 기입되어야 한다. 주기적인 교육은 비상시에 보다 적극적이고 원활한 대처가 가능하게 만든다.

Crew members shall be trained to be familiar with the arrangements of the ship as well as the location and operation of any fire-fighting systems and appliances that they may be called upon to use.

선원은 그들이 사용하게 될 모든 소화장치 및 설비의 위치와 조작은 물론 선박의 배치에도 익숙하도록 훈련을 받아야 한다.

선원은 선박에 승선하면서부터 선박 친숙화 교육을 포함하여 지속적으로 소화훈련을 수행하게 된다.

Training in the use of the emergency escape breathing devices shall be considered as part of on board training.

비상탈출용 호흡구의 사용교육은 선상교육의 일부분으로 간주되어야 한다.

Performance of crew members assigned fire-fighting duties shall be periodically evaluated by conducting onboard training and drills to identify areas in need of improvement, to ensure competency in fire-fighting skills is maintained, and to ensure the operational readiness of the fire-fighting organization.

개선이 필요한 분야를 식별하고, 소화기능에서의 적합성이 유지되는지를 확인하고 그리고 소화조직의 운영상의 준비성을 확인하기 위한 선내 교육 및 훈련을 수행함으로써 소화임무가 부여된 선원의 능력이 정기적으로 평가되어야 한다.

소화훈련은 통상 소화훈련에 앞서 훈련 시나리오를 작성하여 사전공지 및 교육을 통하여 숙지시키고, 관련 장비의 사용법이나 훈련시의 임무에 대한 교육도 동시에 이루어지게 된다. 이를 바탕으로 소화훈련을 수행하고, 훈련에 대한 평가를 시행하고, 미비점이나 개선사항들을 파악하여 다음 훈련시에 반영될 수 있도록 하고 있다.

Fire drills shall be conducted and recorded.

화재훈련은 수행되고 기록되어야 한다.

An onboard means of recharging breathing apparatus cylinders used during drills shall be provided or a suitable number of spare cylinders shall be carried onboard to replace those used.

훈련 중에 사용되는 호흡구의 실린더를 재충전할 수 있는 본선의 수단 또는 사용된 실린더를 대체할 적절한 수의 예비실린더가 본선에 제공되어야 한다.

훈련 중에 호흡구의 실린더를 사용하게 되면 규정 압력을 유지할 수 없기 때문에 본선에 충전설비(공기압축기)가 되어 있다면 바로 재충전하면 되지만, 그렇지 못한 경우는 예비실린더로 교체하여 비치를 한 후에 추후 정박시 사용된 실린더는 육상에 양륙하여 재충전하도록 한다.

A training manual shall be provided in each crew mess room and recreation room or in each crew cabin.

훈련지침서는 각 선원식당과 휴게실 또는 각 선원실에 제공되어야 한다. 일반적으로 훈련지침서는 선원식당과 휴게실에 각각 비치되어 있다.

[그림 3-62] 훈련지침서

The training manual shall be written in the working language of the ship.

훈련지침서는 선박의 통용 언어로 씌어져야 한다.

선박의 통용 언어로 선박의 구성원들이 의사소통을 위해 사용하는 언어로 이러한 언어로 작성이 되어 있어서 선원들이 훈련지침서의 내용을 이해할 수 있기 때문이다.

The training manual shall explain the following in detail

(a) general fire safety practice and precautions related to the dangers of smoking, electrical hazards, flammable liquids and similar common shipboard hazards.

(b) general instructions on fire-fighting activities and fire-fighting procedures including procedures for notification of a fire and use of manually operated call points.

(c) meanings of the ship's alarms.

(d) operation and use of fire-fighting systems and appliances.

(e) operation and use of fire doors.

(f) operation and use of fire and smoke dampers.

(g) escape systems and appliances.

훈련지침서는 다음 사항을 상세하게 설명하여야 한다.

(a) 흡연의 위험성, 전기적 위험성, 가연성 액체 및 유사한 통상적인 선상 위험성에 관련

된 일반 화재안전실무 및 주의사항

(b) 화재를 알리기 위한 절차 및 수동작동 콜포인트의 사용을 포함한 소화활동과 소화 절차에 대한 일반지침

(c) 선내 경보의 의미

(d) 소화장치와 설비의 조작 및 사용

(e) 방화문의 조작 및 사용

(f) 화재 및 연기댐퍼의 조작 및 사용

(g) 탈출장치와 설비

Fire control plans shall be permanently exhibited for the guidance of the ship's officers, showing clearly for each deck the control stations and one copy shall at all times be available on board in an accessible position. Plans and booklets shall be kept up to date.

화재제어도는 선박 사관의 참조를 위하여 항구적으로 게시되어야 하며, 사본 1부는 선내의 접근이 가능한 장소에서 항상 이용이 가능하여야 한다. 일반배치도 및 소책자는 최신화가 유지되어야 한다.

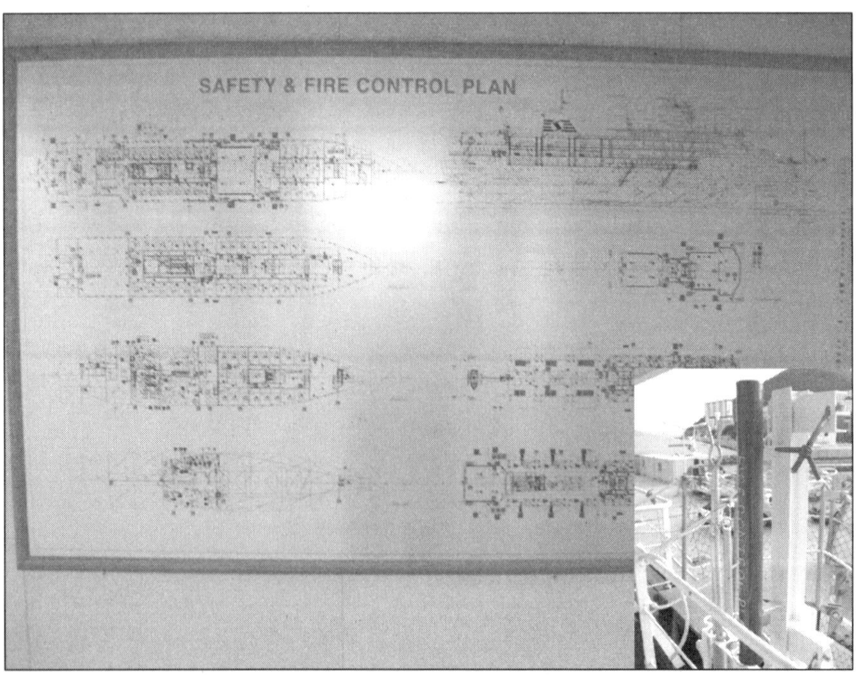

[그림 3-63] 화재제어도 및 갑판실 외측에 비치된 사본 보관함

A duplicate set of fire control plans or a booklet containing such plans shall be permanently stored in a prominently marked weathertight enclosure outside the deckhouse for the assistance of shore-side fire-fighting personnel.

육상소방요원의 지원을 받기 위하여 화재제어도의 복사판 또는 그러한 도면을 포함한 소책자가 갑판실 외측에 명확히 표시된 비바람막이 폐위물에 넣어 항구적으로 보관되어야 한다.

10) Regulation 16 : Operations

The required fire safety operational booklet shall contain the necessary information and instructions for the safe operation of the ship and cargo handling operations in relation to fire safety. The booklet shall include information concerning the crew's responsibilities for the general fire safety of the ship while loading and discharging cargo and while underway. Necessary fire safety precautions for handling general cargoes shall be explained. For ships carrying dangerous goods and flammable bulk cargoes, the fire safety operational booklet shall also provide reference to the pertinent fire-fighting and emergency cargo handling instructions contained in the International Maritime Solid Bulk Cargoes(IMSBC) Code, the International Bulk Chemical Code, the International Gas Carrier Code and the International Maritime Dangerous Goods Code, as appropriate.

요구되는 화재안전 작업용 소책자는 화재안전에 관련된 선박과 화물취급작업의 안전한 조작에 관한 필요한 정보와 지침이 포함되어야 한다. 이 소책자는 화물의 적하 및 양하 중 및 항해 중 선박의 일반적인 화재안전에 대한 선원의 책임에 관한 정보를 포함하여야 한다. 일반 화물취급에 대하여 필요한 화재안전 예방조치가 기술되어야 한다.

위험물 및 인화성 산적화물을 운송하는 선박에 대하여 화재안전 작업용 소책자는 국제해상 고체 산적화물 코드, 국제산적화학품 코드, 국제가스운반선 코드 및 국제해상위험물코드에 포함된 적절한 소화 및 비상 화물취급 지침에 대한 참조가 또한 제공되어야 한다.

The fire safety operational booklet shall be provided in each crew mess room and recreation room or in each crew cabin.

화재안전 작업용 소책자는 각 선원식당 및 휴게실 또는 각 선원실에 제공되어야 한다.

화재안전 작업용 소책자도 훈련지침서와 마찬가지로 선원식당 및 휴게실에 각각 비치되어 있는 것이 일반적이다.

The fire safety operational booklet shall be written in the working language of the ship.

화재안전 작업용 소책자는 선박내 상용 언어로 씌어져야 한다. 이 또한 훈련지침서와 동일한 이유이다.

The fire safety operational booklet may be combined with the training manuals.

화재안전 작업용 소책자는 훈련지침서와 결합될 수 있다. 상당수의 선박에 비치되어 있는 화재안전 작업용 소책자는 훈련지침서와 결합되어 있다.

5. 구명설비 및 장치(Chapter III)의 주요 내용

1) Regulation 1 : Application

The expression all ships means ships constructed before, on or after 1 July 1998; the expressions all passenger ships and all cargo ships shall be construed accordingly.

모든 선박이라 함은 1998년 7월 1일 전이나 이후에 건조된 선박을 말하며 모든 여객선과 모든 화물선이란 표현도 이와 같다.

2) Regulation 3 : Definitions

[그림 3-64] 노출보호복

Anti-exposure suit is a protective suit designed for use by rescue boat crews and marine evacuation system parties.

노출보호복은 구조정의 선원과 해상탈출설비 당사자들이 사용할 수 있도록 설계된 방호복이다. 방수복과 흡사하게 생겼으나 해상에서의 활동성을 좀 더 가미한 방호복이다.

Free-fall launching is that method of launching a survival craft whereby the craft with its complement of persons and equipment on board is released and allowed to fall into the sea without any restraining apparatus.

자유낙하진수라 함은 인원 및 장비를 실은 생존정이 이탈되어 어떠한 억제장치가 없이 바다로 낙하될 수 있도록 하는 생존정 진수방식을 말한다.

[그림 3-65] 자유낙하진수(Free-fall launching)

Immersion suit is a protective suit which reduces the body heatless of a person wearing it in cold water.

방수복이라 함은 찬 수중에서 사람의 체열을 감소시키지 않도록 보온해주는 보호복을 말한다.

해상으로의 탈출시 가능하면 수중으로 탈출하는 것을 지양해야 하지만, 부득이 수중으로 들어가야 하는 경우는 체온저하를 최소화하기 위하여 방수복을 착용하는 것이 최선이다.

Marine evacuation system is an appliance for the rapid transfer of persons from the embarkation deck of a ship to a floating survival craft.

[그림 3-66] 방수복(Immersion suit)

해상탈출설비(Marine Evacuation System)라 함은 선박의 승정갑판부터 물위에 떠 있는 생존정으로 사람을 신속히 옮길 수 있게 하는 장치를 말한다.

일반적으로 여객선에서 많은 수의 승무원 및 여객을 신속하게 탈출시키기 위하여 설치하고 있다.

[그림 3-67] 해상탈출설비(Marine evacuation system)

Thermal protective aid is a bag or suit made of waterproof material with low thermal conductances.

보온복이라 함은 낮은 열전도성을 가진 방수물질로 만들어진 포대기 또는 옷을 말한다. 보온복은 해상에서 구출된 사람이나 저체온 상태에 있는 사람의 체온손실을 막기 위하여 입히는 옷이다.

보온복의 형태는 옷 모양으로 된 것과 포대기 형태로 되어 있는 것들이 있다.

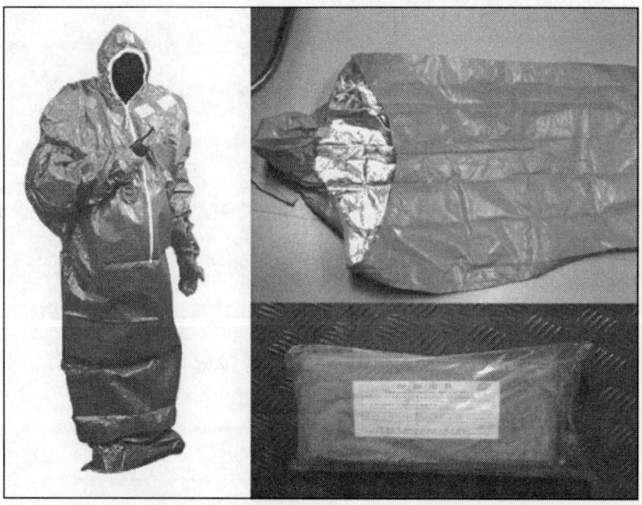

[그림 3-68] 보온복(Thermal protective aid)

3) Regulation 6 : Communications

At least three two-way VHF radiotelephone apparatus shall be provided on every passenger ship and on every cargo ship of 500 gross tonnage and upwards.

모든 여객선 및 총톤수 500톤 이상의 모든 화물선에는 최소한 3대의 쌍방향 VHF 무선전화장치를 비치하여야 한다.

일부 한척의 구명정을 설치하고 있는 제외하고 대다수의 화물선과 여객선에는 2척 이상의 구명정을 설비하고 있는데, 퇴선시에 이러한 구명정간의 통신이나 구명정과 본선 그리고 주위 구조선 및 통항선들과의 교신을 위하여 쌍방향 VHF를 구비해야 한다.

[그림 3-69] 쌍방향 VHF

At least one search and rescue locating device shall be carried on each side of every passenger ship and of every cargo ship of 500 gross tonnage and upwards.

최소한 1대의 수색 및 구조 위치장치를 모든 여객선 및 총톤수 500톤 이상의 모든 화물선의 양현에 탑재하여야 한다.

규정에 따라 선박의 양현에 각각 1대씩 설비되어 있는데, 장치의 안전한 유지 보수를 위하여 그리고 퇴선시 신속한 접근을 위하여 선교의 외부로 통하는 문 인근에 장비를 비치하고 있다.

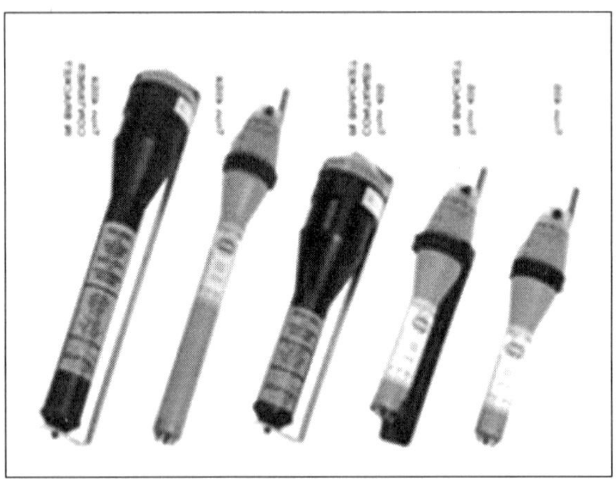

[그림 3-70] 수색 및 구조 위치장치

Not less than 12 rocket parachute flares shall be carried and be stowed on or near the navigation bridge.

12개 이상의 로켓 낙하산 화염신호가 항해선교 또는 그 가까이에 비치되고 보관되어야 한다.

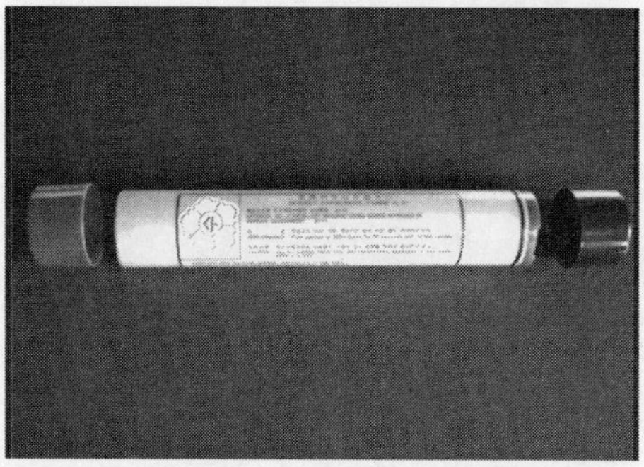

[그림 3-71] 로켓 낙하산 화염신호(Rocket parachute flares)

An emergency means comprised of either fixed or portable equipment or both shall be provided for two-way communications between emergency control stations, muster and embarkation stations and strategic positions on board.

고정식이거나 휴대식인 설비 또는 양쪽 모두를 포함하고 있는 어떤 비상수단이 비상통제위치, 소집 및 승정위치 그리고 기타 필요한 위치 사이에서 쌍방 통신이 가능하도록 비치되어야 한다.

A general emergency alarm system shall be used for summoning passengers and crew to muster stations and to initiate the actions included in the muster list. The system shall be supplemented by either a public address system or other suitable means of communication. Entertainment sound systems shall automatically be turned off when the general emergency alarm system is activated.

총비상경보장치는 여객과 선원을 소집위치로 소집하여 비상배치표상의 행동을 취하도록 하기 위하여 사용되어야 한다. 그 장치는 선내방송장치나 다른 적당한 통신수단에 의하여 보충되어야 한다. 총비상경보장치가 작동되면 오락방송 시스템은 자동적으로 정지되어야 한다.

The general emergency alarm system shall be audible throughout all the accommodation and normal crew working spaces. On passenger ships, the system shall also be audible on all open decks.

총비상경보장치는 모든 거주구역 및 모든 통상의 선원 작업장소에서 들을 수 있어야 한다. 여객선에서는 총비상경보장치를 모든 개방갑판상에서도 들을 수 있어야 한다.

4) Regulation 7 : Personal life-saving appliances

Lifebuoys shall be :

(a) so distributed as to be readily available on both sides of the ship and as far as practicable on all open decks extending to the ship's side. At least one shall be placed in the vicinity of the stern.

(b) so stowed as to be capable of being rapidly cast loose, and not permanently secured in any way. At least one lifebuoy on each side of the ship shall be fitted with a buoyant lifeline in length to not less than twice the height at which it is stowed above the waterline in the lightest seagoing condition, or 30 m, whichever is the greater.

(c) not less than one half of the total number of lifebuoys shall be provided with lifebuoy self-igniting lights, not less than two of these shall also be provided with lifebuoy self-activating smoke signals and be capable of quick release from the navigation bridge ; lifebuoys with lights and those with lights and smoke signals shall be equally distributed on both sides of the ship and shall not be the lifebuoys provided with lifelines

(d) Each lifebuoy shall be marked in block capitals of the Roman alphabet with the name and port of registry of the ship on which it is carried.

(a) 선박의 양현에서 쉽게 이용될 수 있고 가능한 한 선박의 현측에 이르는 모든 개방갑판에서 이용될 수 있도록 분산 배치되어 있어야 한다. 적어도 하나는 선미부근에 있어야 한다.

(b) 신속히 떼어내어 던질 수 있도록 보관되고 어떤 방법으로든 고박되어 있지 아니하여야 한다. 선박의 각 현에 적어도 한 개의 구명부환은 최대경하상태의 흘수선상에서 보관위치까지의 높이의 2배 이상 또는 30미터 중 큰 쪽의 길이와 같은 부양성 구명줄이 부착되어 있어야 한다.

(c) 구명부환 총수량의 반 이상은 자기점화등을 갖추어야 하며 이들 중 둘은 자기발연신호를 갖추고 항해선교로부터 신속히 이탈시킬 수 있는 것이어야 한다. 자기점화등을 갖춘 구명부환 및 자기점화등과 자기발연신호를 갖춘 구명부환은 선박의 양현에 균등히 배치되어야 하며 구명줄을 부착한 구명부환이어서는 안된다.

(d) 각 구명부환에는 그 선박의 선명 및 선적항을 로마식 알파벳의 고형문자로서 표시하여야 한다.

구명부환은 규정에 맞게 선박건조 시에 비치게 되어 나오는데 이후 관리가 적절하게 유지되어야 하기 때문에 주기적으로 검사를 하여야 한다.

선박에서 시행하는 구명부환의 검사는 외관상태를 확인하여 구명부환에 표기된 선명이나 선적항이 명료하지 않으면 다시 표시하여야 하며, 파손된 부분이 있으면 곧 바로 신환을 하여야 한다. 이와 더불어 자기점화등이 같이 부착되어 있는 구명부환은 반드시 자기점화등의 작동상태를 확인하여야 하며, 작동불가시는 반드시 수리를 하거나 수리불가 시는 신환을 하여야 한다.

[그림 3-72] 구명부환(Lifebuoy)

A sufficient number of lifejackets shall be carried for persons on watch and for use at remotely located survival craft stations. The lifejackets carried for persons on watch should be stowed on the bridge, in the engine control room and at any other manned watch station.

당직요원 및 멀리 떨어진 생존정 위치에서 이용할 수 있는 충분한 수의 구명동의를 배치하여야 한다. 당직선원용 구명동의는 선교, 기관실 등 당직이 필요한 모든 개소에 비치되어야 한다. 선교 및 기관실 이외에 대형 선박에서는 선수부에도 일반적인 작업 선원용으로 구명동의를 비치하고 있다.

5) Regulation 8 : Muster list and emergency instructions

Clear instructions to be followed in the event of an emergency shall be provided for every person on board.

Muster lists and emergency instructions shall be exhibited in conspicuous places throughout the ship including the navigation bridge, engine-room and crew accommodation spaces.

비상시 지켜야 할 명확한 지침이 모든 승선자에게 제공되어야 한다.

비상배치표와 비상지침서는 항해선교, 기관실과 선원 거주구역을 포함한 눈에 잘 띄는 장소에 게시되어야 한다.

일반적으로 선박에서는 새로 승선하는 승선자에게 승선하는 즉시 개인별 비상배치표(Personal Muster list)를 제공하고 있으며, 선교, 기관실 및 거주구역에 각각 비상배치표 및 비상지침서를 게시하고 있다.

6) Regulation 9 : Operating instructions

Posters or signs shall be provided on or in the vicinity of survival craft and their launching controls and shall

(a) illustrate the purpose of controls and the procedures for operating the appliance and give relevant instructions or warnings

(b) be easily seen under emergency lighting conditions

(c) use symbols

포스터나 표식이 생존정이나 진수 통제위치 또는 근처에 비치되어야 하며 또한 다음 요건 만족하여야 한다.

(a) 설비를 작동하기 위한 통제의 목적과 절차를 예시하고 해당 지침 또는 경고를 주어야 한다.

(b) 비상전등에 의하여 쉽게 식별이 되어야 한다.

(c) 기호를 사용하여야 한다.

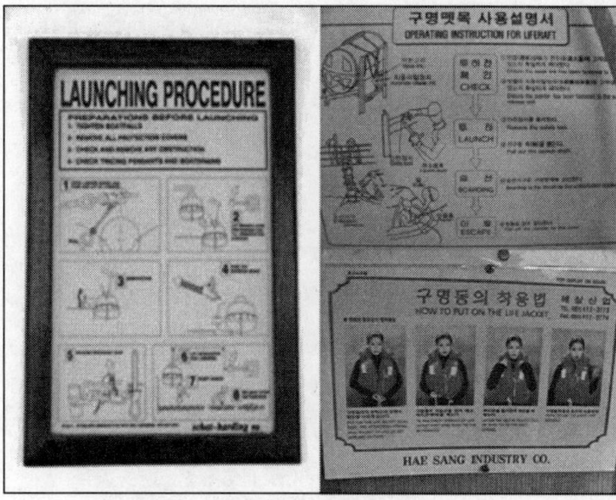

[그림 3-73] 비상지침서

7) Regulation 11 : Survival craft muster and embarkation arrangements

Lifeboats and liferafts shall be stowed as close to accommodation and service spaces as possible.

구명정 및 구명뗏목은 거주구역과 업무구역에 가능한 한 가깝게 탑재되어야 한다. 이는 비상시에 신속히 퇴선을 할 수 있게 하기 위함이다.

Muster stations shall be provided close to the embarkation stations. Each muster station shall have sufficient clear deck space to accommodate all persons assigned to muster at that station.

소집위치는 승정위치에 가까이 배치되어야 한다. 각 소집위치에는 그 위치에 소집하도록 되어 있는 모든 사람을 수용할 수 있을 정도의 충분한 공간이 있어야 한다.

Muster and embarkation stations shall be readily accessible from accommodation and work areas.

소집 및 승정위치는 거주구역과 작업구역으로부터 쉽게 출입할 수 있어야 한다.

Muster and embarkation stations shall be adequately illuminated by lighting supplied from the emergency source of electrical power.

소집 및 승정위치는 비상전원으로부터 공급되는 불빛에 의하여 충분히 조명되어야 한다.

[그림 3-74] 퇴선을 위해 소집된 모습

Davit-launched and free-fall launched survival craft muster and embarkation stations shall be so arranged as to enable stretcher cases to be placed in survival craft.

데비트로 진수되거나 자유낙하로 진수되는 생존정의 소집 및 승정위치에서는 생존정에 구급용 들것을 옮길 수 있어야 한다.

8) Regulation 13 : Stowage of survival craft

Each survival craft shall be stowed in a state of continuous readiness so that two crew members can carry out preparations for embarkation and launching in less than 5 min.

생존정은 두 사람의 선원이 5분내에 승정 및 진수를 할 수 있도록 항상 준비상태에 있어야 한다.

Every liferaft shall be stowed with its painter permanently attached to the ship.

모든 구명뗏목에는 선박에 영구적으로 부착된 페인터가 있어야 한다.

[그림 3-75] 구명뗏목(Liferaft)

9) Regulation 14 : Stowage of rescue boats

Rescue boats shall be stowed in a state of continuous readiness for launching in not more than 5 min, and if the inflated type, in a fully inflated condition at all times, and in a position suitable for launching and recovery.

5분 이내에 진수할 수 있도록 계속 준비된 상태이어야 하고, 팽창형인 경우 항상 완전히 팽창된 상태이어야 하며, 진수 및 회수하기에 적당한 위치에 있어야 한다.

[그림 3-76] 구조정(Rescue boat)

10) Regulation 17 : Rescue boat embarkation, launching and recovery arrangements

The rescue boat embarkation and launching arrangements shall be such that the rescue boat can be boarded and launched in the shortest possible time.

구조정의 승정 및 진수장치는 가능한 한 최단시간 내에 승정되고 진수될 수 있는 것이어야 한다.

구조정은 생존정과 달리 가능한 빠른 시간 내에 진수되어 구조작업을 투입이 되어야 하기 때문이다.

All rescue boats shall be capable of being launched, where necessary utilizing painters, with the ship making headway at speeds up to 5 knots in calm water.

모든 구조정은 페인터를 이용하는 것이 필요한 때에는 잔잔한 수면에서 5노트까지의 속력으로 항진하고 있을 때에도 진수될 수 있는 능력을 가져야 한다.

Recovery time of the rescue boat shall be not more than 5 min in moderate sea conditions when loaded with its full complement of persons and equipment. If the rescue boat is also a lifeboat, this recovery time shall be possible when loaded with its lifeboat equipment and the approved rescue boat complement of at least six persons.

구조정은 그 총정원과 모든 의장품을 실은 채로 보통의 해상상태에서 5분 이내에 회수될 수 있어야 한다. 만일 구조정이 또한 구명정이라면 구명정의 의장품과 적어도 6명의 승인된 구조정 요원을 태운 채 이 회수시간 내에 회수될 수 있어야 한다.

이러한 구조정의 회수시간을 단축시키기 위하여 구조정을 올릴 수 있는 데비트 기어박스에는 속도조절장치가 일반적으로 설비되어 있어 구조정을 올릴 때에 필요에 따라 빠른 속도로 올릴 수 있다.

[그림 3-77] 구조정 데비트 기어박스 속도조절장치

Rescue boat embarkation and recovery arrangements shall allow for safe and efficient handling of a stretcher case. Foul weather recovery strops shall be provided for safety if heavy fall blocks constitute a danger.

구조정 승정과 회수장치는 들것 상자를 안전하고 효과적으로 취급할 수 있어야 한다. 무거운 폴 블록 때문에 위험할 수 있을 경우에는 안전을 위하여 악천후용 회수 환삭이 선박에 비치되어야 한다.

구조정을 회수하는데 있어 가장 시간소요가 많고, 위험한 작업이 무거운 폴 블록을 구조정에 연결하는 작업이다. 만약 날씨마저 좋지 않다면 상황은 더욱 악화되어 구조정의 회수를 지연시키는 것은 물론이고 위험상황을 초래할 수 있는 바 이에 대한 대안으로 악천후용 회수 환삭을 선박에 비치하도록 하여 이 장치를 이용하여 구조정의 안전한 회수가 이루어질 수 있도록 하고 있다.

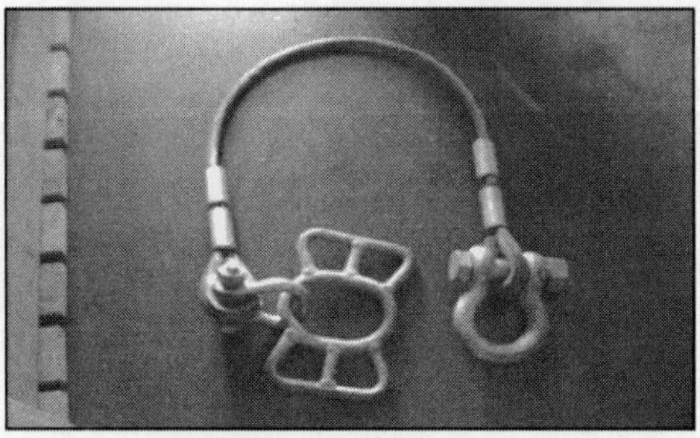

[그림 3-78] 악천후용 회수 환삭

11) Regulation 18 : Line-throwing appliances

A line-throwing appliance shall be provided

구명줄 발사기를 갖추어야 한다.

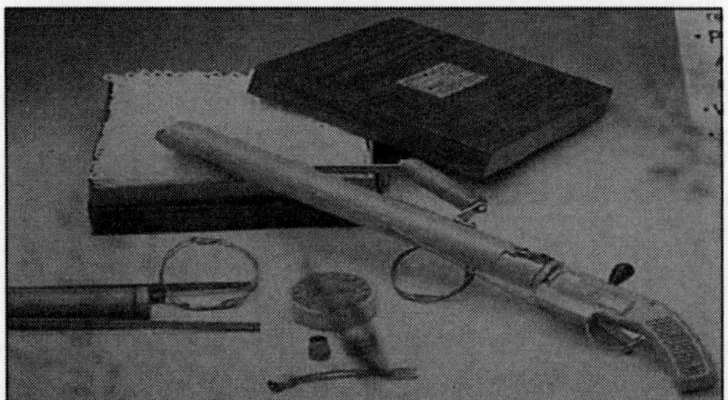

[그림 3-79] 구명줄 발사기

12) Regulation 19 : Emergency training and drills

Every crew member with assigned emergency duties shall be familiar with these duties before the voyage begins.

비상임무를 할당받은 모든 선원은 항해를 시작하기 전에 이런 일들에 익숙하여야 한다.

지속적인 교육훈련을 통한 비상임무에 대한 숙지를 하고 있으며, 새로 승선한 선원도 규정에 따라 항해를 시작하기 전까지는 비상임무를 숙지하여야 한다.

Every crew member shall participate in at least one abandon ship drill and one fire drill every month. The drills of the crew shall take place within 24 h of the ship leaving a port if more than 25% of the crew have not participated in abandon ship and fire drills on board that particular ship in the previous month. When a ship enters service for the first time, after modification of a major character or when a new crew is engaged, these drills shall be held before sailing.

모든 선원은 매달 1회의 퇴선훈련과 1회의 소화훈련에 참가하여야 한다. 총 선원의 25%를 초과하는 선원이 그 전월에 당해 선박에서 선상퇴선과 소화훈련에 참가하지 아니하였다면 선원의 훈련은 출항후 24시간 이내에 행해져야 한다. 선박이 처음으로 운항을 개시할 경우, 주요 개조가 이루어진 후 또는 새 선원이 승선한 경우 이러한 훈련을 항해전에 실시하여야 한다.

소화 및 퇴선 훈련은 매달 시행을 하여야 하며, 전달에 수행했던 훈련에 참가했던 총 선원을 기준으로 25%를 초과하는 선원이 지난달에 훈련에 참가하지 않았다면 출항 후 24시간 내에 훈련을 시행하여야 한다. 이러한 규정은 단순히 선원의 교대뿐만 아니라 부득이한 이유로 선원들 중에 일부가 전월 훈련에 참가하지 못한 경우도 있을 수 있기 때문에 이러한 숫자를 모두 포함한다.

Each abandon ship drill shall include

(a) ensuring that they are made aware of the order to abandon ship

(b) reporting to stations and preparing for the duties described in the muster list

(c) checking that passengers and crew are suitably dressed

(d) checking that lifejackets are correctly donned

(e) lowering of at least one lifeboat after any necessary preparation for launching

(f) starting and operating the lifeboat engine

(g) operation of davits used for launching liferafts

(h) instruction in the use of radio life-saving appliances.

매 퇴선훈련은 다음을 포함하여야 한다.

(a) 퇴선절차를 숙지하고 있는지를 확인한다.

(b) 지휘소에 보고하고 비상배치표에 명기된 바에 따라 소집위치에 집합하여 맡은 바 임무를 준비하도록 한다.

(c) 여객과 선원이 적절한 복장을 취하고 있는지를 점검한다.

(d) 구명동의를 정확히 착용하고 있는가를 점검한다.

(e) 필요한 진수준비 후에 적어도 하나의 구명정을 강하해 보아야 한다.

(f) 구명정 엔진을 시동하고 작동시켜야 한다.

(g) 구명뗏목 진수용 데비트를 작동하여야 한다.

(h) 무선통신구명설비의 사용법을 교육한다.

[그림 3-80] 퇴선 훈련모습

Different lifeboats shall, as far as practicable, be lowered at successive drills.

전회 연습시 내렸던 구명정이 아닌 다른 구명정을 내려야 한다. 이는 구명정이 비상시에 고착되어 강하가 되지 않는 경우에 대비해서 번갈아가면서 사용하도록 규정하고 있는 것이다.

[그림 3-81] 구명정 강하모습

Each lifeboat shall be launched, and manoeuvred in the water by its assigned operating crew, at least once every three months during an abandon ship drill.

각 구명정은 적어도 3개월에 한번 퇴선훈련 시행 중 진수되어 수면위에서 지정된 작동요원에 의해 조종되어야 한다.

[그림 3-82] 진수 후 조종모습

In the case of a lifeboat arranged for free-fall launching, at least once every three months during an abandon ship drill the crew shall board the lifeboat, properly secure themselves in their seats and commence launch procedures up to but not including the actual release of the lifeboat (i.e., the release hook shall not be released). The lifeboat shall then either be free-fall launched with only the required operating crew on board, or lowered into the water by means of the secondary means of launching with or without the operating crew on board. In both cases the lifeboat shall thereafter be manoeuvred in the water by the operating crew. At intervals of not more than six months, the lifeboat shall either be launched by free-fall with only the operating crew on board, or simulated launching shall be carried out.

자유낙하 진수하도록 배치된 구명정의 경우, 퇴선훈련 중 3개월마다 적어도 한번은 선원들이 구명정에 승정하여 각자 좌석에서 적절하게 고정하고, 실제 구명정의 이탈 없이(이탈 혹이 이탈되어서는 아니됨) 진수절차를 시작한다. 그 후 구명정에 요구되는 작동인원만을 태우고 자유낙하 진수하거나, 또는 작동인원을 태우거나 작동인원 없이 2차 진수수단에 의해 수면상으로 강하시켜야 한다.

상기의 모든 경우 이후에 구명정은 수면 위에서 지정된 작동요원에 의해 조종되어야 한다. 6개월을 넘지 않는 간격으로 지정된 작동요원만을 태우고 자유낙하 진수를 시행하여야 한다.

자유낙하 진수방식의 구명정은 실질적인 진수훈련을 위해 진수를 하게 되면 위험성이 너무 크고, 회수시에도 상당히 어려움이 많기 때문에 진수훈련의 주기를 다소 완화하여 6개월에 1번 시행하는 것으로 규정하고 있다.

[그림 3-83] 자유낙하 진수훈련

Fire drills should be planned in such a way that due consideration is given to regular practice in the various emergencies that may occur depending on the type of ships and the cargo.

소화훈련은 선박과 그 선박에 적재하는 화물의 형태에 따라 발생할 수 있는 다양한 비상사태에 대한 정기적인 훈련에 충분한 고려를 하여 계획하여야 한다.

이러한 규정으로 인해 선박에서는 기관실, 스위치 보드룸, 비상발전기실, 주방, 기관실 및 화물구역 등 여러 구역에서 발생할 수 있는 화재에 대비한 소화훈련을 시행하고 있다.

Each fire drill shall include

(a) reporting to stations and preparing for the duties described in the muster list

(b) starting of a fire pump, using at least the two required jets of water to show that the system is in proper working order

(c) checking of fireman's outfit and other personal rescue equipment

(d) checking of relevant communication equipment

(e) checking the operation of watertight doors, fire doors, fire dampers and main inlets and outlets of ventilation systems in the drill area

(f) checking the necessary arrangements for subsequent abandoning of the ship.

각 소화훈련에는 다음 사항이 포함되어야 한다.

(a) 비상배치표에 명기된 바에 따라 소집위치에 집합하고 맡은 바 임무를 준비하도록 한다.

(b) 소화장치가 정상작동 상태에 있는지를 확인하기 위하여 소화펌프를 시동하여 규칙에서 요구하는 적어도 2개의 물분사기를 사용하여 본다.

(c) 소방원 장구 및 개인 구조장비를 점검한다.

(d) 관련 통신장비를 점검한다.

(e) 수밀문, 방화문, 방화 댐퍼와 연습구역 내의 주요 통풍장치의 흡입구와 배기구를 점검한다.

(f) 퇴선에 필요한 장치를 점검한다.

[그림 3-84] 소화 훈련모습

The equipment used during drills shall immediately be brought back to its fully operational condition and any faults and defects discovered during the drills shall be remedied as soon as possible

훈련에 사용된 장비는 즉시 원래의 정상작동 상태로 복귀되어야 하며 훈련 중 발견된 결함은 가능한 조속히 시정되어야 한다.

실질적인 화재가 발생했을 경우에 언제든지 관련 장비를 사용할 수 있도록 준비가 되어 있어야 하기 때문에 훈련 후에는 즉시 관련 장비를 원래 상태로 복귀시켜야 하고, 결함사항은 조속히 시정하여야 한다.

On-board training in the use of the ship's life-saving appliances, including survival craft equipment, and in the use of the ship's fire-extinguishing appliances shall be given as soon as possible but not later than 2 weeks after a crew member joins the ship. However, if the crew member is on a regularly scheduled rotating assignment to the ship, such training shall be given not later than 2 weeks after the time of first joining the ship. Instructions in the use of the ship's fire- extinguishing appliances, life-saving appliances, and in survival at sea shall be given at the same interval as the drills. Individual instruction may cover different parts of the ship's life-saving and fire-extinguishing appliances, but all the ship's life-saving and fire-extinguishing appliances shall be covered within any period of 2 months.

생존정의 의장품을 포함한 선박의 구명설비의 사용에 관한 선상훈련과 소화설비의 사용에 관한 훈련은 선원이 승선한 후 가능한 빨리 늦어도 2주일 이내에 행하여져야 한다. 그러나 만일 선원이 규칙적인 교대, 배치계획에 따라 배선된 경우 그러한 훈련은 그 선박에 최초로 승선한 때부터 2주일 이내에 행해져야 한다.

선박의 소화장치, 구명설비 사용에 관한 교육과 해상에서 생존에 관한 교육은 같은 간격으로 훈련되어야 한다. 이 경우 선박의 구명 및 소화설비의 다른 부분에 대해서는 개별적인 교육을 실시할 수 있지만, 본선의 모든 구명설비 및 소화설비에 대한 교육은 2개월 이내에 행해져야 한다.

신규 승선자에게는 소화 및 퇴선에 있어서 필수적인 설비에 있어서는 2주내에 교육 및 훈련을 수행하여 습득할 수 있도록 하고 있으며, 그 외에 구명 및 소화에 관련된 모든 정비나 설비에 대한 교육은 2개월 내에 수행되도록 계획을 수립하여 진행하고 있다.

Every crew member shall be given instructions which shall include but not necessarily be limited to

(a) operation and use of the ship's inflatable liferafts;

(b) problems of hypothermia, first-aid treatment for hypothermia and other appropriate first-aid procedures

(c) special instructions necessary for use of the ship's life-saving appliances in severe weather and severe sea conditions

(d) operation and use of fire-extinguishing appliances

모든 선원은 최소한 다음 사항을 포함한 교육을 받아야 한다.

(a) 팽창식 구명뗏목의 작동과 사용방법

(b) 체온저하의 문제와 이를 위한 구급처치법 및 기타 적절한 구급절차

(c) 악천후 및 악천후 항해시 선박의 구명설비를 사용하는데 필요한 특별교육

(d) 소화설비의 작동과 사용방법

The date when musters are held, details of abandon ship drills and fire drills, enclosed space entry and rescue drills, drills of other life-saving appliances and on board training shall be recorded in such log-book. If a full muster, drill or training session is not held at the appointed time, an entry shall be made in the log-book stating the circumstances and the extent of the muster, drill or training session held.

소집을 한 일자, 퇴선훈련과 소화훈련, 밀폐구역 진입 및 구조훈련의 상세, 다른 구명설비의 연습과 선내 훈련은 항해일지에 기록되어야 한다. 만일 소집, 연습 또는 훈련시간이 지정된 때에 충분히 행하여지지 아니한다면 실행된 소집, 연습 또는 훈련시간의 정도와 당시의 주위환경을 항해일지에 기록하여야 한다.

일반적으로 선박에서는 선내 훈련과 관련하여 연간 계획을 수립하여 진행을 하고 있다. 하지만 여러 사유로 인하여 계획된 일정대로 훈련을 시행할 수 없는 경우가 발생하는데 그럴 경우는 계획에 따른 훈련을 시행하지 못한 연유를 항해일지에 기록한다.

13) Regulation 20 : Operational readiness, maintenance and inspections

Before the ship leaves port and at all times during the voyage, all life-saving appliances shall be in working order and ready for immediate use.

선박이 출항 전 그리고 항해 중에는 모든 구명설비가 작동 가능하고 또한 즉시 사용될 수 있는 상태에 있어야 한다.

선박에서의 비상상황은 언제든지 발생할 수 있기 때문에 이에 대한 대비가 항상 이루어져 있어야 한다.

Falls used in launching shall be inspected periodically with special regard for areas passing through sheaves, and renewed when necessary due to deterioration of the falls or at intervals of not more than 5 years, whichever is the earlier.

진수를 위하여 사용되는 폴은 활차를 통해 지나는 부위에 대한 특별한 주의와 함께 주기적으로 점검하여야 하며, 폴의 마모 등으로 필요한 경우 또는 5년을 넘지 아니하는 간격 중 빠른 쪽의 기간 내에 신환하여야 한다.

폴은 구명정을 진수하는데 있어 구명정을 잡고 있는 일정 굵기의 와이어 줄로 'Boat fall'이라고 한다. 이러한 폴은 구명정이 해상에 안전하게 진수될 수 있도록 하는 기능을 수행하는 것으로 주기적인 점검과 신환이 이루어지지 않는다면 자칫 절단으로 인한 인명사고가 발생할 수 있다. 일반적으로 선박에서는 이러한 폴의 신환일을 구명정 부근에 표기하고 있다. 물론 신환일 전이라도 주기적인 점검을 통하여 마모상태가 불량하다면 신환을 하게 된다.

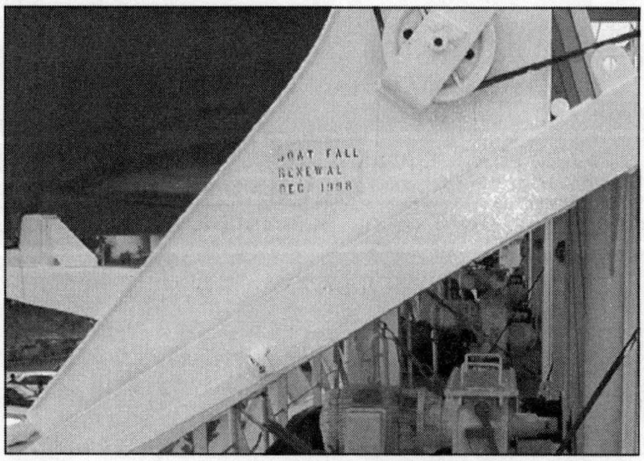

[그림 3-85] 구명정 폴(신환일 표기)

The following tests and inspections shall be carried out weekly and a report of the inspection shall be entered in the log-book

(a) all survival craft, rescue boats and launching appliances shall be visually inspected to ensure that they are ready for use.

(b) all engines in lifeboats and rescue boats shall be run for a total period of not less than 3 min.

(c) lifeboats, except free-fall lifeboats, on cargo ships shall be moved from their stowed position, without any persons on board, to the extent necessary to demonstrate satisfactory operation of launching appliances, if weather and sea conditions so allow.

(d) the general emergency alarm shall be tested.

다음의 시험과 검사가 매주 시행하여야 하며 점검기록이 항해일지에 기록되어야 한다.

(a) 모든 생존정, 구조정 그리고 진수설비가 사용 준비상태에 있음을 확인하기 위하여 육안으로 검사되어야 한다.

(b) 구명정과 구조정의 모든 엔진은 주위온도가 엔진을 시동하고 작동하는데 요구되는 최저온도를 상회하는 조건으로서 3분 이상 작동하여 보아야 한다.

(c) 자유낙하 구명정을 제외한 화물선의 구명정은 날씨와 해상상태가 허락된다면 승선인원 없이 적재 위치로부터 진수장치가 만족스럽게 작동한다는 것을 시험할 수 있는 범위까지 움직여 보아야 한다.

(d) 총비상경보장치를 시험하여야 한다.

The following tests and inspections shall be carried out monthly and a report of the inspection shall be entered in the log-book

(a) All lifeboats, except free-fall lifeboats, shall be turned out from their stowed position, without any persons on board if weather and sea conditions so allow.

(b) Inspection of the life-saving appliances, including lifeboat equipment, shall be carried out monthly.

다음의 시험과 검사가 매월 시행하여야 하며, 점검기록이 항해일지에 기록되어야 한다.

(a) 자유낙하 구명정을 제외한 모든 구명정은 날씨와 해상상태가 허락된다면 승선인원 없이 적재 위치로부터 밖으로 향하도록 움직여 보아야 한다.

(b) 구명정의 의장품을 포함한 구명설비의 검사가 시행되어야 한다.

실질적으로 선박에서 이루어지는 주간 및 월간 검사는 점검표 일명, 'Maintenance &

Inspection log'라는 서류를 바탕으로 시행되며, 관련 사항은 당연히 항해일지에도 기록이 된다.

Every inflatable liferaft, inflatable lifejacket, and marine evacuation system shall be serviced at intervals not exceeding 12 months.

모든 팽창식 구명뗏목과 팽창식 구명동의, 해상탈출설비는 12개월을 초과하지 아니하는 간격으로 정비되어야 한다.

관련하여 해당 설비는 선박 자체적으로 점검하거나 정비할 수 없기 때문에 통상 육상 전문업체에 양륙하여 점검 및 정비를 시행하고, 관련 증빙 서류를 발급받아 본선에 비치하고 있다.

Hydrostatic release units, other than disposable hydrostatic release units, shall be serviced at intervals not exceeding 12 months.

일회용 수압이탈장치 이외의 수압이탈장치도 12개월을 초과하지 아니한 간격으로 정비되어야 한다.

수압이탈장치도 앞서 언급한 구명뗏목을 점검할 때에 같이 점검이 이루어지도록 하고 있다.

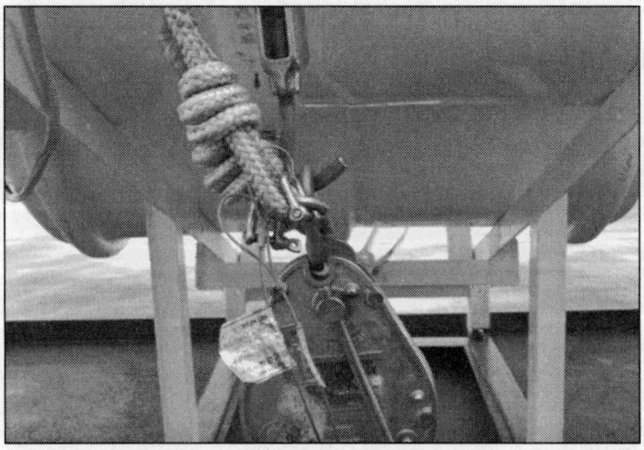

[그림 3-86] 수압이탈장치

14) Regulation 31 : Survival craft and rescue boats

Cargo ships shall carry one or more totally enclosed lifeboats on each side of the ship than can accommodate the total number of persons on board.

화물선의 각 현에 총승선인원을 수용할 수 있는 1대 이상의 전폐형 구명정을 설비하여야 한다.

규정에 따르면 각 현에 1개 이상의 전폐형 구명정을 설비하여야 하기 때문에 최소한 2대 이상의 구명정이 설비되어야 한다. 통상 화물선에서는 양현에 각각 1대의 구명정을 설비하고 있다.

[그림 3-87] 진폐형 구명정 설비모습

Cargo ships may carry one or more free-fall lifeboats being free-fall launched over the stern of the ship that can accommodate the total number of persons on board.

화물선은 총승선인원을 수용할 수 있으며 선박의 선미쪽으로 자유낙하 진수할 수 있는 하나 이상의 자유낙하 구명정을 설비할 수 있다.

앞서 언급한 양현에 진폐형 구명정을 각각 1대씩 설비하는 것을 대신하여 선미에 1대의 자유낙하 구명정을 설비하는 것도 허용하고 있어, 상당수의 화물선에서 선미에 1대의 자유낙하 구명정을 설비하고 있다.

[그림 3-88] 자유낙하 구명정 설비모습

Cargo ships where the horizontal distance from the extreme end of the stem or stern of the ship to the nearest end of the closest survival craft is more than 100 m shall carry a liferaft stowed as far forward of aft, or one as far forward and another as far aft, as is reasonable and practicable. Such liferaft or liferafts may be securely fastened so as to permit manual release.

화물선에 있어서 선수 또는 선미 최끝단으로부터 가장 가까이에 있는 생존정의 최끝단까지 측정한 수평거리가 100m를 넘는 경우에는 합리적이고 실행가능한 한 앞쪽이나 뒤쪽 멀리 1개의 구명뗏목을 한척 더 탑재하거나, 또는 하나는 최대한 앞쪽으로, 다른 하나는 최대한 뒤쪽으로 구명뗏목을 비치하여야 한다. 그러한 구명뗏목은 단단히 고박되어 수동으로만 진수되게 할 수 있도록 되어 있어야 한다.

선박이 대형화 되면서 거주구 주변에만 생존정을 설치함으로 인해 이러한 생존정에서 상당한 거리에 떨어져 있는 곳에서 선원들이 작업 등으로 인해 머물고 있을 때에 갑작스런 비상상황으로 부득이 퇴선을 해야 하는 경우에도 대비를 하고자 추가적인 구명뗏목을 비치하도록 규정하고 있다. 통상 대다수의 화물선이 선미 거주구 형태라 선수부에 추가적으로 구명뗏목을 설비하고 있다.

All survival craft required to provide for abandonment by the total number of persons on board shall be capable of being launched with their full complement of persons and equipment within a period of 10 min from the time the abandon ship signal is given.

총승선인원을 퇴선시키도록 되어 있는 모든 생존정은 인원과 의장품을 만재한 상태로 퇴선신호가 울린 후 10분 이내에 진수될 수 있는 능력을 가져야 한다.

규정된 시간내에 퇴선이 이루어질 수 있도록 생존정은 언제든지 사용 가능하도록 준비가 되어야 있어야 하고, 선원들은 반복적인 훈련을 통하여 신속한 퇴선이 이루어질 수 있도록 하고 있다.

Chemical tankers and gas carriers carrying cargoes emitting toxic vapours or gases shall carry lifeboats with a self-contained air support system.

유독성 증기 또는 가스를 분출하는 케미칼탱커와 가스캐리어는 자장식 공기공급장치를 갖춘 구명정을 설비하여야 한다.

퇴선시 구명정이 유독성 증기 또는 가스가 분출되어 있는 구역을 통과할 수 있기 때문에 이러한 가스가 구명정내로 유입이 되지 않게 그리고 구명정 내의 탈출선원들이 호흡을 하기 위하여 자장식 공기공급장치가 설비되어 있어야 한다.

[그림 3-89] 구명정 자장식 공기공급장치

Oil tankers, chemical tankers and gas carriers carrying cargoes having a flash point not exceeding 60°C(closed cup test) shall carry fire-protected lifeboats.

인화점이 60℃를 초과하지 아니하는 화물을 운송하는 유탱커, 케미칼탱커 및 가스캐리어는 내화구명정을 비치하여야 한다.

유출된 화물에 화재가 발생했을 때에 이러한 구역을 부득이 통과하여 퇴선을 하는 경우가 발생할 수 있기 때문에 이에 대한 대비책으로 불에 견딜 수 있는 내화구명정을 설비하도록 규정하고 있는데, 통상 관련 선박에 설비되어 있는 내화구명정은 불로부터 구명정을 보호할 수 있도록 구명정 주변으로 물을 분사할 수 있는 장치를 설비하고 있다.

[그림 3-90] 구명정 물분사모습

15) Regulation 32 : Personal life-saving appliances

An immersion suit of an appropriate size shall be provided for every person on board the ship.

선내의 모든 인원에게 적절한 크기의 방수복을 제공하여야 한다.

16) Regulation 35 : Training manual and on-board training aids

A training manual shall be provided in each crew mess room and recreation room or in each crew cabin.

훈련지침서는 선원식당과 휴게실 또는 선실에 비치되어야 한다.

일반적으로 훈련지침서는 선원식당 및 휴게실에 비치를 하고 있으며, 선실에는 비치를 하고 있지 않다.

The training manual shall include the following in detail

(a) donning of lifejackets, immersion suits and anti-exposure suits, as appropriate
(b) muster at the assigned stations
(c) boarding, launching, and clearing the survival craft and rescue boats
(d) method of launching from within the survival craft
(e) release from launching appliances
(f) methods and use of devices for protection in launching areas, where appropriate
(g) illumination in launching areas
(h) use of all survival equipment
(i) use of all detection equipment
(j) with the assistance of illustrations, the use of radio life-saving appliances
(k) use of drogues;
(l) use of engine and accessories
(m) recovery of survival craft and rescue boats including stowage and securing
(n) hazards of exposure and the need for warm clothing
(o) best use of the survival craft facilities in order to survive
(p) methods of retrieval, including the use of helicopter rescue gear(slings, baskets,

stretchers), breeches-buoy and shore life-saving apparatus and ship's line-throwing apparatus;

(q) all other functions contained in the muster list and emergency instructions

(r) instructions for emergency repair of the life-saving appliances."

훈련지침서는 다음 사항을 상세히 포함하여야 한다.

(a) 구명동의, 방수복 및 노출보호복의 적절한 착용법

(b) 지정된 장소에의 소집

(c) 생존정 및 구조정 탑승, 진수 및 이탈

(d) 생존정 내부에서의 진수조작방법

(e) 진수장치로부터의 이탈

(f) 해당 진수장소에서의 보호장비의 사용법

(g) 진수장소의 조명

(h) 모든 생존정설비의 사용법

(i) 모든 탐지설비의 사용법

(j) 도해를 포함한 무선구명설비의 사용법

(k) 해묘의 사용법

(l) 기관 및 부속품의 사용법

(m) 탑재 및 고박을 포함한 생존정 및 구조정의 회복

(n) 노출의 위험 및 따뜻한 의복의 필요성

(o) 생존을 위한 생존정 시설의 최선의 사용법

(p) 헬리콥터 구조장치(멜빵, 광주리, 들것), 반바지형 부환 그리고 육상구명장치 및 선박의 구명줄 발사기의 사용법을 포함한 구출방법

(q) 비상배치표 및 비상지침에 기재된 모든 다른 임무

(r) 구명설비의 비상수리를 위한 지침

훈련지침서는 일반적으로 선사에서 일괄적으로 제작하여 보급하는 경우가 많은데 보급된 훈련지침서에는 규정에서 요구하는 사항을 모두 포함하도록 제작되어 있고, 일반적인 장비들을 기준으로 설명이 되어 있기 때문에 관련 사항을 본선에서 선박에 맞게 수정 보완하여야 한다.

17) Regulation 37 : Muster list and emergency instructions

The muster list shall specify details of the general emergency alarm and public address system and also action to be taken by crew and passengers when this alarm is sounded. The muster list shall also specify how the order to abandon ship will be given.

비상배치표에는 총비상경보 또는 선내방송장치의 상세와 이런 경보가 울릴 때 승객 및 승무원이 취하여야 할 행동이 자세히 명시되어야 한다. 선내 비상배치표에는 퇴선명령이 어떻게 내려지는지 명시하여야 한다.

The muster list shall show the duties assigned to the different members of the crew including

(a) closing of the watertight doors, fire doors, valves, scuppers, side scuttles, skylights, portholes and other similar openings in the ship
(b) equipping of the survival craft and other life-saving appliances
(c) preparation and launching of survival craft
(d) general preparations of other life-saving appliances
(e) muster of passengers
(f) use of communication equipment
(g) manning of fire parties assigned to deal with fires
(h) special duties assigned in respect to the use of fire-fighting equipment and installations.

비상배치표는 다음에 명기하는 사항을 포함하여 각 승무원에게 부여된 임무를 표시하여야 한다.

(a) 선박에 있는 수밀문, 방화문, 밸브, 배수구, 현창, 공기창 겸용 출입구 및 기타 유사한 개구의 폐쇄
(b) 생존정 및 다른 구명설비의 의장품
(c) 생존정의 준비 및 진수
(d) 다른 구명설비의 일반적인 준비
(e) 여객의 소집
(f) 통신장비의 사용

(g) 화재를 진압하도록 지정된 소화분대의 인원구성

(h) 소화장비의 사용 및 설치에 관하여 지정된 특별한 임무

비상배치표에는 전승무원의 임무와 소지품이 명시되어 있고, 이외는 별도로 개인별 비상배치표를 배부하여 좀 더 임무 및 소지품 숙지를 잘 할 수 있도록 하고 있다.

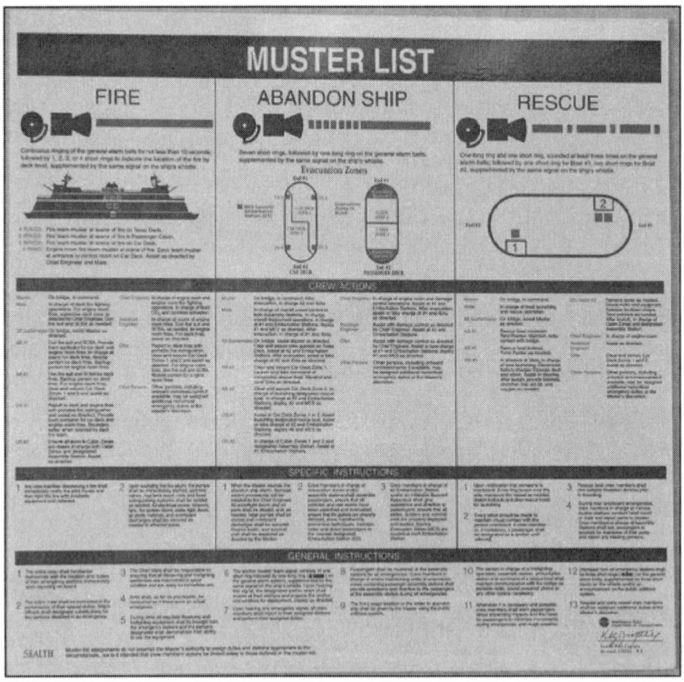

[그림 3-91] MUSTER LIST

The muster list shall specify which officers are assigned to ensure that life-saving and fire appliances are maintained in good condition and are ready for immediate use.

비상배치표는 구명 및 소화설비가 양호한 상태로 유지되고 즉시 사용될 수 있게 하기 위하여 어느 사관이 지정되어 있는지를 명시하여야 한다.

The muster list shall be prepared before the ship proceeds to sea. After the muster list has been prepared, if any change takes place in the crew which necessitates an alteration in the muster list, the master shall either revise the list or prepare a new list.

비상배치표는 선박이 출항하기 전에 작성되어야 한다. 비상배치표가 작성된 후 만약

비상배치표의 변경을 요하는 승무원의 변화가 있다면 선장은 배치표를 수정하든지 또는 새로운 배치표를 작성하여야 한다.

통상 선박에서는 정박 중에 승무원의 교대가 있거나 승무원의 직급 변경이 있다면 출항 전에 비상배치표를 수정하여 게시한다. 이러한 비상배치표의 수정은 3등항해사가 하고, 최종적으로 선장의 승인을 얻는다.

3.2 FSS code

3.2.1 FSS code의 도입

SOLAS(International Convention for the Safety Of Life At Sea, 1974)이 지속적으로 개정되면서 SOLAS 협약 제II-2장에서 요구하는 화재안전장치에 대한 구체적인 기술적 사항에 관한 국제표준을 제공하고자 도입되었다.

3.2.2 FSS code의 적용과 구성

1. FSS code의 적용

SOLAS(International Convention for the Safety Of Life At Sea, 1974) 협약 제 II-2장의 개정사항에 대한 발효와 더불어 FSS code(Fire Safety System code)의 발효를 2002년 7월 1일부로 발효하게 되었다. 이에 따라 별도의 명문규정이 없는 한 FSS code는 2002년 7월 1일 이후에 용골이 거치되거나 이와 동등한 건조단계에 있는 선박에 적용된다.

2. FSS code의 구성

앞서 언급한 바와 같이 FSS code는 SOLAS 협약 제II-2장 규정에 대한 기술사항을 담고 있는 협약으로 다음과 같이 총 15개의 장으로 구성되어 있다.

Chapter 1	General	일반사항
Chapter 2	International shore connections	국제육상시설 연결구
Chapter 3	Personnel protection	개인보호
Chapter 4	Fire extinguishers	소화기
chapter 5	Fixed gas fire-extinguishing systems	고정식 가스 소화장치
Chapter 6	Fixed foam fire-extinguishing systems	고정식 포말 소화장치
Chapter 7	Fixed pressure water-spraying and water-mist fire-extinguishing systems	고정식 가압식 분사 및 물분무 소화장치
Chapter 8	Automatic sprinkler, fire detection and fire alarm systems	자동스프링클러 장치
Chapter 9	Fixed fire detection and fire alarm systems	고정식 화재탐지 및 경보장치
Chapter 10	Sample extraction smoke detection systems	시료채취 연기탐지장치
Chapter 11	Low-location lighting systems	낮은 위치의 조명장치
Chapter 12	Fixed emergency fire pumps	고정식 비상소화펌프
Chapter 13	Arrangement of means of escape	비상탈출설비의 배치
Chapter 14	Fixed deck foam systems	고정식 갑판 포말 소화장치
Chapter 15	Inert gas systems	이너트가스 장치
Chapter 16	Fixed hydrocarbon gas detection systems	고정식 탄화수소 감지장치

3.2.3 FSS code 주요 규정

1. Chapter 1 General

1) 정 의

Administration means the Government of the State whose flag the ship is entitled to fly.

주관청은 선박이 게양하는 국기의 정부를 말한다.

Convention means the 1974 SOLAS Convention, as amended.

협약은 개정된 1974년 SOLAS 협약을 말한다.

2. Chapter 2 International shore connections

International shore connections shall be of steel or other equivalent material. The flange shall have a flat face on one side and, on the other side, it shall be permanently attached to a coupling that will fit the ship's hydrant and hose. The connection shall be kept aboard the ship together with a gasket, together with four bolts of 16 mm diameter and 50 mm in length, four 16 mm nuts, and eight washers.

국제 육상시설 연결구는 강 또는 기타 적절한 재료의 것이어야 한다. 플랜지는 그 1면을 1평면으로 하고, 다른 면에는 선박의 소화전 및 소화호스에 적합한 이음매를 항구적으로 부착하여야 한다. 국제 육상시설 연결구는 가스켓 그리고 직경 16밀리미터, 길이 50밀리미터의 4개의 볼트와 16밀리미터 너트 4개 그리고 8개의 와셔와 함께 선내에 비치하여야 한다.

규정에서 언급하고 있는 부속품으로 구성되어 선박에는 별도의 보관함을 제작하여 비치하고 있으며, 보관함은 식별이 용이하게 표시가 되어 있다.

[그림 3-92] International shore connections

3. Chapter 3 Personnel protection

1) Fire-fighter's outfit

A fire-fighter's outfit shall consist of a set of personal equipment and a breathing apparatus.

소방원 장구는 1조의 개인장구와 1개의 호흡구로 구성되어야 한다.

Personal equipment shall consist of the following :

(a) protective clothing of material to protect the skin from the heat radiating from the fire and from burns and scalding by steam. The outer surface shall be water-resistant

(b) boots of rubber or other electrically non-conducting material

(c) rigid helmet providing effective protection against impact

(d) electric safety lamp (hand lantern) of an approved type with a minimum burning period of 3 hours. Electric safety lamps on tankers and those intended to be used in hazardous areas shall be of an explosion-proof type

(e) axe with a handle provided with high-voltage insulation

개인장구는 다음과 같이 구성되어야 한다.

(a) 화재의 방사열 및 증기 등에 의한 화상으로부터 피부를 보호하는 재료로 제조된 보호천으로써, 그 표면은 방수성이어야 한다.

(b) 고무 또는 전기적으로 절연성재료로 제조된 장화 및 장갑

(c) 충격에 효과적으로 보호되는 견고한 헬멧

(d) 3시간 이상 동안 점등할 수 있는 승인된 형식의 전기안전등(휴대식 손전등), 탱거 및 위험구역에서 사용될 전기안전등은 방폭형의 것이어야 한다.

(e) 고압에 대한 절연성을 가진 손잡이가 달린 도끼

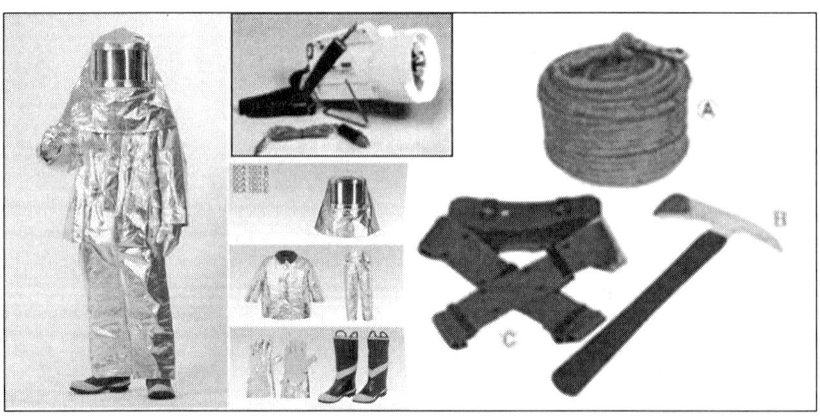

[그림 3-93] 개인장구

Breathing apparatus shall be a self-contained compressed air breathing apparatus for which the volume of air contained in the cylinders shall be at least 1,200ltrs or other self-contained breathing apparatus which shall be capable of functioning for at least 30 min. All air cylinders for breathing apparatus shall be interchangeable.

호흡구는 실린더 내에 저장되어 있는 공기용적이 1200리터 이상인 자장식 압축공기 호흡구 또는 30분 이상 동안 기능을 발휘할 수 있는 다른 자장식 호흡구이어야 한다.

호흡구의 모든 공기실린더는 상호교환이 가능하여야 한다.

선박에는 몇 세트의 자장식 압축공기 호흡구와 더불어 각 세트당 2개의 예비 압축공기 실린더를 같이 비치하도록 규정하고 있기 때문에 이러한 압축공기 실린더들은 상호 교환이 가능하도록 설비하여야 한다.

Compressed air breathing apparatus shall be fitted with an audible alarm and a visual or other device which will alert the user before the volume of the air in the cylinder has been reduced to no less than 200 ltrs.

압축공기 호흡구는 실린더 내의 공기용적이 200리터 미만으로 감소되기 전에 사용자에게 경고를 주는 가시가청경보 또는 다른 장치를 부착하여야 한다.

자장식 호흡구를 장착하고 화재진압을 수행할 경우 공기압력을 원활하게 점검할 수 없는 상황이 있을 수 있기 때문에 가시가청경보를 울릴 수 있는 장치를 부착하도록 규정하고 있으며, 경보가 울리게 되면 즉시 현장을 탈출하여야 한다.

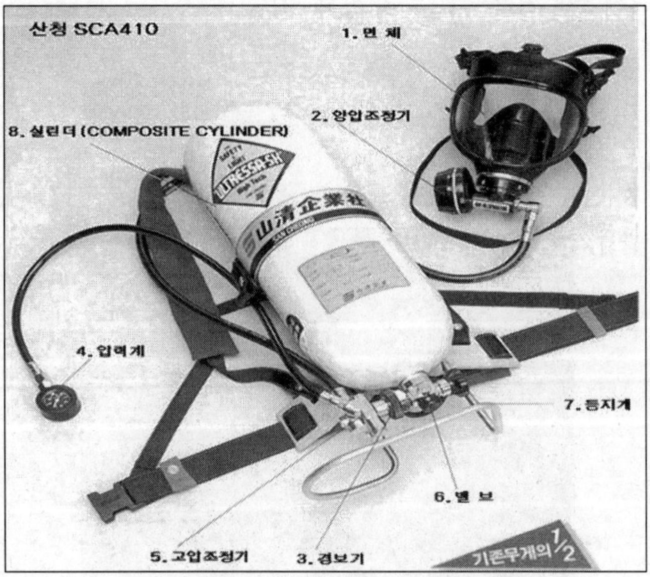

[그림 3-94] 자장식 압축공기 호흡구

For each breathing apparatus a fireproof lifeline of at least 30 m in length shall be provided.

호흡구에는 길이 30미터 이상의 내화성 구명줄을 비치하여야 한다.

2) Emergency Escape Breathing Devices(EEBD)

An EEBD is a supplied-air or oxygen device only used for escape from a compartment that has a hazardous atmosphere.

비상탈출용 호흡구(EEBD)는 위험한 대기를 가진 구역으로부터 탈출용으로만 사용되는 공기공급 또는 산소장치이다.

The EEBD shall have a service duration at least 10 minutes.

비상탈출용 호흡구는 10분 이상동안 지속적으로 사용할 수 있어야 한다.

Brief instructions or diagrams clearly illustrating their use shall be clearly printed on the EEBD. The donning procedures shall be quick and easy to allow for situations where there is little time to seek safety from a hazardous atmosphere.

비상탈출용 호흡구 표면에는 이들의 사용을 명확하게 설명하는 간단한 지침서 또는 도해가 명확하게 인쇄되어야 한다. 위험 대기로부터 안전하게 탈출하는 데 시간이 없는 상황에 대비하여 착용절차는 쉽고 빨라야 한다.

화재발생시 가장 위험한 요소는 연기로 인한 질식이다. 이러한 점을 고려하여 화재발생시 신속하게 사용할 수 있도록 간단한 착용법으로 이루어져 있고, 관련하여 통상 호흡구 전면에 착용법이 설명되어 있다.

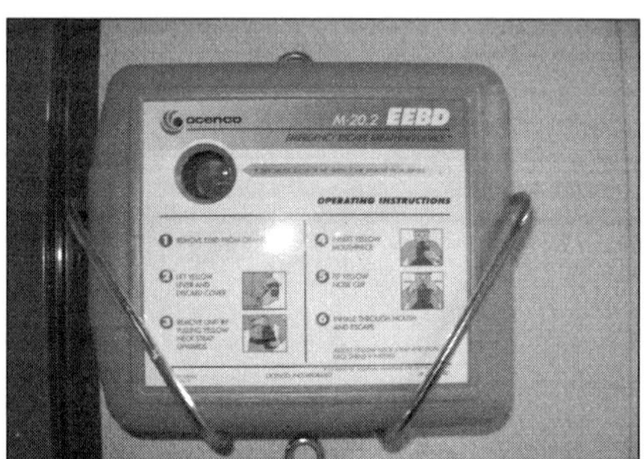

[그림 3-95] 비상탈출용 호흡구(EEBD)

4. Chapter 4 Fire extinguishers

1) Fire extinguisher

Each powder or carbon dioxide extinguisher shall have a capacity of at least 5kg, and each foam extinguisher shall have a capacity of at least 9 *l* . The mass of all portable fire extinguishers shall not exceed 23 kg.

각 분말 또는 이산화탄소 소화기 용량은 5kg 이상이어야 하며, 각 포말소화기 용량은 9리터 이상이어야 한다. 모든 휴대식 소화기의 무게는 23kg을 초과해서는 아니된다.

휴대식 소화기의 무게가 23kg을 초과하지 않는 것으로 규정되어 있어, 통상 기관실에 비치되는 45리터 및 135리터 포말소화기는 휴대식 소화기에 속하지 않는다고 할 수 있다.

2) Portable foam applicators

A portable foam applicator unit shall consist of a foam nozzle/branch pipe, either of a self-inducing type or in combination with a separate inductor, capable of being connected to the fire main by a fire hose, together with a portable tank containing at least 20 liters of foam concentrate and at least one spare tank of foam concentrate of the same capacity.

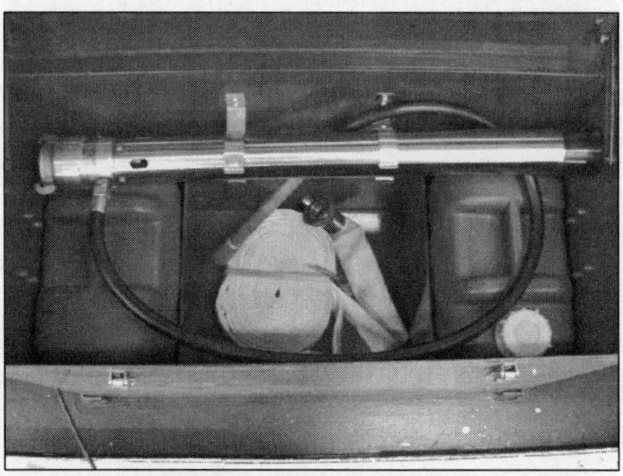

[그림 3-96] 휴대식 포말방사기

휴대식 포말방사기는 자체 유도형 또는 분리 인덕터와 조합되는 것으로서 소화호스로 소화주관에 연결할 수 있는 포말노즐/지관, 20리터 이상의 포말원액을 담은 휴대식 탱크 및 1개 이상의 동등한 용량의 포말원액 예비탱크로 구성되어야 한다.

휴대식 포말방사기는 기관실에 주로 비치가 되어 있으며, 식별이 용이한 보관함안에는 규정에서 요구하는 구성품을 모두 담고 있다.

5. Chapter 5 Fixed gas fire-extinguishing systems

The means of control of any fixed gas fire-extinguishing system shall be readily accessible, simple to operate and shall be grouped together in as few locations as possible at positions not likely to be cut off by a fire in a protected space. At each location there shall be clear instructions relating to the operation of the system having regard to the safety of personnel.

고정식 가스소화장치의 제어장치는 신속히 접근할 수 있고, 간단히 조작할 수 있는 것이어야 하며, 보호되는 구역의 화재로 인하여 차단될 우려가 없는 최소한의 위치에 가능한 한 일괄 배치되어야 한다. 각 위치에는 인원의 안전을 고려하여 장치조작에 관한 분명한 설명서를 비치하여야 한다.

고정식 가스소화장치는 여러 구획에 나눠 분출될 수 있도록 되어 있는 경우가 많기 때문에 각각의 구획에 가스소화장치를 작동시키기 위해서는 각각의 제어장치를 설비하고 있어야 한다. 일반적으로 이러한 연유로 인해 고정식 가스소화장치의 가스가 충전되어 있는 실린더가 배치되어 있는 장소에 설비되어 있다.

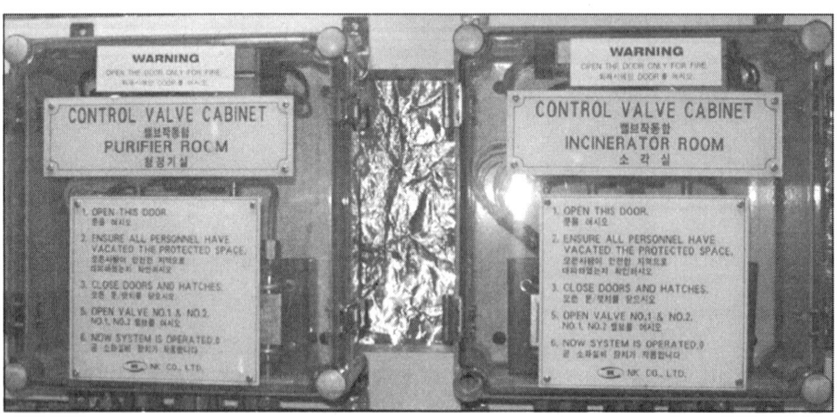

[그림 3-97] 고정식 가스소화장치의 제어장치

For machinery spaces the fixed piping system shall be such that 85% of the gas can be discharged into the space within 2 min.

기관구역을 보호하기 위하여 고정된 관장치는 가스의 85%가 2분 이내에 기관구역에 방출될 수 있도록 하여야 한다.

기관구역의 경우 화재의 위험성이 큰 장소이면서 유류를 취급하는 많은 장비들이 있기 때문에, 자칫 화재발생시 초기 진압이 불가할 경우는 화재확산이 빨리 이루어지기 때문에 가능한 빠른 시간 내에 화재를 진압할 필요가 있다.

고정식 가스소화장치를 사용하여 화재를 진압하고자 할 경우 화재의 확산을 최소화하고, 질식효과를 극대화하기 위하여 많은 양의 가스가 빠른 시간 내에 방출되어야 한다.

6. Chapter 9 Fixed fire detection and fire alarm systems

Any required fixed fire detection and fire alarm system with manually operated call points shall be capable of immediate operation at all times.

수동작동 콜 포인트를 부착하도록 요구된 고정식 화재탐지 및 경보장치는 항상 즉시 사용할 수 있어야 한다.

There shall be not less than two sources of power supply for the electrical equipment used in the operation of the fixed fire detection and fire alarm system, one of which shall be an emergency source of power.

고정식 화재탐지 및 경보장치의 작동시 사용되는 전기설비에 대한 동력공급원은 2개 이상이어야 하며 그 중 1개는 비상전원이어야 한다. 여러 연유로 인하여 선내 전원이 꺼지더라도 화재탐지는 가능하도록 비상전원을 통한 동 장치들은 작동이 되어야 한다.

Detectors shall be operated by heat, smoke or other products of combustion, flame, or any combination of these factors.

탐지기는 열, 연소 연기 또는 기타 연소물, 화염 또는 이들의 혼합된 요인에 의해서 작동하여야 한다.

선박은 여러 재질로 이루어져 있기 때문에 화재가 발생하면 장소에 따라 열, 연기 또는 화염을 달리 발산할 수 있기 때문에 탐지기는 주변 특성에 맞게 설치되어야 한다.

[그림 3-98] 연기탐지기

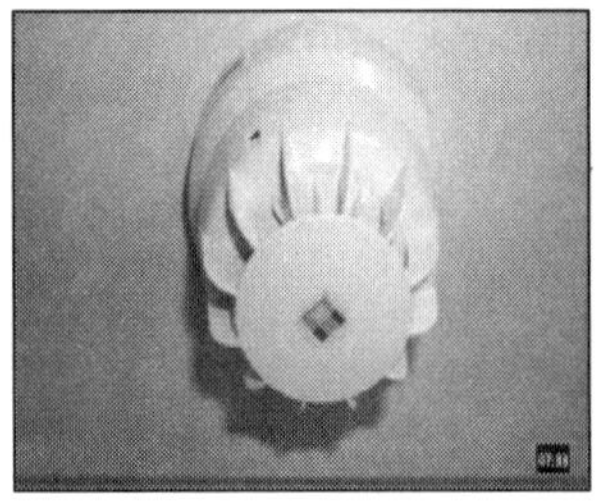

[그림 3-99] 열탐지기

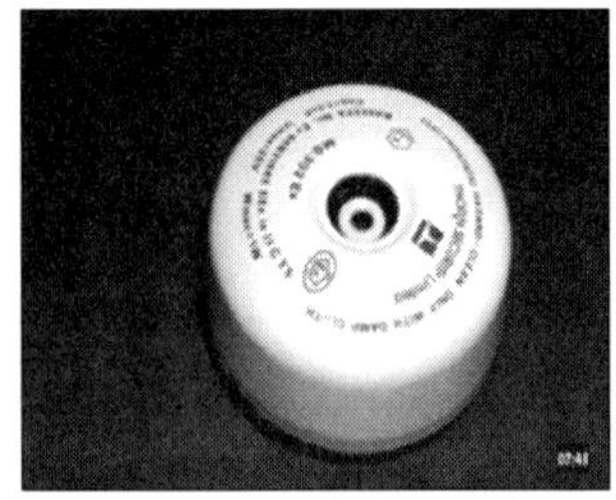

[그림 3-100] 화염탐지기

The activation of any detector or manually operated call point shall initiate a visual and audible fire detection alarm signal at the control panel and indicating units.

탐지기 또는 수동작동 콜포인트가 작동할 경우에도 제어반 및 표시판에 가시 가청경보장치가 발하여져야 한다.

제어반 또는 표시판에는 탐지기 및 수동 작동 콜포인트가 작동된 위치표시와 더불어 가시 가청경보가 발하게 되어 있다.

7. Chapter 12 Fixed emergency fire pumps

The emergency fire pump shall be of a fixed independently driven power-operated pump.

비상소화펌프는 독립된 구동원으로부터 구동되는 고정식의 펌프이어야 한다.

비상소화펌프는 기관구역과 격리되어 있는 곳에 설치하게 규정되어 있으며, 선내 주전원외에 비상전원으로도 구동이 될 수 있어야 한다. 만약, 기관실이 침수되어 발전기와 더불어 기관실에 설치되어 있는 소화펌프를 작동시킬 수 없는 상황에서도 비상소화펌프는 구동될 수 있어야 한다.

8. Chapter 14 Fixed deck foam systems

The arrangements for providing foam shall be capable of delivering foam to the entire cargo tanks deck area as well as into any cargo tank the deck of which has been ruptured.

포말공급장치는 화물탱크 갑판구역의 전역뿐만 아니라 갑판이 파손되어 있는 화물탱크 내부로 포말을 방출하는 능력을 가진 것이어야 한다.

포말공급장치에 의한 화물 전 구역에서 발생할 수 있는 화재에 대비하여 외부 갑판상에는 일정한 간격으로 포말을 방출할 수 있는 모니터가 여러 개 설치되어 있다.

Operation of a deck foam system at its required output shall permit the simultaneous use of the minimum required number of jets of water at the required pressure from the fire main. The simultaneous use of the minimum required jets of water shall be possible on deck over the full length of the ship, in the accommodation, service spaces, control stations and machinery spaces.

고정식 갑판 포말장치가 규정된 방사율로 작동하는 경우에 소화주관으로부터 규정된 압력으로 규정된 최소 수의 물줄기가 동시에 방수될 수 있어야 한다. 요구되는 최소 수의 물줄기는 선박의 전장에 걸쳐 갑판상, 거주구역, 업무구역, 제어장소 및 기관구역에서 동시에 사용할 수 있어야 한다.

화재의 확산 및 여러 군데 다발적으로 발생할 수 있는 화재에 대비하여 일정 이상의 압력을 유지하여 규정수의 물줄기를 방수할 수 있도록 설비되어 있다.

Foam from the fixed foam system shall be supplied by means of monitors and foam applicators.

고정식 갑판 포말장치는 모니터 및 포말방사기에 의하여 포말을 방출하여야 한다.

[그림 3-101] 고정식 갑판 포말장치의 모니터

9. Chapter 15 Inert gas systems

The IG system shall be capable of

(a) inerting empty cargo tanks and maintaining the atmosphere in any part of the tank with an oxygen content not exceeding 8% by volume and at a positive pressure in port and at sea except when it is necessary for such a tank to be gas-free

(b) eliminating the need for air to enter a tank during normal operations except when it is necessary for such a tank to be gas-free

(c) purging empty cargo tanks of hydrocarbon or other flammable vapours, so that subsequent gas-freeing operations will at no time create a flammable atmosphere within the tank;

(d) delivering inert gas to the cargo tanks at a rate of at least 125% of the maximum rate of discharge capacity of the ship expressed as a volume.

(e) delivering inert gas with an oxygen content of not more than 5% by volume to the cargo tanks at any required rate of flow.

이너트가스 장치는 다음의 기능을 가져야 한다.

(a) 빈 화물탱크를 불활성화하여야 한다. 그리고 화물탱크의 모든 부분에서 공기 중 산소농도가 8%를 초과하지 아니하도록 유지하여야 한다. 다만, 당해 화물탱크를 가스프리할 필요가 있는 경우를 제외하고 항내 및 해상에서 정압을 유지하여야 한다.

(b) 당해 화물탱크를 가스프리할 필요가 있는 경우를 제외하고 정상 작동 중에는 공기가 탱크에 유입되지 아니하도록 하여야 한다.

(c) 빈 화물탱크에 탄화수소 또는 다른 가연성 증기를 제거하여 연속된 가스프리 작업 중 어떠한 경우에도 탱크 내부의 가연성 분위기를 형성하지 아니하여야 한다.

(d) 선박의 최대양하용적의 125 % 이상의 비율로 화물탱크에 이너트가스를 공급할 수 있어야 한다.

(e) 화물탱크에 산소농도 5% 이하의 이너트가스를 요구되는 유입 비율로 공급할 수 있어야 한다.

빈 화물탱크에 불활성화되어 있지 않은 상태에서 화물을 선적하게 되면 폭발에 위험성이 있기 때문에 불활성화되어 있어야 하며, 화물선적이 아니더라도 화물탱크에 남아 있는 잔량의 화물에서 나오는 유증기가 자칫 외부에서 유입되는 공기와 만나는 경우 또한 폭발에 위험성이 있기 때문에 가스프리할 경우를 제외하고는 불활성시켜야 하며, 화물탱크는 항상 정압상태를 유지하여 외부로부터 공기의 유입을 차단하여야 한다.

가스프리 작업이란 화물탱크에 가연성 증기를 제거하고 산도농도를 21%로 유지하는 것으로 오선 화물탱크에 남아 있는 가연성 증기를 제거하기 위하여 불활성가스를 불어넣어 치환시켜야 한다.

선박에서 양하작업을 수행하면 화물탱크로부터 화물이 육상으로 이송되기 때문에 화물탱크의 압력이 화물이 빠져나가는 만큼 떨어지게 되는데 이를 보상시키지 않으면 화물탱크 압력 하강으로 자칫 외부의 공기가 유입될 수 있기 때문에 화물이 빠져나간 만큼의 공간에 불활성 가스를 공급하여 화물탱크의 정압을 유지시켜야 한다.

화물탱크는 가스프리 작업을 제외하고는 화물의 존재 여부와 상관없이 항상 산소농도를 5% 이하로 유지하여 폭발을 방지하여야 한다.

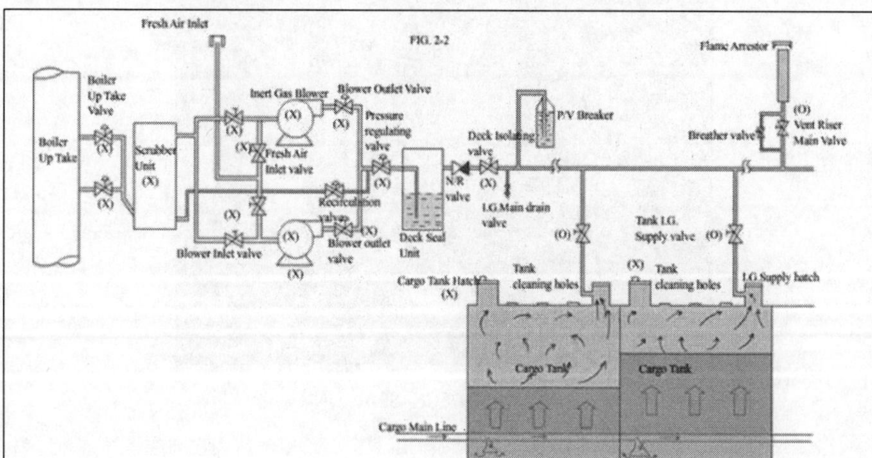

[그림 3-102] IG 장치 계통도

The system shall be designed to ensure that if the oxygen content exceeds 5% by volume, the inert gas shall be automatically vented to atmosphere.

이너트가스 장치는 산소농도가 5%를 초과하는 경우에 이너트가스를 대기 중으로 자동배출되도록 설계되어야 한다.

화물탱크의 폭발을 막기 위하여 화물탱크내의 산소농도를 5% 이하로 유지하기 위해서는 이러한 화물탱크에 공급되는 이너트가스의 산소농도 또한 5% 이하여야 하기 때문에 이너트가스 장치에서 생성되는 가스의 산소농도가 5%를 초과하는 경우는 화물탱크로 공급이 차단되어야 하고, 대기 중으로 배출되어야 한다.

At least two non-return devices shall be fitted in order to prevent the return of vapour and liquid to the inert gas plant, or to any gas-safe spaces.

이너트가스 설비 또는 가스안전구역에 증기 및 액체가 역류하는 것을 방지하기 위하여 2개 이상의 역류방지장치를 부착하여야 한다.

일반적으로 증기나 액체의 역류를 막기 위한 이중 설비로 'Deck seal unit(Deck water seal 이라고도 함)' 및 'Non-return valve'가 갑판상 화물구역 내에 설치되어 있다.

[그림 3-103] DECK SEAL UNIT

The operation status of the inert gas system shall be indicated in a control panel.

이너트가스 장치의 작동상태는 제어반에 표시되어야 한다.

이너트가스의 작동상태를 확인하고 필요에 따라 제어할 수 있는 제어반이 설치되어 있어야 하며, 통상 이너트가스 장치를 화물 적양하 작업시에 가장 많이 활용하기 때문에 화물제어실에 설치하고 있다.

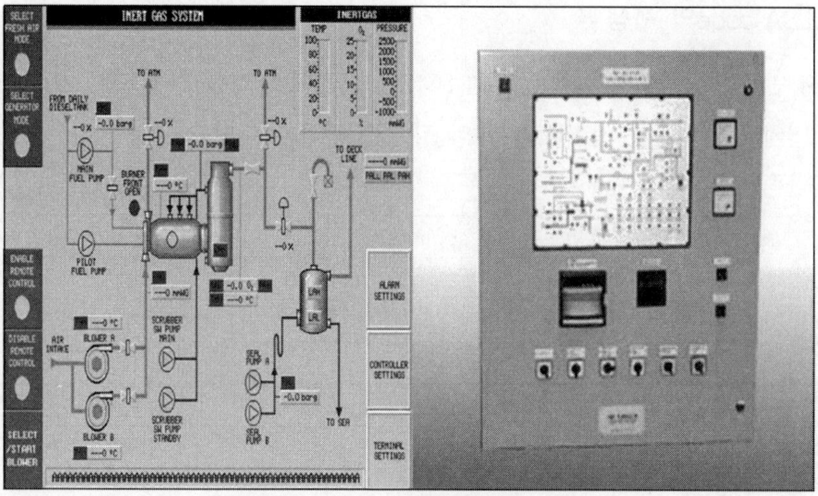

[그림 3-104] 이너트가스 장치 제어반

3.3 LSA code

3.3.1 LSA code의 도입

LSA code는 SOLAS(International Convention for the Safety Of Life At Sea, 1974) 제 III 장에 의하여 요구되는 구명설비에 대한 국제적인 기준을 제공하기 위하여 도입되었다.

3.3.2 LSA code의 적용과 구성

1. LSA code의 적용

SOLAS(International Convention for the Safety Of Life At Sea, 1974) 협약 제 III 장의 개정사항의 발효 하에서 1998년 7월 1일 이후 강제화되었다.

2. LSA code의 구성

앞서 언급한 바와 같이 LSA code는 SOLAS 협약 제 III 장 규정에 대한 기술사항을 담고 있는 협약으로 다음과 같이 총 7개의 장으로 구성되어 있다.

Chapter 1	General	일반사항
Chapter 2	Personal life-saving appliances	개인용 구명설비
Chapter 3	Visual signals	가시신호
Chapter 4	Survival craft	생존정
Chapter 5	Rescue boats	구조정
Chapter 6	Launching and embarkation appliances	진수 및 승정설비
Chapter 7	Other life-saving appliances	기타의 구명설비

3.3.3 LSA code 주요 규정

1. Chapter 2 Personal life-saving appliances

1) Lifebuoys

Every lifebuoy shall have a mass of not less than 2.5 kg and if it is intended to operate the quick release arrangement provided for the self-activated smoke signals and self-igniting lights, have a mass of not less than 4 kg

It shall be fitted with a grabline not less than 9.5 mm in diameter and not less than 4 times the outside diameter of the body of the buoy in length. The grabline shall be secured at four equidistant points around the circumference of the buoy to form four equal loops.

모든 구명부환은 2.5kg 이상의 무게를 가져야 한다. 만약 자기발연부신호 및 자기점화 등을 위하여 장치된 순간이탈장치를 작동시키게 되어 있는 경우 4kg 이상의 무게를 가져야 한다.

직경 9.5mm 이상이고, 부환 본체 외경의 4배 이상의 길이를 가진 손잡이 줄을 부착할 것. 손잡이 줄은 4개의 동등한 크기의 고리를 형성하도록 부환의 원주에 같은 거리를 두고 4곳에 부착되어야 한다.

일반적으로 윙브릿지(Wing bridge)에 설비되어 있는 구명부환의 경우 자기발연부신화 및 자기점화등과 함께 이탈되어야 하기 때문에 다른 구명부환에 비해서는 무게가 무거워야 장착된 다른 장치들과 이탈이 가능하다.

또한 구명부환 주변에 손잡이 줄은 구명부환의 도움을 받고자 하는 사람들이 용이하게 잡고 사용할 수 있도록 장착이 되어야 한다.

[그림 3-105] 윙브릿지 구명부환

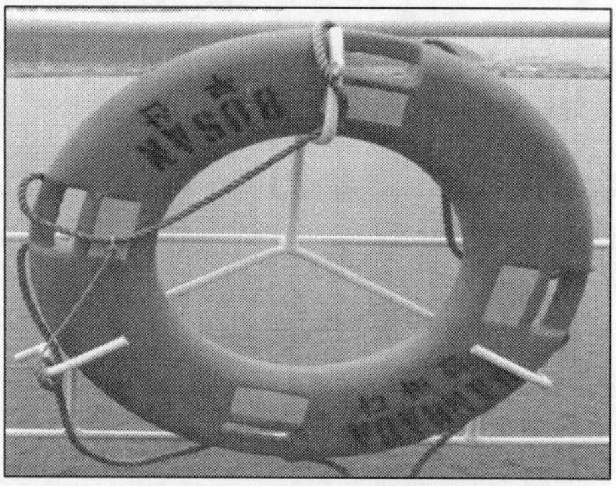

[그림 3-106] 일반 구명부환

Lifebuoy self-igniting lights shall be such that they cannot be extinguished by water and be provided with a source of energy for a period of at least 2 h.

구명부환 자기점화등은 물에 의하여 꺼지지 아니하여야 하며, 최소 2시간 이상 동안의 에너지원을 구비하여야 한다.

구명부환에 설치되어 있는 자기점화등은 야간에 구명부환의 식별을 용이하게 하기 위하여 부착된 것으로 일정 시간 이상의 에너지원을 구비하여야 한다.

[그림 3-107] 자기점화등이 달린 구명부환

아울러 이러한 자기점화등이 외부에 설치되어 있는 관계로 주기적인 작동상태 점검을 통하여 언제든지 사용할 수 있는 상태로 유지하여야 한다.

Lifebuoy self-activating smoke signals shall emit smoke of a highly visible colour at a uniform rate for a period of at least 15 min when floating in calm water and not be swamped in a seaway and continue to emit smoke when fully submerged in water for a period of at least 10s.

구명부환 자기발연부신호는 잔잔한 수면에 떠 있을 때 최소한 15분 동안 동일한 비율로 잘 보이는 색의 연기를 발해야 하고, 거친 해상에서 물속에 잠기지 않아야 하며, 만약에 최소한 10초 이상동안 물속에 완전히 잠겼어도 계속해서 연기를 낼 수 있어야 한다.

자기발연부신호는 주간에 식별이 용이하게 하기 위하여 장착된 것으로 구명부환 자체는 날씨가 좋지 않을 경우는 더욱 식별이 용이하지 않기 때문에 이러한 환경에서도 자기발연부신호는 계속적으로 연기를 발할 수 있도록 제작되어야 한다.

[그림 3-108] 자기발연부신호 작동모습

2) Lifejackets

An adult lifejacket shall be so constructed that at least 75% of persons who are completely unfamiliar with the lifejacket can correctly don it within a period of 1 min without assistance, guidance or prior demonstration and after demonstration, all persons can correctly don it within a period of 1 min without assistance.

성인용 구명동의는 구명동의에 대하여 전혀 익숙하지 못한 사람의 75 % 이상이 도움,

지침 또는 사전 시범 없이 1분 이내에 정확히 입을 수 있어야 하고, 착용 시범 후에는 모든 사람이 도움 없이 1분 이내에 정확히 착용할 수 있어야 한다.

구명동의는 일반적인 사람들이 누구나 손쉽게 착용할 수 있도록 제작이 되어 비치되어야 한다. 구명동의를 착용해야 하는 상황이란 통상 아주 급박한 상황이 많기 때문에 구명동의를 착용하는데 시간이 지체된다면 그 만큼 더 위험한 상황에 직면할 수 있다하겠다.

An adult lifejacket shall be so constructed that it allows the wearer to jump into the water from a height of at least 4.5 m while holding on to the lifejacket, and from a height of at least 1m with arms held overhead, without injury and without dislodging or damaging the lifejacket or its attachments.

착용자가 구명동의를 잡은 상태에서 4.5m 이상의 높이에서 수면으로 뛰어들고, 손으로 머리를 잡은 상태에서 1m 이상의 높이에서 수면으로 뛰어들더라도 부상당하지 아니하고 구명동의가 벗겨지거나 손상되지 아니하여야 한다.

구명동의 착용자가 수면으로 들어갈 경우는 통상 일정 높이를 있는 장소에서 입수를 하게 된다. 이러한 점을 감안하여 일정 높이에서 수면으로 뛰어들어도 잘 벗겨지거나 손상이 되지 않는 형태로 제작되어야 한다. 일단 수면으로 뛰어든 후에 구명동의를 다시 착용한다는 것은 상당히 어렵기 때문이다.

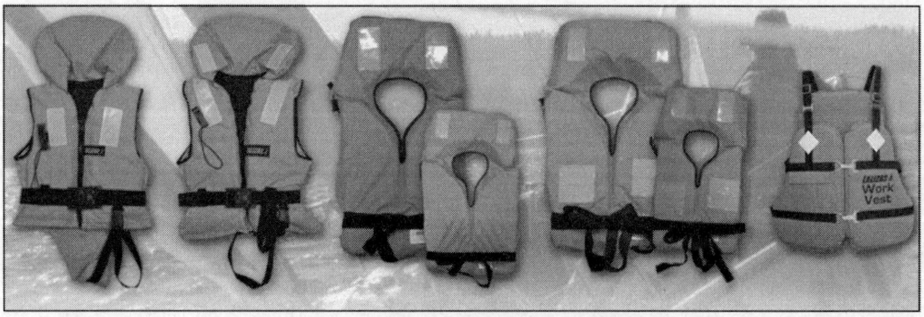

[그림 3-109] 구명동의(Lifejacket)

An adult lifejacket shall allow the person wearing it to swim a short distance and to board a survival craft.

성인용 구명동의는 이를 착용한 사람이 짧은 거리를 헤엄쳐서 생존정에 오를 수 있어야 한다.

Each lifejacket shall be fitted with a whistle firmly secured by a lanyard.

각 구명동의에는 끈으로 연결된 호각이 단단히 부착되어야 한다.

Each lifejacket light shall have a source of energy for a period of at least 8 h.

구명동의 표시등은 최소한 8시간 동안 유지할 수 있는 에너지원을 가져야 한다.

3) Immersion suits

An immersion suit can be unpacked and donned without assistance within 2 min.

2분 내에 외부의 도움 없이 방수복을 펴서 입을 수 있어야 한다.

An immersion suit shall cover the whole body with the exception of the face.

얼굴을 제외한 몸 전체를 덮어야 한다. 방수복은 해상에서의 체온저하를 가능한 최소화하고자 비치하는 것으로 노출된 신체부분이 가능한 적게 하여야 한다.

An immersion suit on its own, or worn in conjunction with a lifejacket shall have sufficient buoyancy and stability in calm fresh water.

방수복 자체로, 또는 필요시 구명동의와 함께 착용하였을 경우 잔잔한 청수에서 충분한 부력과 복원성을 가져야 한다.

해상에서 체온저하를 막기 위해서는 가능한 움직임을 최소화하여야 하는데 부력과 복원성을 가지고 있지 않다면 부득이한 움직임이 필요하고 이로 인해 체온저하를 유발할 수 있기 때문이다.

An immersion suit shall permit the person wearing it, and also wearing a lifejacket if the immersion suit is to be worn in conjunction with a lifejacket to swim a short distance through the water and board a survival craft.

방수복을 입었을 때, 그리고 구명동의와 함께 입는 것이라면 구명동의를 입은 채로 짧은 거리를 헤엄쳐서 구명정에 오를 수 있어야 한다. 방수복의 주요 기능이 해상에서의 체온저하를 최소화하고자 제작된 것으로 앞서 언급한 바와 같이 얼굴만 노출이 되어 있으려 일정한 부력과 복원성을 갖고 있어야 한다.

하지만 필요에 따라 구명정으로 오르기 위해 수영을 해야 하는 경우와 같이 일정거리를 이동할 수 있을 정도의 활동성은 확보되어야 한다.

[그림 3-110] 해상에서 방수복 착용자 모습

An immersion suit shall provide the wearer with sufficient thermal insulation to ensure that the wearer's body core temperature does not fall more than 2°C after a period of 6 h immersion in calm circulating water at a temperature of between 0°C and 2°C.

방수복은 0℃~2℃의 온도의 조용히 유동하는 물에서 6시간 동안 있었을 때 체온이 2℃ 이상 떨어지지 아니할 정도의 충분한 보온성을 제공하여야 한다. 방수복의 주 기능이 해상에서의 체온저하를 막기 위한 것으로 일정 이상의 보온성을 유지할 수 있도록 제작되어야 한다.

4) Anti-exposure suits

An anti-exposure suit shall be constructed with waterproof materials such that it covers the whole body(covering for the hands and head may be provided by separate gloves and a hood, both of which shall be permanently attached to the suit) and can be unpacked and donned without assistance within 2 min.

노출보호복은 방수재료로 제작된 것으로 전신을 감쌀 수 있어야 하고(손과 머리는 별도의 장갑 및 모자에 의하여 보호될 수 있으며, 장갑 및 모자는 보호복에 영구적으로 부착되어야 함), 도움이 없이 2분 이내에 풀어서 입을 수 있어야 한다.

An anti-exposure suit shall permit the person wearing it to perform all duties associated with abandonment, assist others and operate a rescue boat.

노출보호복은 이를 착용한 사람이 퇴선과 관련된 모든 임무를 행하며 남을 도우고 구조정을 운전할 수 있어야 한다.

노출보호복은 구조정 및 해상탈출장비를 담당하는 사람들이 착용하고 그 업무를 수행하고자 비치하는 보호복으로 관련 업무를 수행하기 위해서는 담당자의 체온저하도 가능한 막으면서 구조업무도 수행할 수 있도록 방수복에 비해서는 활동성이 더 가미되어야 한다.

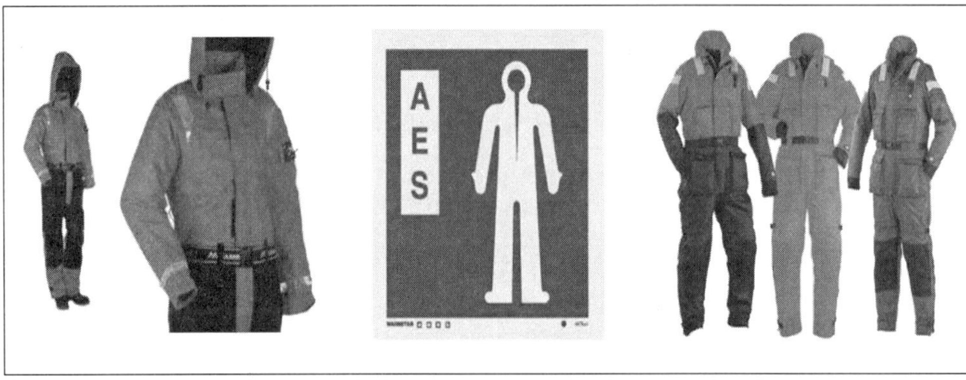

[그림 3-111] 노출보호복

5) Thermal protective aids

The thermal protective aid shall cover the whole body of persons of all sizes wearing a lifejacket with the exception of the face. Hands shall also be covered unless permanently attached gloves are provided and be capable of being unpacked and easily donned without assistance in a survival craft or rescue boat, and permit the wearer to remove it in the water in not more than 2 min, if it impairs ability to swim.

보온복은 모든 신장에 대하여 구명동의를 착용한 사람의 얼굴을 제외한 전신을 덮을 수 있어야 한다. 영구 부착된 장갑이 붙어 있지 않는 한 양손도 덮을 수 있어야 하고, 구명정이나 구조정에서 도움없이 풀어서 쉽게 착용할 수 있어야 하며, 만약 수영하는데 지장이 있다면 착용자가 2분 이내에 수중에서 벗어 버릴 수 있어야 한다.

보온복은 착용자의 몸으로부터 대류 및 증발로 인한 체온저하를 감소시키기 위하여 제작된 것으로 필요에 따라서는 손쉽게 착용과 더불어 벗어 버릴 수 있어야 한다.

[그림 3-112] 보온복

2. Chapter 3 Visual signals

1) Rocket parachute flares

The rocket parachute flare shall

(a) be contained in a water-resistant casing

(b) have brief instructions or diagrams clearly illustrating the use of the rocket parachute flare printed on its casing;

(c) have integral means of ignition

(d) be so designed as not to cause discomfort to the person holding the casing when used in accordance with the manufacturer's operating instructions.

(e) burn with a bright red colour

(f) have a burning period of not less than 40 s

로켓 낙하산 화염신호는 다음에 따라야 한다.

(a) 방수용 기내에 수납될 것

(b) 로켓 낙하산 화염신호의 사용법을 명확히 나타내는 간단한 설명문이나 그림이 용기상에 인쇄될 것

(c) 일체형의 점화장치를 보유하고 있을 것

(d) 제조자의 작동지시에 따라 사용할 때 용기를 들고 있는 사람에게 불편을 주지 아니하도록 설계될 것

(e) 밝은 적색으로 탈 것

(f) 40초 이상의 연소시간을 가질 것

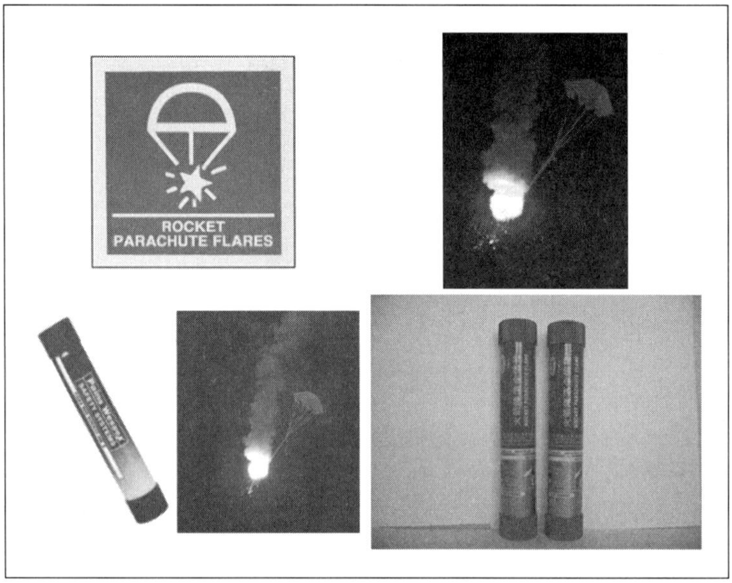

[그림 3-113] Rocket parachute flares

모든 시각신호는 작동시키는 사람이 누구나 쉽게 사용할 수 있도록 사용법이 용기에 표시되어 있으며, 다른 도구의 도움 없이 자체적으로 작동이 가능하도록 되어 있다. 로켓 낙하산 화염신호도 이에 해당이 된다.

또한 조난위치를 보다 용이하게 식별할 수 있도록 일정 이상의 조도와 연소시간을 유지할 수 있도록 제작되어야 한다.

2) Hand flares

The hand flare shall

(a) be contained in a water-resistant casing;

(b) have brief instructions or diagrams clearly illustrating the use of the hand flare printed on its casing

(c) have a self-contained means of ignition

(d) be so designed as not to cause discomfort to the person holding the casing and not endanger the survival craft by burning or glowing residues when used in accordance with the manufacturer's operating instructions.

(e) burn with a bright red colour;

(f) have a burning period of not less than 1 min

신호홍염은 다음에 따라야 한다.

(a) 방수용기내에 수납될 것

(b) 신호홍염의 사용법을 명확히 나타내는 간단한 설명문이나 그림이 그 용기상에 인쇄될 것

(c) 자체 수납된 점화장치를 보유하고 있을 것

(d) 제조자의 작동지시에 따라 사용할 때, 요기를 들고 있는 사람이 불편을 느끼지 아니하고, 화염의 잔재에 의하여 생존정에 위험을 끼치지 않도록 설계될 것

(e) 밝은 적색으로 연소할 것

(f) 1분 이상의 연소시간을 가질 것

[그림 3-114] 신호홍염(Hand flares)

3) Buoyant smoke signals

The buoyant smoke signal shall

(a) be contained in a water-resistant casing

(b) not ignite explosively when used in accordance with the manufacturer's operating instructions

(c) have brief instructions or diagrams clearly illustrating the use of the buoyant smoke signal printed on its casing

(d) emit smoke of a highly visible colour at a uniform rate for a period of not less than 3 min when floating in calm water

(e) not be swamped in a seaway

발연부 신호는 다음에 따라야 한다.

(a) 방수용기내에 수납될 것

(b) 제조자의 작동 지시에 따라 사용하였을 때 폭발적으로 점화하지 아니할 것

(c) 발연부 신호의 사용법을 명확히 나타내는 간단한 설명문이나 그림이 용기상에 인쇄될 것

(d) 조용한 수면에 떠있을 때 3분 이상 동안 일정한 비율로 잘 보이는 색깔의 연기를 분출할 것

(e) 거친 해상에서 잠기지 않을 것

[그림 3-115] Buoyant smoke signals

3. Chapter 4 Suvival craft

1) General requirements for liferafts

Every liferaft shall be so constructed as to be capable of withstanding exposure for 30 days afloat in all sea conditions.

모든 구명뗏목은 모든 해상상태에서 30일 동안 떠 있어도 견딜 수 있도록 제작되어야 한다.

The liferaft shall have a canopy to protect the occupants from exposure which is automatically set in place when the liferaft is launched and waterborne.

구명뗏목은 진수되어 물위에 떠 있을 때 자동적으로 설치되어 승무원들을 보호할 수

있는 천정덮개를 가져야 한다. 구명뗏목에 있는 승무원들의 탈수를 가능한 줄일 수 방법 중에 하나는 직사광선을 피하는 것인데 이를 위해 천정덮개가 필요하다.

The liferaft painter system shall provide a connection between the ship and the liferaft and shall be so arranged as to ensure that the liferaft when released and, in the case of an inflatable liferaft, inflated is not dragged under by the sinking ship.

구명뗏목 페인터장치는 배와 구명뗏목 사이를 연결하는 것이어야 하고 구명뗏목이 이탈될 때 또는 팽창식 구명뗏목인 경우, 팽창되었을 때, 침몰하는 배에 의하여 끌려 들어가지 않도록 배치되어야 한다.

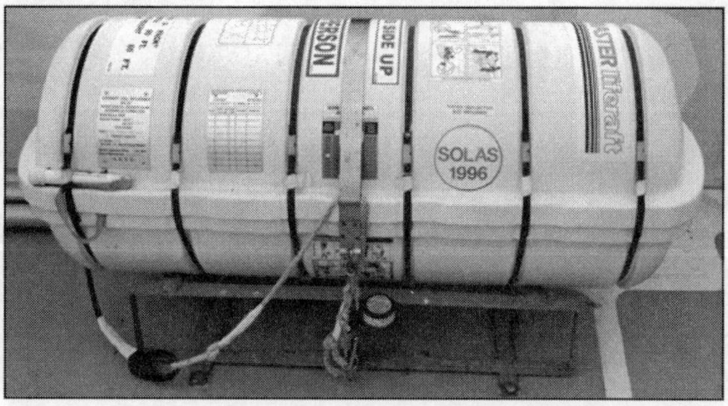

[그림 3-116] 구명뗏목에 설치된 페인터

If a weak link is used in the float-free arrangement, it shall not be broken by the force required to pull the painter from the liferaft container.

자유부양장치에 사용되는 위크링크는 구명뗏목 수납용기에서부터 페인터를 끌어당기는 힘에 의하여서 떨어지지 아니하여야 한다.

위크링크와 페인터는 서로 연결이 되어 있어 페인터는 구명뗏목이 본선으로부터 일정 조건하에서 이탈되는 것을 방지하면서 구명뗏목이 팽창할 수 있도록 팽창장치를 작동시켜 주는 트리거 역할도 수행한다. 만약 이러한 팽창장치가 작동시키기 전에 위크링크가 떨어지면 구명뗏목이 팽창하지 않기 때문에 일정 이상의 강도는 유지를 하여야 한다. 물론, 너무 강하면 팽창된 구명뗏목이 선박침몰시에 같이 끌려 들어갈 수 있기 때문에 일정한 강도가 유지되어야 한다.

If a hydrostatic release unit is used in the float-free arrangements, it shall automatically release the liferaft at a depth of not more than 4 m.

만약 자유부양장치에 수압이탈장치가 사용되는 경우 4 m 이하의 수심에서 구명뗏목을 자동적으로 이탈시킬 수 있어야 한다.

구명뗏목을 작동시킬 수 없는 급박한 상황이라도 구명뗏목은 자동으로 부양하여 작동될 수 있도록 설비되어야 하는데, 수압이탈장치는 구명뗏목이 구명뗏목을 설비하고 있는 선박과 동시에 침몰할 경우 수압에 의해 자동으로 구명뗏목을 선박으로부터 이탈시켜주는 역할을 하게 된다.

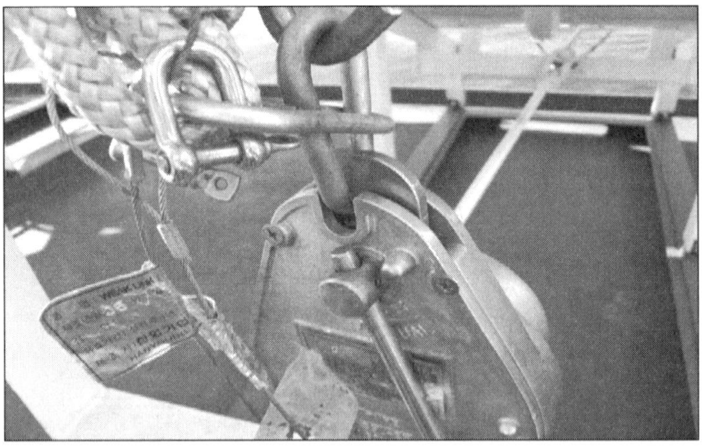

[그림 3-117] 구명뗏목의 수압이탈장치 및 위크링크

2) Inflatable liferafts

The main buoyancy chamber shall be divided into not less than two separate compartments.

주부력실은 2 개 이상의 분리된 구획으로 나누어져야 한다.

주부력실이 파공이나 가스주입이 잘 이루어지지 않는다면 구명뗏목이 부양하지 않기 때문에 이에 대비하여 2개 이상으로 분리하여 어느 한 구획에 손상이 발생하더라도 구명뗏목이 부양할 수 있어야 한다.

The liferaft shall be capable of being inflated by one person.

구명뗏목은 1명에 의해 팽창 가능해야 한다.

The stability of the liferaft when in the inverted position shall be such that it can be righted in a seaway and in calm water by one person.

구명뗏목의 복원성은 뒤집힌 상태에 있을 때, 해상 및 조용한 수면위에서 한사람의 힘으로 복원될 수 있어야 한다. 구명뗏목이 팽창하는 과정에서 또는 다른 이유로 인해 뒤집히더라도 이를 손쉽게 복원시킬 수 있도록 제작되어야 한다.

[그림 3-118] 구명뗏목 복원

3) General requirements for lifeboats

All lifeboats shall be of sufficient strength to be capable of being launched and towed when the ship is making headway at a speed of 5 knots in calm water.

모든 구명정은 본선이 잔잔한 수면에서 5노트의 속력으로 전진하고 있을 때, 진수 되고 예인될 수 있기 위한 충분한 강도를 가져야 한다.

퇴선시에 구명정을 이용하고자 할 경우 선박이 움직이고 있을 상황도 있을 수 있기 때문에 이러한 경우에 대한 강도를 유지가 필요하다.

Every cargo ship lifeboat shall be so arranged that it can be boarded by its full complement of persons in not more than 3 min from the time the instruction to board is given.

모든 화물선의 구명정은 탑승지시가 내려진 시각부터 3분 이내에 정원 모두가 승정할 수 있도록 배치되어야 한다.

3분 이내의 승정이 가능하기 위해서는 구명정의 구조적인 측면에서는 적절한 크기의 출입구가 확보되어야 하고, 반복적인 훈련을 통한 숙달이 필요하다.

The engine shall be provided with either a manual starting system, or a power starting system with two independent rechargeable energy sources.

기관은 수동시동장치나 두 개의 독립된 충전 가능한 에너지원을 가진 동력시동장치를 가져야 한다.

일반적으로 선박에 장착되어 있는 구명정 기관의 시동동력은 배터리로 기동을 하게 되고, 만약 이러한 배터리 기동이 되지 않는다면 수동시동을 하게 된다. 하지만, 별도의 동력 배터리를 추가로 설비하고 있다면 수동시동장치를 설비하지 않아도 되기 때문에 일부 선박의 구명정에는 2세트의 배터리를 장착하여 기관 수동장치를 장착하고 있지 않다.

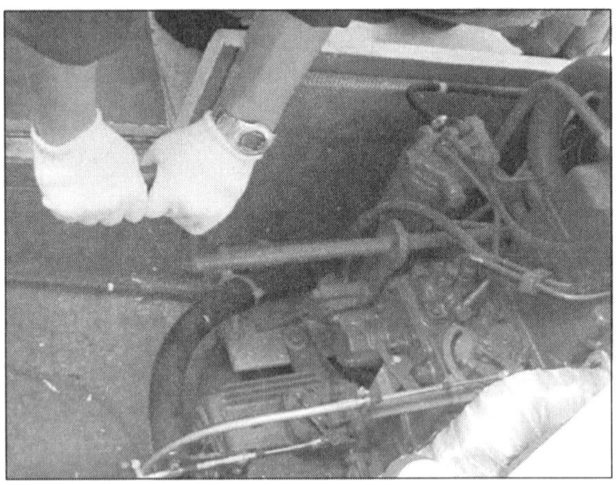

[그림 3-119] 구명정 수동시동장치

The speed of a lifeboat when proceeding ahead in calm water, when loaded with its full complement of persons and equipment and with all engine powered auxiliary equipment in operation, shall be at least 6knots and the lifeboat shall be provided to run the fully loaded lifeboat at 6knots for a period of not less than 24 h.

구명정은 잔잔한 수면에서 정원 및 의장품을 만재하고 기관으로 구동되는 모든 보조설비가 작동 중일 때 최소한 6노트의 속력으로 전진할 수 있어야 하고, 구명정은 6노트의 속력으로 24시간 이상 달릴 수 있어야 한다.

구명정은 위험한 수역을 벗어나기 위해 일정한 속력을 유지할 수 있어야 하며, 필요에 따라 이동해야 할 상황에 대비하여 어느 정도의 연료를 갖추고 있어야 한다.

The mechanism shall only open when the release mechanism is operated with the boat fully waterborne or, if the boat is not waterborne, by multiple, deliberate and sustained action which shall include the removal or bypassing of safety interlocks designed to prevent premature or inadvertent release.

이탈장치는 구명정이 완전히 물에 뜬 상태에서 해당 이탈장치가 작동될 때에만 열리거나, 또는 구명정이 물에 뜨지 않은 경우에는 조기이탈, 또는 의도하지 않은 이탈을 방지하기 위하여 설계된 안전장치를 제거하거나, 바이패스하는 것을 포함하는 다중적이고, 고의적이며 지속적인 동작에 의해서 열려야 한다.

일반적으로 구명정의 이탈장치는 수압이 형성되어야 작동이 되도록 설비되어 있기 때문에, 구명정이 수면에 안착했을 경우에 자연스럽게 수업이 형성되어 이탈장치를 작동시킬 수 있다. 하지만 필요에 따라 구명정이 수면에 도달하기 전에 이탈해야 하는 경우도 있을 수 있는데, 이러한 경우는 안전장치를 제거하여야만 이탈장치를 작동시킬 수 있다. 이탈장치의 안전장치는 잘못된 오작동을 막는 역할도 한다.

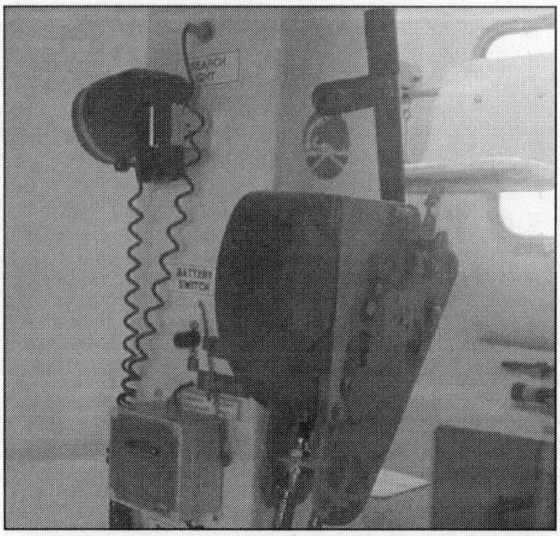

[그림 3-120] 이탈장치의 안전장치(수압 인터록)

모든 구명정은 선수 근처에 있는 페인터를 고박시킬 수 있는 장치를 갖추어야 한다. 이 장치는 본선이 조용한 해상에서 5노트까지의 속력으로 전진하면서 예인하고 있을 때, 구명정이 불안전상태에 있거나 또는 불안정형태를 보이지 않도록 하는 것이어야 한다.

구명정의 선수에는 페이터를 연결하는 장치가 설비되어 있어 퇴선시 본선에서 이탈하고자 할 때에 구명정 내부의 페인터 이탈장치를 작동시켜 페이터를 제거한다. 더불어 구명정에 연결되어 있는 페인터의 반대쪽은 본선에 결부되어 있어 구명정이 본선으로부터 이탈되는 것을 막아준다.

[그림 3-121] 구명정 페인터 설치모습

4) Totally enclosed lifeboats

Access to Every totally enclosed lifeboat shall be provided by hatches which can be closed to make the lifeboat watertight and it is possible to row the lifeboat.

모든 전폐형 구명정의 출입구는 구명정이 수밀이 되도록 닫을 수 있는 해치를 갖추어야 하고, 모든 구정정은 노를 저을 수 있어야 한다.

전폐형 구명정은 외부로부터의 해수의 침입을 막을 수 있어야 함과 동시에 퇴선시 승무원이 승선할 수 있는 해치가 마련되어 있어야 하기 때문에 설치된 해치는 수밀이 되어야 한다.

또한 전폐형 구명정에 노를 저을 수 있도록 장치를 설치하기 위해서는 별도의 개구부가 필요한데 일반적으로 해치 주변에 노를 설치할 수 있도록 구조적으로 제작하고 있다.

[그림 3-122] 노 설치를 위한 크러치

The engine and transmission shall be controlled from the helmsman's position.

기관 및 전달장치는 조정자(정장)의 위치에서 조종될 수 있어야 한다.

일반적으로 구명정은 정장이 기관 및 조타까지 수행하기 때문에 정장이 위치하는 곳에 관련 설비를 설비하고 있다.

[그림 3-123] 구명정 정장석 전면

5) Fee-fall lifeboats

Each free-fall lifeboat shall be of sufficient strength to withstand, when loaded with its full complement of persons and equipment, a free-fall launch from a height of at least 1.3 times the free-fall certification height.

자유낙하식 구명정은 구명정의 정원과 의장품을 만재하였을 때 승인된 자유낙하식 높이의 1.3 배 이상의 높이에서 자유낙하 진수를 견디어내기에 충분한 강도를 가져야 한다.

자유낙하식 구명정은 높은 위치에서 승정원과 의장품을 실은 상태에서 중력에 의해 해상으로 직접적으로 낙하하기 때문에 상당한 충격이 있을 수 밖에 없다. 그렇기 때문에 이러한 충격에 충분히 견딜 수 있는 강도로 제작이 되어야 한다.

Each free-fall lifeboat shall have two independent activation systems for the release mechanisms which may only be operated from inside the lifeboat and be marked in a colour that contrasts with its surroundings and be designed to test the release system without launching the lifeboat.

자유낙하식 구명정의 이탈장치는 오직 구명정 내부에서만 작동되고, 주위의 색과 대조되는 색으로 표시를 한 2개의 독립된 이탈장치, 작동장치를 갖추어야 하고, 구명정을 진수하지 않고 이탈장치 시험이 가능하도록 설계되어야 한다.

자유낙하식 구명정은 본선에서 이탈시 상당한 위험성이 있기 때문에 반드시 구명정 내부에서만 구명정을 이탈시킬 수 있는 장치를 갖추고 있어야 한다. 아울러 같은 장치는 식별이 용이하게 되어 있어야 한다.

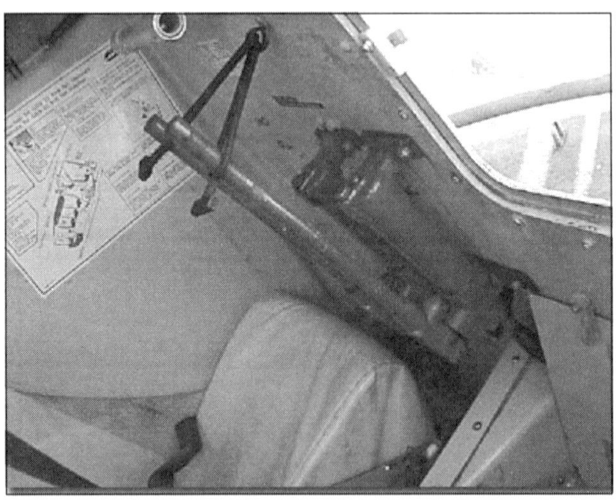

[그림 3-124] 자유낙하식 구명정 내부 이탈장치

6) Lifeboats with a self-contained air support system

A lifeboat with a self-contained air support system shall be so arranged that, when proceeding with all entrances and openings closed, the air in the lifeboat remains safe and breathable and the engine runs normally for a period of not less than 10 min. During this period the atmospheric pressure inside the lifeboat shall never fall below the outside atmospheric pressure. The system shall have visual indicators to indicate the pressure of the air supply at all times.

자장식 공기공급장치를 가진 구명정은 모든 입구 및 개구를 닫고서 항진할 때, 구명정의 공기가 안정하게 그리고 호흡할 수 있게 유지되고, 기관이 10분 이상 정상적으로 운전될 수 있도록 배치되어야 한다. 이 시간동안 구명정내의 대기압은 바깥 대기압 이하로 떨어져서는 아니된다. 이 장치는 공급공기의 압력을 항상 표시할 수 있는 가시지시계를 가져야 한다.

공기공급장치는 구명정 내부의 승무원의 호흡 및 외부로부터의 유해가스의 유입을 막는 역할을 하는 것으로 일정한 압력으로 공기가 공급될 수 있도록 하는 장치가 같이 설비되어 있다.

[그림 3-125] 구명정 자장식 공기공급장치

7) Fire-protected lifeboats

A fire-protected lifeboat when waterborne shall be capable of protecting the number of persons it is permitted to accommodate when subjected to a continuous oil fire that envelops the lifeboat for a period of not less than 8 min.

물에 떠 있을 때 내화 구명정은 8분 이상 구명정을 감싸는 계속적인 기름 화재를 당하였을 때 그 수용인원을 보호할 수 있어야 한다.

구명정을 화재로부터 보호하기 위해서는 구명정을 불에 타지 않는 재질로 제작하거나 아니면 구명정을 불로부터 직접적인 접촉을 막는 방법이 있는데, 일반적으로 구명정은 불로부터 직접적인 접촉을 막는 방법을 채택하고 있다. 관련하여 가장 대표적인 방법이 구명정 주변으로 해수를 살수하여 구명정을 불로부터 직접적인 접촉이 이루어지지 않도록 보호막을 형성하는 것이다.

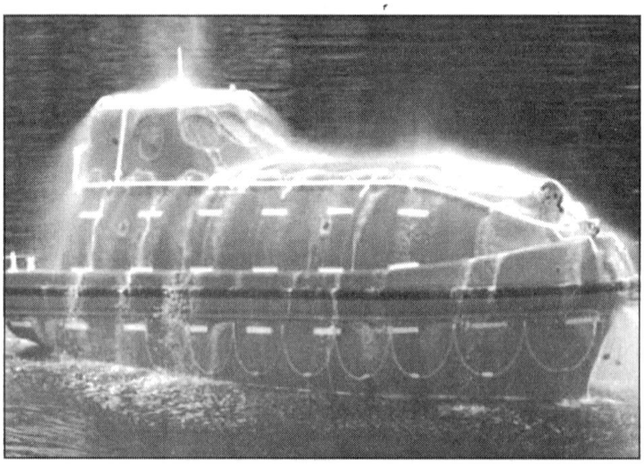

[그림 3-126] 구명정 해수 살수모습

On a lifeboat which has a water spray fire-protection system, the water for the system shall be drawn from the sea by a self-priming motor pump and the system shall be arranged for flushing with fresh water and allowing complete drainage.

살수 방화장치를 가진 구명정에 있어 살수장치에 사용되는 물은 자체주입식 전동펌프에 의하여 바다로부터 끌어올려져야 하고, 이 장치는 청수로 소제할 수 있고 완전한 배수가 될 수 있도록 배치되어야 한다.

일반적으로 살수 방화장치는 해수를 이용하여 살수를 하기 때문에 관련 장치들의 관리가 잘 이루어지지 않는다면 파공 등과 같은 손상이 발생할 수 있기 때문에 주기적인 관리와 더불어 훈련 등으로 사용 후에는 반드시 청수로 소제를 하여야 한다. 이러한 이유 때문에 청소 소제를 할 수 있도록 제작되어야 한다.

[그림 3-127] 구명정 청소 소제 연결구

4. Chapter 5 Rescue boats

Rescue boats shall be capable of carrying at least five seated persons and a person lying on a stretcher all wearing immersion suits, and lifejackets if required.

구조정은 모두 방수복과 요구되는 경우 구명동의를 입고 앉아 있는 사람 5명 이상과 들것에 누운 사람 1명을 수용할 수 있어야 한다. 구조정은 해상의 조난자를 구출하기 위한 용도로 사용되기 때문에 구조업무에 종사하는 사람 및 구조자를 수용할 수 있는 용적을 가지고 있어야 한다.

Rescue boats shall have sufficient mobility and manoeuvrability in a seaway to enable persons to be retrieved from the water.

구조정은 충분한 이동성 및 조종성을 가지고 있어서 해상에서 물에 빠진 사람을 구조할 수 있어야 한다.

앞서 언급한 바와 같이 구조정은 해상에 조난자를 구조하기 위한 것으로 이동성 및 조종성이 갖추어져 있지 않다면 원활한 구조업무가 수행될 수 없다하겠다.

Every rescue boat shall be so arranged that an adequate view forward, aft and to both sides is provided from the control and steering position for safe launching and manoeuvring.

구조정은 제어 및 조종위치로부터 안전한 진수 및 항해를 위하여 전방, 후방 및 양현의 적절한 시야가 확보되어야 한다.

5. Chapter 6 Launching and embarkation appliances

"With the exception of free-fall lifeboats, each launching appliance shall be so arranged that the fully equipped survival craft or rescue boat it serves can be safely launched against unfavourable conditions of trim of up 10° and list of up to 20° either way."

자유낙하식 구명정을 제외하고, 각 진수설비는 완전히 의장품을 갖춘 생존정 또는 구조정이 10°까지의 트림 및 20°까지의 횡경사의 불리한 조건에서 안전하게 진수될 수 있어야 한다.

생존정이나 구조정을 본선에서 이탈시키고자 할 경우가 항상 선박이 평형상태에 있는 것은 아니기 때문에, 일정 이상 선박에 경사가 있더라도 충분히 생존정과 구조정을 본선에서 이탈시킬 수 있도록 설비되어야 한다.

Each launching appliance shall, as far as practicable, remain effective under conditions of icing.

모든 진수설비는 실행 가능한 한 결빙상태에서도 효과적으로 사용될 수 있어야 한다.

A lifeboat launching appliance shall be capable of recovering the lifeboat with its crew.

구명정 진수장치는 승무원을 태운 상태에서 구명정을 회수할 수 있어야 한다.

Rescue boat launching appliances shall be provided with foul weather recovery strops for recovery where heavy fall blocks constitute a danger.

구조정 진수설비는 무거운 폴 블록 때문에 위험할 수 있을 경우에 악천후용 회수 환삭이 선박에 비치되어야 한다.

구조정은 조난자의 신속한 구출도 중요하지만 신속하게 구조정을 회수하여 조난자의 응급처치를 수행하는 것도 중요하다. 이를 위해서는 신속한 구조정의 회수가 이루어져야 하는데 구조정 회수장치의 일부인 폴 블록의 경우는 다루기가 너무 어렵고 위험하다. 만약 날씨마저 좋지 않다면 더욱 구명정의 회수가 어려워지기 때문에 이에 대비하여 악천후에도 폴 블록을 잘 다룰 수 있도록 하는 회수 환삭이 비치되어야 한다. 이러한 악천후 회수 환삭은 슬링, 와이어 등 여러 형태로 제작되어 있다.

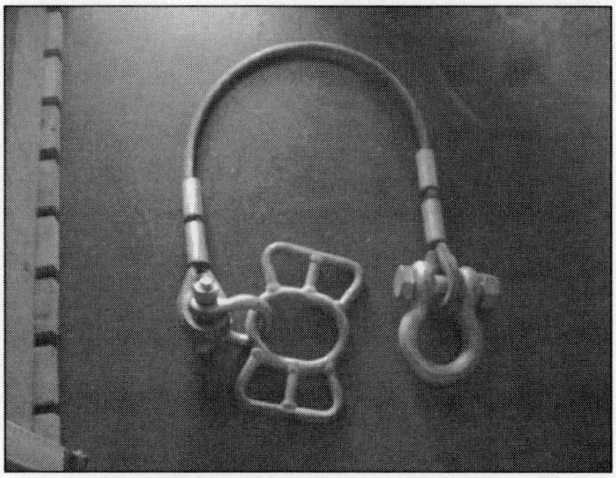

[그림 3-128] 악천후용 와이어 회수 환삭

Except for free-fall lifeboats, the launching mechanism shall be so arranged that it may be actuated by one person from a position on the ship's deck and from a position within the survival craft or rescue boat.

자유낙하식 구명정을 제외하고, 진수용 기계장치는 본선의 갑판상에서, 생존정 또는 구조정내의 어떤 위치에서 한사람에 의하여 조작될 수 있도록 배치되어야 한다.

An efficient hand gear shall be provided for recovery of each survival craft and rescue boat.

생존정 및 구조정을 회수하기 위하여 유효한 수동기어가 설비되어야 한다. 일반적으로 생존정 및 구조정의 회수는 전기나 압축공기로 기동되는 모터에 의해 회수되는데, 만약 이러한 장치들이 작동되지 않는 경우에도 회수가 이루어질 수 있도록 수동기어를 설비하여야 한다.

Every launching appliance shall be fitted with brakes capable of stopping the descent of the survival craft or rescue boat and holding it securely when loaded with its full complement of persons and equipment.

각 진수설비는 생존정 또는 구조정이 정원 및 의장품을 만재하였을 때 그들의 하강을 정지시키고, 그것을 안전하게 붙잡아 둘 수 있는 브레이크를 부착하여야 한다.

[그림 3-129] 구명정 브레이크 장치

6. Chapter 7 Other life-saving appliances

1) Line-throwing appliances

Every line-throwing appliance shall

(a) be capable of throwing a line with reasonable accuracy

(b) include not less than four projectiles each capable of carrying the line at least 230 m in calm weather

(c) include not less than four lines

(d) have brief instructions or diagrams clearly illustrating the use of the line-throwing appliance.

모든 구명줄 발사기는 다음의 조건을 만족하여야 한다.

(a) 어느 정도의 정확성을 가지고 줄을 발사할 수 있을 것

(b) 조용한 날씨에 길이 230m 이상의 줄을 운반할 수 있는 4개 이상의 발사체를 가질 것

(c) 4개 이상의 줄을 가질 것

(d) 구명줄 발사기의 사용법을 분명히 예시하는 간단한 지시문이나 도해를 가질 것

구명줄 발사기는 일정한 정확성을 가지고 줄이 발사될 수 있어야 하는데, 로켓 원리로 작동이 되면서 발사 당시의 주변의 여러 외력에 의해 원하는 방향으로 발사하기 어려운 경우가 발생할 수 있기 때문에 이에 대비하여 최소한 4번을 발사할 수 있도록 구비되어야 한다.

[그림 3-130] 구명줄 발사기

2) General alarm and public address system

“The general emergency alarm system shall be capable of sounding the general emergency alarm signal consisting of seven or more short blasts followed by one long blast on the ship's whistle or siren and additionally on an electrically operated bell or klaxon or other equivalent warning system, which shall be powered from the ship's main supply and the emergency source of electrical power. The system shall be capable of operation from the navigation bridge and, except for the ship's whistle, also from other strategic points. The system shall be audible throughout all the accommodation and normal crew working spaces. The alarm shall continue to function after it has been triggered until it is manually turned off or is temporarily interrupted by a message on the public address system.”

총비상경보장치는 선박의 기적 또는 사이렌 그리고 이에 추가하여 선박의 주전원 및 비상전원으로부터 공급하는 전기작동 벨 또는 클랙슨 또는 다른 동등한 경고장치에 의한 7회 이상의 단음에 이어 1회의 장음으로 구성된 총 비상경보신호를 발할 수 있어야 한다. 그 장치는 항해선교에서 작동할 수 있어야 하며, 선박의 기적을 제외하고는, 또한 다른 중요 위치로부터도 작동할 수 있어야 한다.

이 시스템은 거주구 및 통상적인 선원작업구역의 전역에 걸쳐 들을 수 있어야 한다. 이 경보는 작동이 된 후 수동으로 정지시킬 때까지 또는 선내방송으로 인하여 일시 중단될 때까지 계속 작동하여야 한다.

총비상경보장치는 비상시에 선내외 전구역에 비상경보를 발할 수 있는 장치로 일반적으로 자동 및 수동장치로 이루어져 있다. 수동장치의 경우는 장치를 작동시키는 동안 즉, 버튼을 누르고 있는 동안 경보가 발하게 되어 있고, 자동장치의 경우는 장치를 작동시키는 동안, 즉 버튼을 한번 누르면 단음 7회에 이어 1회의 장음으로 이루어진 경보신호를 발하게 되며, 이 경보는 수동으로 정지시킬 때까지 즉, 버튼을 다시 한번 누를 때까지 작동이 지속되게 구성되어 있다.

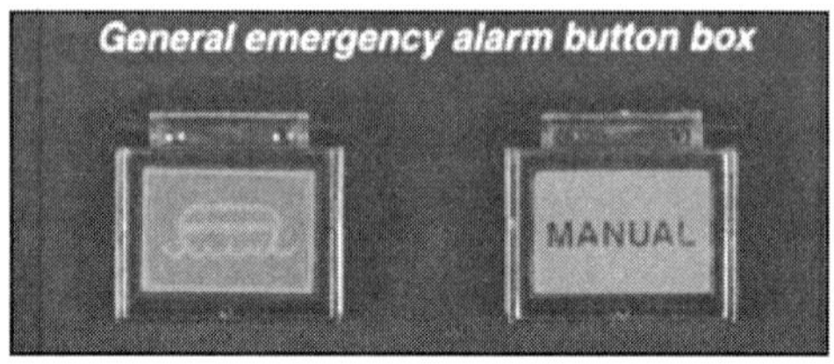

[그림 3-131] 총비상경보장치

The public address system shall be a loudspeaker installation enabling the broadcast of messages into all spaces where crew members or passengers, or both, are normally present, and to muster stations. It shall allow for the broadcast of messages from the navigation bridge and such other places on board the ship as the Administration deems necessary.

선내방송장치는 승무원, 승객 또는 선원과 승객이 통상적으로 거주하는 모든 장소와 소집장소에 메시지를 방송할 수 있는 확성기 시스템이어야 한다. 선내방송장치는 선교 및 주관청이 필요하다고 간주하는 선내 다른 구역에서부터 방송할 수 있어야 한다.

일반적으로 선내방송장치는 선교에 설비되어 있으며, 통상 거주구 및 비상소집장소에 관련 방송내용이 잘 청취될 수 있도록 설비되어야 있다.

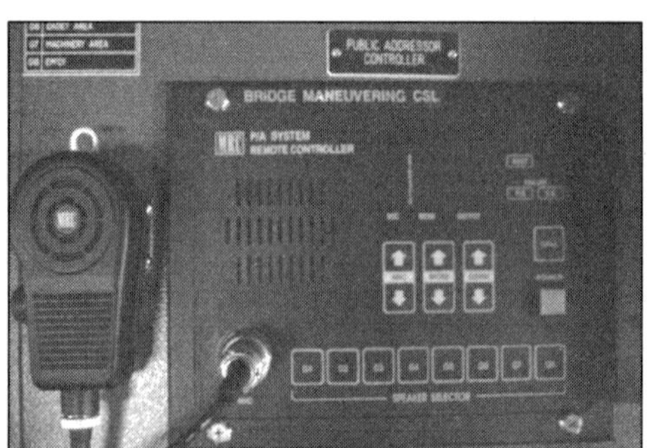

[그림 3-132] 선내방송장치

참고 문헌

- 윤수원, SOLAS Training Manual, 해인출판사, 2005
- 윤여정, 전승환, 천문항해학, 한국해양대학교, 2013
- 윤여정, 지문항해학, 해인출판사, 1969
- 윤여정, 전승환, 문성배, 지문항해학, 한국해양대학교, 2013
- 윤점동, 국제해상충돌예방규칙, 세종출판사, 2012
- 이윤철, 국제해사협약, 다솜출판사, 2013
- 이형기 외, 항해실습(I), 상학당, 2008
- ICS, Bridge Procedure Guide, 해인출판사, 1999
- Nathaniel Bowditch, The American Practical Navigator, National Imagery and Mapping Agency, 2002
- IMO SOLAS(International Convention for the Safety of Life at Sea), 2015 Consolidation Edition
- IMO FSS Code(International Code for Fire Safety Systems), 2015 Consolidation Edition
- IMO LSA Code(International Life-Saving Appliance Code), 2015 Consolidation Edition

찾아보기

□ 저자 약력 □

■ **윤 귀 호** (주저자)

- 한국해양대학교 항해학과 졸업(공학사)
- 현대상선(주) 일등항해사
- 한국해양대학교 실습선 선장
- 한국해양대학교 교수

■ **배 병 덕**

- 한국해양대학교 항해학과 졸업(공학사)
- 현대상선(주) 선장
- 한국해양대학교 실습선 선장
- 한국해양대학교 교수

■ **김 종 성**

- 한국해양대학교 항해학과 졸업(공학사)
- 목포해양대학교 대학원 공학석사/공학박사
- 한진해운(주) 수석항해사
- 한국해양대학교 실습선 한바다호 선장
- 한국해양대학교 교수

■ **조 익 순** (교신저자)

- 한국해양대학교 항해학과 졸업(공학사)
- 일본 고베대학 대학원 공학박사
- 선박안전기술공단 해사안전연구센터장
- 한국해양대학교 실습선 교학부장
- 한국해양대학교 교수

■ **이 윤 석**

- 한국해양대학교 항해학과 졸업(공학사)
- 일본 고베대학 대학원 공학석사/공학박사
- 한국해양대학교 실습선 한나라호 선장
- 한국해양대학교 교수

■ **전 해 동**

- 한국해양대학교 항해학과 졸업(공학사)
- 영국 사우스햄튼 대학교 법학석사/법학박사
- 현대상선(주) 항해사
- 한국해양대학교 실습선 교학부장
- 한국해양대학교 교수

최신항해실습(I)

2016년 3월 10일 1판 1쇄 발행

저 자 윤 귀 호 외
발행인 남 승 우

발행처 도서출판 상尙 학學 당堂
서울시 동작구 사당로9가길 6(사당동)
TEL : 02) 595-1692~4
FAX : 02) 595-1394
E-mail : namdoo6@unitel.co.kr
신고 : 2-155호 (1968. 11. 29)

정가 29,000원

ISBN 978-89-6587-141-5 93500
ISBN 978-89-6587-140-8 (전5권)

〈디자인〉 편집 엄해숙 / 표지 토틀컴